AGRICULTURAL DIMENSIONS of GLOBAL CLIMATE CHANGE

Edited by
Harry M. Kaiser, Ph.D.
Thomas E. Drennen, Ph.D.
Department of Agricultural Economics
Cornell University, Ithaca

SᵗL

St. Lucie Press
Delray Beach, Florida

Printed and bound in the U.S.A. Printed on acid-free paper.

Library of Congress Cataloging-in-Publication Data

Agricultural dimensions of global climate change / edited by Harry M. Kaiser and Thomas E. Drennen.

p. cm.

Includes bibliographic references and index.

ISBN 0-9634030-3-6 : $59.95

1. Climate changes. 2. Crops and climate. I. Kaiser, Harry Mason, 1956– II. Drennen, Thomas E.

QC981.8.C5A35 1993

338. 1'4—dc20

93-28519

CIP

Direct all inquiries to St. Lucie Press, Inc., 100 E. Linton, Blvd., Suite 403B, Delray Beach Florida, 33483.

Phone: (407) 274-9906

Fax: (407) 274-9927

StL

Published by

St. Lucie Press

100 E. Linton Blvd., Suite 403B

Delray Beach, FL 33483

TABLE OF CONTENTS

PREFACE

This book explores the social and scientific issues concerning agriculture and global climate change. This is a large and important topic that crosses the boundaries of many disciplines. To deal with the multidisciplinary nature of the problem, we assembled a group of experts from a variety of fields for a two-day conference in October 1992. Attendees discussed such topics as: the current scientific evidence of global climate change; the potential impacts on world and regional agriculture; the contribution of agriculture to greenhouse gas emissions; the status of international negotiations to formulate a response to the risk of climate change; ethical issues to guide the policy debate; and resolving communication problems between policy makers and scientists. This book is an outgrowth of that conference.

The primary aim of the book is to give the reader a thorough understanding of the central issues surrounding this complex problem. There are several other objectives as well. A second goal is to examine important issues critical to global climate change research. The book summarizes relevant findings on potential agricultural impacts and contributions and points out key areas that need further research to clear up uncertainties. The third objective is to provide information to help policy makers take a rational stance on climate change policies. The Framework Convention on Climate Change, signed this past summer (June, 1992) by 153 countries, begins a process of targeting sources of greenhouse gases for emission reductions. This book should help guide that discussion, clearly showing the issues relating to agricultural emissions. A fourth goal is to present a multidisciplinary view of global climate change and agriculture. The topics should appeal to a broad range of readers, including scientists, policy makers, economists, agronomists, climatologists, and lawyers. The last objective is to discuss why there are communication problems between scientists and policy makers on climate change research and policy and suggest possible ways to reduce this communication gap.

The book is organized into six sections. Section 1 is an introduction by the editors on the agricultural dimensions of global climate change. The purpose of this section is to summarize some of the main points and relevant issues addressed in the book. While this chapter is similar to an executive summary, one should turn to the individual chapters for substantially more detail. Also, the material in this chapter does not cover all chapters of the book.

Section 2 includes two chapters that examine the scientific evidence of global climate change. In the first chapter, Phil Jones presents scientific evidence of global warming from instrumental record data. Jones reports that the earth's surface temperature has increased by about 0.5°C since the middle of the nineteenth century. Jones indicates that while the warming has not been equal everywhere, most areas have warmed, and the 1980s are the warmest decade recorded. Jones notes that while it is still not possible to absolutely attribute the warming to the greenhouse effect, it is certainly likely that it is the case. In the other chapter of this section, Ben Santer discusses some of the problems associated with the use of coupled atmosphere-ocean general circulation models in studies which attempt to detect the greenhouse gas-induced climate change signal predicted by a model with observed records of surface temperature changes. Santer shows that the magnitude of model uncertainties affect both the predictions of how climate might change in response to changes in greenhouse gases, and the estimates of the decadal- to century-time scale natural variability (the "noise") of the climate system. Santer concludes that the only way to detect a climate change signal and attribute it convincingly to changes in greenhouse gases is by reducing the scientific uncertainties associated with the use of models to define the greenhouse gas signal and natural variability. He notes, however, that such uncertainty does not relieve policy makers of the responsibility of taking reasonable precautions.

Section 3 contains six chapters that address various aspects of how agriculture will be potentially affected by global climate change. In order to put the impacts of future climate on agriculture into perspective, the first chapter by Rene Gommes discusses current climate and population constraints for agriculture. Gommes reports that rainfall remains the dominant factor behind the large scale fluctuations in food supply. He notes that while technical advances have increased productivity in many areas, they have also led to farming systems which are less adaptable and more dependent on rainfall. Population pressures have made many agricultural systems more fragile, resulting in an increased variability in food production.

The second chapter in this section by Cynthia Rosenzweig and Martin Parry examines the potential impacts of climate change on world food supply. The authors look at several climate change and adaptation scenarios and predict that climate change will lead to a slight decrease in global food supply, larger disparities in grain production between developed and developing countries, and higher grain prices. While more major forms of adaptation (e.g., installation of irrigation) may offset some of the decrease in global food supply, the adaptation strategies do little to reduce the disparity between developing and developed countries. This is followed by a chapter by Pierre Crosson that examines the impacts of climate change based on the hot and dry climate of the 1930s on current agriculture in four Midwestern states: Missouri, Iowa, Nebraska, and Kansas (MINK). Crosson finds that crop yields and production would decrease from 15-20% under the occurrence of the 1930s hot and dry climate assuming farmers could not make management adjustments and there was no fertilizing effect of increased CO_2 in the atmosphere. However, most of this decline is eliminated assuming there is a CO_2 fertilizer effect and farmers could make management adjustments in response to the change in climate. Some contend that a CO_2 "fertilizer effect" will accompany global warming because elevated atmospheric CO_2 increases rates of net photosynthesis and it reduces stomatal openings, which results in increased water use efficiency by the plant and therefore may enhance crop yields.

The fourth chapter in this section by Harry Kaiser, Susan Riha, Dan Wilks, and Radha Sampath deals with farm-level adaptation issues in Minnesota and Nebraska. These authors investigate whether fairly minor farm management adaptation strategies (without a CO_2 fertilizer effect) are sufficient to keep Minnesota and Nebraska farmers' profits from falling under two climate change scenarios (a mild warmer and wetter and a hotter and drier scenario). The results suggest that farmers in Minnesota are actually better off due to climate change while Nebraska producers are worse off under the more severe climate change scenario. The effects of climate change on agriculture predicted from the studies in this section depend critically upon whether or not a CO_2 fertilizer effect is assumed. Dave Wolfe and Jon Erickson tackle the issue of whether CO_2 fertilization will benefit agriculture in the fifth chapter of this section. The authors contend that previous studies have overstated the benefits of the CO_2 fertilizer effect because they are based on short term laboratory experiments under optimal conditions with water and nutrients non-limiting, temperatures near optimum, and weed, diseases, and insect pressures nonexistent. Consequently, this may overstate the benefits from CO_2 enrichment, especially in developing countries where irrigation, fertilizer, herbicides, and pesticides are either not available, or are prohibitively expensive. In the last chapter of this section, Steve Kyle looks at the possible distributional effects of global warming and emission abatement. Kyle concludes that while the effects of global warming are highly uncertain, it is clear that countries which rely heavily on agricultural production and employment will suffer more, as will countries less able to invest in adaptation to change. The effects of greenhouse gas abatement programs are clearer, falling principally on energy-using countries and sectors.

Section 4 of the book deals explicitly with international climate change negotiations and includes two chapters. In the first chapter, Tom Drennen discusses the status of international negotiations to formulate a response to climate change since the 1992 Rio Earth Summit. He reviews the basic obligations for various countries under the Framework Convention on Climate Change. He then provides an analysis which indicates that while the Convention will have little impact on slowing projected temperature increases in the short term, it should have a larger impact in the long run. This optimism stems from the notion that the Convention starts a process, which will require countries to continue talking and providing data on emission levels, and that this flow of information will ultimately lead to more substantial commitments.

The second chapter in this section is by Henry Shue, who examines four different questions of justice regarding international global warming abatement agreements. The four questions are: (1) What is a fair allocation of the costs of preventing the global warming that is still avoidable?; (2) What is a fair allocation of the costs of coping with the social consequences of the global warming that will not in fact be avoided?; (3) What background allocation of wealth would allow international bargaining (about issues like (1) and (2)) to be a fair process?; and (4) What is a fair allocation of emissions of greenhouse gases over the long term and during the transition to the long term allocation?

The purpose of Section 5 is to address the issue of whether agriculture contributes significantly to climate change and how should agriculture be treated in the negotiations relative to other sectors. In the first of two chapters, John Duxbury and Arvin Mosier review the sources and sinks of greenhouse gases, with a particular emphasis on agricultural emissions. They conclude that agricultural emissions account for between a quarter and a third of all anthropogenic emissions, if one assumes that CO_2 released from deforestation is all agriculturally related. They provide examples of where these emissions could be reduced. Nevertheless, the authors contend that the greatest potential for emission reductions are from the energy sectors.

Owen Greene and Julian Salt examine the realistic potential for monitoring compliance with any agreements for reducing emissions from the agricultural sector. They discuss the range of uncertainty in emission estimates for the different agricultural practices and provide methods for verifying country level emissions. They conclude that the potential for verification is quite limited and that it would make more sense for negotiations to focus not on actual emissions, but on specific agricultural practices, such as the management practices for animal wastes.

The last section of the book deals with bridging the communications gap between scientists and policy makers. Susan Offutt acknowledges that a definite "gap" exists, but that the Intergovernmental Panel on Climate Change process has gone a long way towards closing this gap. She notes, however, that much climate change research still takes place in a vacuum, with little of the results of good research reaching the policy makers. Some of this is due to institutional barriers, some to unwillingness of researchers to get involved with the policy debate. She urges researchers to talk about their results and problems during all phases of their research projects, not just during the grant writing phase.

In the final chapter, J. Christopher Bernabo, Peter Eglinton, and Chester Cooper report on the results of the Joint Climate Project to Address Decision Makers' Uncertainties. They note that both decision makers and researchers are uncomfortable with the existing gap in communication and reaffirm Offutt's conclusion that policy makers want to meet more with researchers both during the course of research projects and after their conclusions are reached. They note that unfortunately, additional research in many areas will not resolve existing uncertainties, and may even add new levels of uncertainty. Decision makers need to work with this uncertainty rather than use it as an excuse for inaction.

One thing we learned from this experience is that editing a book is a mammoth undertaking and our reliance on several individuals for support was crucial for completing the project. Indeed, this book would not have been possible without the support of several people. We would like to thank John Duxbury, Director of the Agricultural Ecosystems Program at Cornell University, for his complete support of this project. John not only provided us with the financial support to develop the book, but also with intellectual insights into the organization of topics and selection of authors. We also owe a debt of gratitude to several individuals who also helped organize the topics and suggest authors. Susan Riha, Dan Wilks, Tim Mount, Steve Kyle, and Duane Chapman spent a lot of time with us from May to October of 1992 outlining this book. In addition, we thank the authors of the chapters that follow for not only writing interesting and important essays, but also for getting their manuscripts and re-drafts back to us on time.

Everyone who edits a book realizes how much work is required in terms of manuscript preparation, and technical and copy editing. We were fortunate to have three extremely talented people assist us in this regard. We owe our gratitude to Jonell Blakeley and Eleanor Smith for carrying out all the tasks involved in manuscript preparation. We are also indebted to Renee Serowski of St. Lucie Press for technical and copy editing the book. Jonell, Eleanor, and Renee did an outstanding job in preparing and laying out this book. Finally, we thank Dennis Buda, publisher of St. Lucie Press, for his enthusiastic support of the book and his efficiency in carrying this project to its completion.

Harry M. Kaiser

Thomas E. Drennen
Cornell University, Ithaca

1

INTRODUCTION

AGRICULTURAL DIMENSIONS of GLOBAL CLIMATE CHANGE

Thomas E. Drennen

Harry M. Kaiser
Department of Agricultural Economics
Cornell University

I. INTRODUCTION

This chapter explores key issues related to global climate change and agriculture. Much of its content draws from subsequent chapters in the book. Hence, it is somewhat of a summary of the highlights of the book. The chapter begins with a discussion of the concept of climate change, noting the causes and effects, the likely rate of change, and the key areas of remaining uncertainty. This is followed by an examination of the implications of these predicted climatic changes for agriculture at the farm, regional, and world levels. The adequacy of the international response to date is then discussed, focussing on the terms of the Framework Convention and obligations for the developed and developing worlds. Next, the importance of the agricultural sector itself as a major source of several of the greenhouse gases is discussed. Finally, we offer our conclusions and recommendations for confronting the threat of climate change.

II. NATURE OF THE PROBLEM

The atmospheric concentrations of greenhouse gases are increasing as a result of human activity. These gases, which include carbon dioxide (CO_2), methane (CH_4), nitrous oxides (N_2O), water vapor, ozone (O_3), and chlorofluorocarbons (CFCs), affect the global energy balance by partially absorbing outgoing infrared radiation, resulting in increased global surface temperatures. This is the basic mechanism commonly referred to as the greenhouse effect and is illustrated in Figure 1-1.

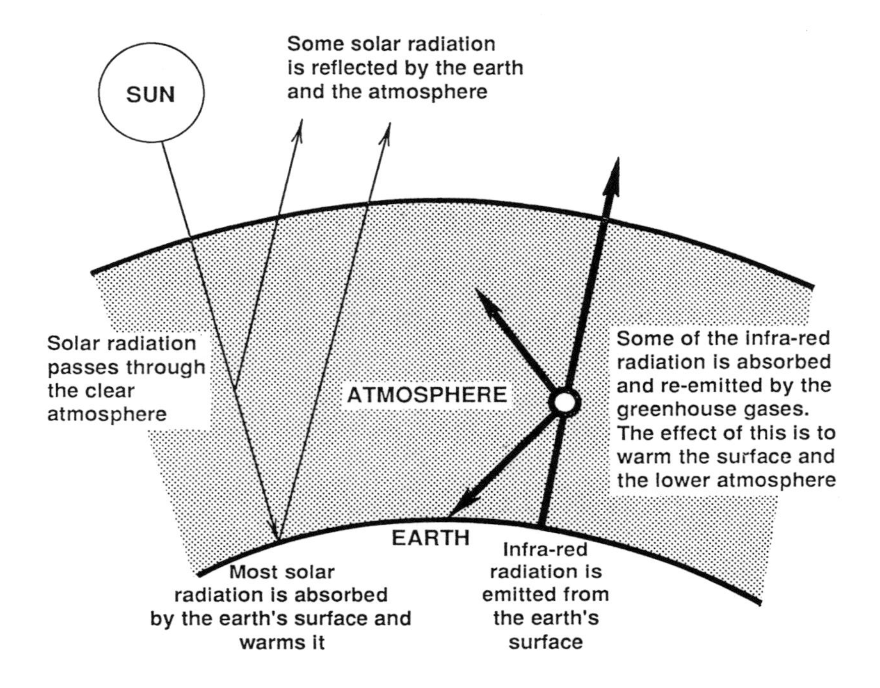

Figure 1-1. The greenhouse effect.
Source: IPCC, 1992.

There is no question whether the greenhouse effect is real. In the absence of carbon dioxide in the atmosphere, the planet would be cold and lifeless with average temperatures 33°C colder than they are today. The real concern is what will happen as concentrations of key greenhouse gases increase. Venus is the extreme example; with an atmosphere containing over 90% CO_2, its surface temperature is 477°C (Hoffert, 1992).

On earth, atmospheric concentrations of CO_2 began increasing with the onset of the Industrial Revolution from a level of 270 parts per million by volume (ppmv) in 1860 to a level of 355 ppmv in 1991 (Intergovernmental Panel on Climate Change (IPCC), 1992). For a doubling of CO_2 over pre-industrial levels (or its equivalence in terms of the other greenhouse gases), scientists predict a warming of 1.5–4.5°C (2.7–8.1°F) (IPCC, 1992). The timing of this change is dependent on many factors, some related to the role of the climate system itself, and others related to increases in global population and the consumption of fossil fuels. The IPCC (1992) predicts a likely rate of change of 0.3°C per decade.

Table 1-1. General circulation model predictions of changes in average global temperature and precipitation due to doubled atmospheric concentrations of CO_2.

Model	Change in surface air temperature (°C)	Change in precipitation (%)
Princeton University Geophysical Fluid Dynamics Laboratory	4.0	8.7
Goddard Institute of Space Studies	4.2	11.0
National Center for Atmospheric Research	3.5	7.1
Oregon State University	2.8	7.8
United Kingdom Meteorological Office	5.2	15.8
Average	**3.9**	**10.1**

Source: Karl et al. (taken from Rosenzweig 1989).

The primary tool for predicting the consequences of increasing levels of greenhouse gases are General Circulation Models (GCMs). GCMs are used to predict the complex response of the climate system as levels of CO_2 are doubled from pre-Industrial Revolution levels (see Chapter 3 for a more complete discussion of GCMs). Table 1-1 summarizes the predictions of five major GCMs in terms of changes in average global surface air temperature and precipitation levels. On average, these models predict a 3.9°C increase in surface temperatures and a 10% increase in global precipitation for a doubling of CO_2. While displaying general agreement as to global averages, there are many important differences in regional predictions. This is evident in Table 1-2, which compares the predictions of two of the GCMs for changes in air temperature and precipitation for eight regions of the U.S. For example, the GISS model predicts an *increase* in precipitation of 7% for the Northern Plains while the GFDL model predicts a *decrease* of 3.4%. Such differences, resulting from the underlying approaches used to model the various physical components of the climate system, frustrate policy makers who find recommending solutions in the face of such uncertainty a daunting task.

Table 1-2. GISS and GFDL model predictions of changes in average temperature and precipitation in the U.S. due to doubled atmospheric concentrations of CO_2.

	Change in annual surface air temperature (°C)		Change in precipitation (%)	
Region	GISS	GFDL	GISS	GFDL
Northwest	4.4	4.5	23.0	2.7
California	4.9	4.9	6.2	1.8
Northern Mountains	4.8	5.5	18.0	1.7
Southern Mountains	4.9	5.1	5.0	-1.4
Northern Plains	4.7	5.9	7.0	-3.4
Southern Plains	4.4	4.5	-7.8	-0.3
Delta	5.3	4.4	2.4	0.3
Southeast	3.5	4.9	10.5	-7.8
Average	**4.6**	**5.0**	**8.0**	**-0.8**

Source: Adams et al. (1988).

Formulating an effective response to climate change is further hindered by the inability of the scientific community to provide compelling evidence of an increasing global temperature. Detection of increasing global atmospheric temperatures is a difficult task, especially since the temperature change is not immediate; there is a lag from when CO_2 is added to the atmosphere to the date that full temperature equilibrium occurs, due largely to the huge thermal mass of the oceans. This expected temperature change is often referred to as a temperature change commitment, meaning that the future global temperature will reach this level as a result of emissions released to date. Estimates of the magnitude of these lags range from one to several decades. Jones (Chapter 2) concludes that the actual temperature increase since the mid 19th century is about 0.5°C. While this trend is consistent with projections of General Circulation Models (GCMs), it is not yet possible to attribute this increase to the greenhouse effect (Jones, Chapter 2; National Academy of Sciences, 1991).

In addition to increases in the global mean surface temperatures, other effects of climatic change will likely include: an increase in the number of extreme weather events; sea-level rise; and changes in soil temperatures and soil moisture levels. Some countries may well adapt to these changes, while for others the changes could cause massive hardship. Unfortunately, those countries with the fewest resources and who have contributed the least to the problem, are the ones who may suffer the most. Countries such as Egypt and Bangladesh will have difficulty escaping the damage of rising sea

Table 1-3. Summary of major greenhouse gases.

	CO_2	CH_4	CFC-11	CFC-12	N_2O
Atmospheric concentration	(ppmv)	(ppmv)	(pptv)	(pptv)	(ppbv)
Pre-industrial	280	0.8	0	0	288
Present (1990)	355	1.72	280	484	310
Current rate of change	1.8 (0.5%)	0.015 (0.9%)	9.5 (4%)	17 (4%)	0.8 (0.25%)
Atmospheric lifetime (yrs)	50-200	10	65	130	150
Radiative effectiveness					
Per molecule	1	21	12400	15800	206
Per unit mass	1	58	3970	5750	206

Source: IPCC (1990, 1992).

levels, while countries with more resources may find it possible to adapt, through relocation or the use of technology. And the higher temperatures combined with lower soil moisture will make it even more difficult to successfully grow crops in areas of Africa, where thousands already die annually from the ravages of drought. Rosenzweig and Parry (Chapter 5) predict that decreases in crop production in developing countries due to climate change could increase the number of people at risk from hunger from 60 to 350 million by 2060.

III. SOURCES OF GREENHOUSE GASES

While CO_2 receives the most attention as a factor in global warming, there are other gases to consider, including CH_4, N_2O, and CFCs. The past and current concentrations of the key greenhouse gases, rates of increase, and atmospheric lifetimes are summarized in Table 1-3.

A. Carbon Dioxide

Future emissions of CO_2 depend largely on the consumption of fossil fuels and rates of deforestation. Current emissions of carbon dioxide from fossil fuel combustion are about 6.0 ± 0.5 Gt C[1] per year (IPCC, 1992). Estimates of CO_2 released as a result of land use changes, mainly deforestation, are in the range of 1.6±1.0 Gt C per year (IPCC, 1992). Total net fluxes to the atmosphere from these two sources are between 6.1 and 9.1 Gt C. However, the atmospheric stock of CO_2 is increasing at just 3.2 to 3.6 Gt C (IPCC, 1990b), implying that the remainder (2.9–5.5 Gt C) is returned to the biosphere and oceans.

The basic mechanism for transferring CO_2 from the atmosphere to the oceans is the differences in partial pressures of the CO_2 in the surface waters and in the atmosphere. Mixing and circulating between the surface waters and the deeper oceans eventually move the CO_2 into deeper waters. There is vast literature devoted to the complex modelling of the oceans and their uptake,[2] but it is sufficient to state that the oceans act as a net sink for carbon.

The land biota is another possible sink for the excess carbon dioxide. Several possible mechanisms have been proposed, including: increased plant productivity due to higher ambient CO_2 levels (referred to as CO_2 fertilization); increased biosphere productivity due to warmer temperatures; increased plant productivity due to increased fertilizer use; and changing forest practices. The IPCC (1992) notes that actual CO_2 emissions from the tropics are less than would be expected from estimates of the extent of deforestation; they hypothesize that remaining forests have increased their carbon sequestration due to the CO_2 fertilizer effect. (This fertilizer effect is discussed more completely in the next section of this chapter.) This finding reinforces the importance of preserving tropical forests to limit climatic change.

B. Methane

The sources and sinks for methane (CH_4) are less well quantified than is the case for CO_2. The largest source is natural wetlands and bogs where methane is continuously formed through anaerobic decomposition of organic matter. Other sources include: rice paddies; enteric fermentation (the intestinal fermentation which occurs in animals such as cows); biomass burning; coal mining; drilling, venting, and transmission of natural gas; and termites. Of the anthropogenic sources, approximately two-thirds (205–245 Tg CH_4 per year) are attributed to agricultural sources (Duxbury and Mosier, Chapter 12). The actual annual increase in CH_4 is 32 Tg CH_4.

C. Nitrous Oxide

Atmospheric concentrations of nitrous oxide are increasing at 3–4.5 Tg N per year. The main sources of these emissions now appear to be the cultivation of soils and denitrification of fertilizers, although vast uncertainty regarding the sources and sinks of N_2O exists (Duxbury and Mosier, Chapter 12; IPCC, 1992). This uncertainty, which currently leaves a missing sink of 1.4–4.8 Tg N unaccounted for (IPCC, 1992), is unlikely to be resolved anytime soon and will make verification of any actual changes in emission levels most difficult.

D. Halogenated Species

Another key group of greenhouse gases include those termed the halogenated species and which include CFCs, hydrochlorofluorocarbons (HCFCs), halon, methyl chloroform, and carbon tetrachloride. These gases are previously regulated under international agreement (Montreal Protocol, 1987) because of their ability to damage the stratospheric ozone layer. Consumption of CFCs 11, 12, and 113 have decreased by 40% from 1986 levels (IPCC, 1992). At a recent meeting of the Parties to the Montreal Protocol, the deadline for the complete phase out of most of these chemicals was advanced from 2000 to 1995 (IER, Dec. 2, 1992). Unfortunately, many of the proposed substitutes for these chemicals are also greenhouse gases. Preliminary reports indicate that certain replacements, such as fluorinated gases like CF_4 and C_2F_6, could have GWPs[3] at least as large as CFC-11 and CFC-12, and possibly significantly higher (Ravishankara et al., 1993).

IV. AGRICULTURAL IMPACT OF GLOBAL WARMING

The unprecedented levels of climatic change predicted by some will have tremendous implications for climate-sensitive systems such as forestry, other natural resources, and agriculture. With respect to agriculture, changes in temperature, precipitation, and solar radiation will have an effect on the productivity of crop and livestock agriculture. Climate change will also have economic effects on agriculture, including changes in farm profitability, prices, supply, demand, trade, and regional comparative advantage. This section examines recent predictions for how global warming may impact world and U.S. agriculture.

A. World-Wide Agronomic Effects

The agronomic impacts of global climate change will depend upon how temperature, precipitation, and solar radiation change over time. Crop yields are critically influenced by these climate factors, as well as by the type of soil and the type of plant that is being grown. Studies that have examined how global climate change may affect agricultural productivity of various regions have generally combined predictions of climate models with crop simulation models to predict yield effects due to a changing climate.

Various regions of the world will be impacted differently under climate change, with some countries gaining in agricultural productivity and others losing. Some current predictions are that countries located in northern latitudes will experience increases in precipitation in addition to increases in temperature. Consequently, this will have the effect of enhancing crop yields in the northern regions of the former Soviet Union, Canada, and Europe

(Kettunen et al., 1988; IPCC, 1990; Rosenzweig and Parry, 1993). The predicted yield increases in these higher latitude regions are due primarily to a lengthening of the growing season and the mitigation of negative cold weather effects (e.g., frost kill) on plant growth (Rosenzweig and Parry, 1993). Other studies predict that an increase in precipitation will also occur in the southern middle latitude countries (e.g., parts of Latin America, Northern Africa, and middle India), which should also enhance crop yields (Pittock, 1989; Pittock and Nix, 1986; Walker et al., 1989; IPCC, 1990). On the other hand, climate change is expected to have negative effects on crop and livestock productivity in northern middle latitude countries like the U.S., Western Europe, and most of Canada's currently productive agricultural regions (Williams et al., 1988; Smit, 1989; U.S. Environmental Protection Agency, 1990; Santer, 1985; IPCC, 1990). This predicted result is due to a shortening of the growing period for the plant caused by an increased temperature;[4] a decrease in water availability for the plant caused by a combination of increased evapotranspiration rates, losses in soil moisture, and (in some cases) decreases in precipitation; and lower vernalization.[5] Two exceptions to these general findings are China and Japan, which may benefit from climate change in terms of enhanced agricultural productivity because much of their agriculture is located near coasts. Consequently, these two countries are not predicted to experience the interior continental drying that is predicted by GCMs for many countries in the northern middle latitudes (Tobey, Reilly, and Kane, 1992). It is important to note that these general results are subject to a high degree of uncertainty due to the current large degree of inaccuracy of climate models, as well as crop simulation models.

The predicted effect of climate change on crop yields depends critically upon whether or not the so called CO_2 "fertilizer effect" is assumed. The CO_2 fertilizer effect refers to an enhancement in crop yields due to elevated atmospheric CO_2, which increases rates of net photosynthesis and reduces stomatal openings, resulting in increased water use efficiency by the plant. For example, in Chapter 5 of this book, Rosenzweig and Parry (1993) report a world average decrease in yields for wheat, rice, maize, and soybeans for three climate scenarios they considered when a CO_2 fertilizer effect is not assumed. World average wheat yields decline by as much as 33%, rice yields by as much as 25%, maize yields by as much as 31%, and soybean yields by as much as 57% under climate change. On the other hand, when a CO_2 fertilizer effect is assumed, world average crop yields do not decline as much, or actually increase in several cases, under the climate change scenarios. In this situation, the impact of three climate change scenarios on world average wheat, rice, maize, and soybean yields range from -13% to 11% (wheat), -5% to -2% (rice), -24% to -15% (maize), and -33% to 16% (soybeans). Other studies (e.g., Crosson, 1993; Adams et al., 1990) produce the same conclusion: that the assumption about CO_2 fertilization has a major influence on the impact of climate change on crop yields. In fact, the general result from these studies is that crop yields decrease under climate warming when there is no CO_2 fertilization, but that these decreases can be largely offset by assuming a positive CO_2 fertilization effect.

Recently, some have questioned the widely held belief that there will be a large CO_2 fertilizer effect accompanying climate change. For instance, Wolfe and Erickson (1993), in Chapter 8 of this book, contend that many studies overstate the benefits of CO_2 fertilization. These authors argue that the magnitude of the CO_2 fertilizer effect used in past studies is based on short term laboratory experiments under optimal conditions with water and nutrients non-limiting, temperatures near optimum, and weed, diseases, and insect pressures nonexistent. Consequently, this may overstate the benefits from CO_2 fertilization, especially in developing countries where irrigation, fertilizer, herbicides, and pesticides are either not available, or are prohibitively expensive. Wolfe and Erickson (1993) also argue that current models do not account for the interactions between CO_2 and other environmental factors. For example, there may be no benefit, or even a negative effect, from increased atmospheric CO_2 at low temperatures (less than 15°C). Also, some studies have indicated that growth stimulation from high CO_2 does not continue with prolonged exposure. We agree with Wolfe's and Erickson's (1993) conclusion that the use of experimental results with elevated CO_2 from controlled experiments overstates the likely beneficial effects of increased CO_2 concentrations. Since there is debate over the magnitude of this potentially beneficial effect, we recommend that research on the magnitude of the CO_2 fertilizer effect be one of the top priorities for future work on agricultural impacts of climate change. Furthermore, we also recommend that until this uncertainty is resolved, further impact analyses should be conducted with and without the CO_2 fertilizer effect in the crop simulation models so that the full range of possibilities are presented.

B. World-Wide Economic Effects

The impact of climate change on crop yields is only one part of the picture of how a changing climate will impact the welfare of countries. In order to get a complete picture, one has to consider the associated economic effects of price, production, and trade changes due to climate change. The effects of climate change on supply, demand, prices, and trade flows have been simulated in two studies by using predicted crop yield effects with world trade models. The results of each study are briefly summarized below.

In the first study, Tobey et al. (1992) use a partial equilibrium world trade model called SWOPSIM (Static World Policy Simulation) to simulate the economic impacts of three sets of scenarios corresponding to different assumed crop yield changes due to climate change. Rather than using specific predictions of climate change-induced yield effects, this study examines a wide range of possible yield changes to determine how sensitive the world market is to crop yield levels. One of the most interesting findings of this study is that the impact of even the most severe yield reduction scenarios on national economies of the world is fairly modest. For example, even under crop yield reductions of 50% for the U.S., Canada, and the European Community, the decrease in world economic welfare is only 3% assuming that there is no change in crop yields in the rest of the world.[6] The authors also find that world economic welfare may actually increase if the decrease in agricultural productivity of the U.S., Canada, and Europe is offset by increases in productivity in other areas such as the former Soviet Union and China. These results, which indicate very modest impacts on economic welfare even under severe yield reductions in the U.S., Canada, and the European Community, are partially due to the predicted change in world agricultural prices found in this study. That is, world prices are actually predicted to decrease in one scenario, remain relatively constant in another, and increase only modestly in the most extreme scenario. A key conclusion of their report, therefore, is that the welfare effects of climate change on individual countries will not only depend upon changes in domestic yields, but also on changes in world prices, and the country's relative strength as an exporter and importer. When even large negative yield effects are considered, there will be interregional adjustments in the world market that will cushion the impacts of climate change on economic welfare.

In the second and more recent study, Rosenzweig and Parry (1993) combine their predicted crop yields with a computable general equilibrium world trade model to simulate potential economic effects of climate change. In terms of production, the authors conclude that climate change will reduce world grain production by up to 20% assuming no CO_2 fertilization effect and that no adaptation will occur. With CO_2 fertilization and modest levels of adaptation (e.g., shifts in planting dates and changes in crop cultivars), global production is predicted to decrease by 5%. However, these global reductions in production may be largely offset if one also assumes the possibility of more major adaptation strategies, such as installing irrigation. The authors also find that climate change will increase the disparities in grain production between developing and developed countries. Their results indicate that developed countries may actually benefit from climate change in terms of production, while production in many developing countries is predicted to decrease. Furthermore, the results show that adaptation strategies do little to reduce these disparate effects. This is due to the fact that developed countries have many more resources to utilize in adaptation strategies than do developing countries. Because Rosenzweig and Parry's (1993) results indicate global production decreases under most scenarios, their economic model predicts that commodity prices will rise. Moreover, the predicted price increases are substantially higher than those predicted by Tobey et al. (1992). The combination of production declines in developing countries and increases in prices due to climate change would increase the number of people at risk of hunger. This result holds even at high levels of assumed adaptation strategies with CO_2 fertilization.

While there are some similarities, there are also some key differences between these two studies, which is not surprising given the level of uncertainty in climate change research. The first study suggests that climate change will not cause a major disruption in world agriculture markets. The second study indicates that climate change may lead to a more inequitable distribution of grain production and cause prices to rise significantly. This divergence in results is due to many factors such as the use of different economic models, different climate change scenarios, different yield effects, and different assumptions. Nevertheless, the lack of consensus between results illustrates the imprecision in modeling efforts to look at potential impacts of climate change.

C. Effects of Climate Change on U.S. Agriculture

Several studies have looked at the potential effects of climate change on U.S. agriculture. Adams (1989) and Adams et al. (1990) conducted a comparative static analysis of the potential effects of a double CO_2 induced change in climate on the regional comparative advantage of U.S. agriculture. Similar to the Rosenzweig and Parry study (1993), Adams et al. (1990) generated separate yield impacts due to climate change with and without CO_2 fertilization. The authors found that there were tremendous differences in predicted yield impacts between the two climate scenarios they investigated,[7] as well as between the CO_2 scenarios. Not surprisingly, crops that were irrigated did not have as severe yield reductions as rainfed crops. Under the CO_2 fertilizer effect scenario, the negative impacts of climate change on crop yields were substantially mitigated and, in some cases, yields actually increased. In terms of regional distribution of yield impacts, regions in the northern U.S. were generally less severely affected while regions in the south suffered the largest yield losses due to climate change. These results are comparable to a study of Midwestern crop agriculture by Kaiser et al. (1993) (see Chapter 7), who found that the cooler northern locations will not have as large a negative yield effects as the southern parts of the midwest.

Adams et al. (1990) combined the yield effects with a sector level mathematical programming model of U.S. agriculture to simulate climate change impacts on prices, regional and national production, consumption, consumer and producer surplus, and other market variables (Cheng and McCarl, 1989). As was true for the yield results, the economic results varied tremendously in magnitude and even direction among the various scenarios.

For example, in the two scenarios without the CO_2 fertilizer effect, the predicted climate change by the less severe climate scenario resulted in a decrease in U.S. crop production of 10%, and an 18% increase in prices; while the more severe climate change scenario resulted in a 39% decrease in production, and a 109% increase in prices. In both cases, society is worse off in terms of economic welfare (change of -$6.5 billion and -$35.9 billion for the two climate change scenarios, respectively). If the potentially positive CO_2 fertilizer effect is included, the predicted changes from both climate models are less severe. In this case, the predicted climate change of the less severe climate change scenario actually results in a 10% increase in crop production and an 18% decrease in prices. In contrast, the more severe climate change scenario results in a 19% decrease in production, and a 28% increase in crop prices. The change in economic surplus associated with the two climate change scenarios in this case is +$9.9 and -$10.5 billion, respectively. The results of this study again point out the importance of resolving the uncertainty over the magnitude of the CO_2 fertilizer effect in climate change impact assessment.

D. Agricultural Adaptation to Climate Change

As has already been alluded to in this section, if there is a change in climate, agriculture will adapt in order to minimize the negative effects of the climatic change. The question then becomes: How successful will agriculture be in adapting strategies to eliminate these negative effects? Agricultural adaptation to a changing climate will occur in several forms, including technical innovations, changes in agricultural land areas, and changes in use of irrigation. Technological innovations include the development of new plant cultivars (varieties) that may be bred to better match the changing climate. Changes in agricultural land areas in response to both changes in demand for agricultural products due to increasing population, as well as changes in supply due to climate change, will be another form of adaptation. Finally, as the climate warms there will likely be a shift towards greater use of irrigation systems to grow crops.

Several of the previously discussed studies have included various adaptation strategies in their simulations to determine how effective such actions will be in mitigating negative effects. Kaiser et al. (1993) considered three types of relatively minor farm-level adaptations to climate change. The first is the possibility of switching cultivars for a particular crop. A second is to alter crop mix in response to a change in climate. If climate change affects the relative yield and profitability of one crop in favor of another, then farmers should respond by making the appropriate change in crop mix. Finally, a third is to make adjustments in scheduling of field operations in response to climate change. The findings of this study indicate that farmers in cooler locations like

Minnesota could successfully adapt, and in fact benefit, from climate warming. On the other hand, producers in warmer locations like Nebraska could be worse off, even using adaptation strategies, under the more severe climate change scenarios. One of the more interesting results of this study is the comparison of crop yields between when adaptation is allowed (adaptation case) and when it is assumed that farmers cannot adapt by changing cultivars, crop selection, and timing of planting and harvesting (no adaptation case). When it is assumed that farmers cannot use adaptive farm management strategies over time, simulated crop yields are drastically lower than in the adaptation case for all three climate warming scenarios considered. For instance, in considering even the mild climate change scenario, the difference in corn yields between the adaptation and no adaptation cases reaches 40 bushels per acre (about one-third of average yields in 1980) by the year 2060.

In another study of the Midwestern U.S., Crosson (1993; see Chapter 6) examines the impacts of the hot and dry climate of the 1930s on current agriculture under a variety of adaptation scenarios. Crosson finds that crop yields and production would decrease from 15–20% under the occurrence of a hot and dry climate (based on 1930s weather patterns) assuming farmers could not make management adjustments and there was no CO_2 fertilizer effect. However, the assumption of farm-level adaptation to this climate eliminates over 50% of the production losses. Hence, it appears that agricultural adaptation may greatly reduce the predicted negative effects of climate change, and, in some cases, lead to positive economic effects for agriculture in some locations.

Rosenzweig and Parry (1993; see Chapter 5) also simulate yield effects under two levels of adaptation to climate change. Level 1 adaptation strategies include shifts in crop planting by one month, additional watering of crops that are already under irrigation, and switches in plant cultivars in response to the new climate. Level 2 adaptation strategies include shifts in crop planting dates by more than one month, increased fertilization, and installation of irrigation systems. The authors note that while some adaptation will take place, farmers, particularly in developing countries, will not be able to use all of these adaptation strategies (e.g., install irrigation, increase fertilizer amounts). Under Level 1 adaptation, negative climate change yield effects are somewhat mitigated in developed countries, but still tend to be negative in the developing countries. In the case of Level 2 adaptation, the negative yield effects are almost all mitigated for two of the three climate change scenarios, but not for the most extreme scenario. For the latter climate change scenario, neither Level 1 nor 2 adaptation can fully overcome the negative yield effects due to climate change, even under the CO_2 fertilization scenario.

V. THE INTERNATIONAL RESPONSE

Global recognition of the likely adverse effects of climate change led to the commencement of international negotiations in February 1991. The process culminated 16 months later with the signing of the Framework Convention on Climate Change by 153 countries at the 1992 United Nations Conference on the Environment and Development (UNCED).

While the ultimate objective of the Convention (Art. 2) is the "stabilization of greenhouse gas concentrations in the atmosphere," the actual commitments (Art. 4) oblige developed countries to attempt to limit their emissions of key greenhouse gases to 1990 levels by the year 2000. This is not a binding commitment. The Clinton Administration recently announced its intention of firmly committing the U.S. to the goal of reducing emissions of all greenhouse gases to 1990 levels by the year 2000 (White House Press Release), a goal which many other countries had expressed a willingness to accept, but which the U.S. had avoided under the previous administration. The U.S. plan to accomplish this task will be presented in August, 1993.

Unfortunately, limiting emissions to 1990 levels will not stabilize atmospheric concentrations, and will have little impact on slowing the onset of future warming. Drennen (Chapter 10) shows that a commitment by all industrialized countries to return emissions of fossil-fuel related CO_2 and CH_4 would have a minimal impact on reducing future temperature change commitments (Figure 1-2). This analysis assumes that countries will either be unwilling, or unlikely, to agree to limits on non-energy related emissions, due to uncertainties regarding the magnitudes and locations of sources and sinks, the selection of an appropriate weighting scheme, uncertainties regarding the costs of taking action, and concerns about equity. The primary reason for this limited effect of a freeze of fossil-fuel related emissions are projections of increasing aggregate emissions from the developing world, including China, due to increasing populations and economic development.

Further reducing the onset of climate change will be an arduous task. The IPCC (1990b) estimates that stabilization of atmospheric concentrations of greenhouse gases would require reductions of 60–80% for CO_2, 15–20% for CH_4, and 70–80% for N_2O.

The initial lack of strong commitments in the Framework Convention led to open criticism of the agreement. Yet, it is possible to argue that the Framework Convention is an important first step in developing an international strategy. Aman (1993) refers to the two phases of negotiation. The most important component of Phase I is to establish widespread dialogue, involving as many players as possible, to design an agreement that is seriously "committed to renewal, re-evaluation and reconsideration, and one that includes the key conflicting perspective that must be resolved if a meaningful

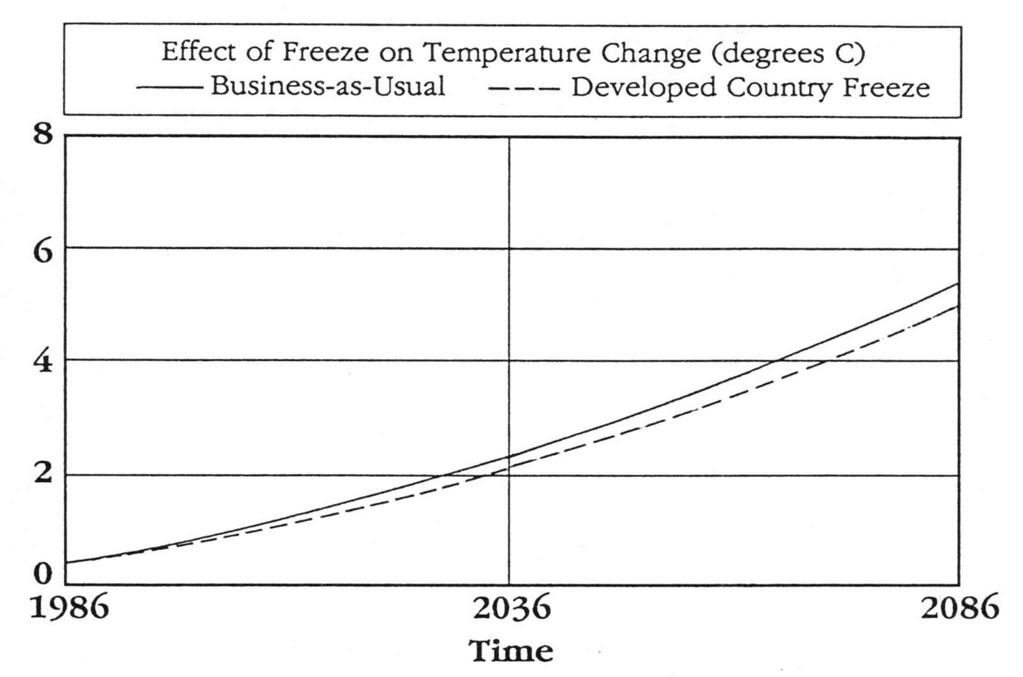

Figure 1-2. Projected temperature change for two scenarios: 1) business-as-usual and 2) industrialized country emissions freeze of industrial-related greenhouse gases.

Phase II is to occur" (Aman, 1993). Looked at in this light, Aman likens the signing of the Framework Convention to the ending of Phase I of negotiations, a phase that "is a necessary, though not sufficient, step towards a meaningful, substantive agreement."

The Framework Convention is a flexible document, designed to keep negotiations moving. While the Convention does not take effect until ratified by 50 countries, negotiations are continuing on such issues as clarifying financial mechanisms, and establishing methodologies for countries to use in preparing national inventories of sources and sinks. After the Framework Convention officially enters into force, a Conference of the Parties (CoP) becomes the mechanism for continuing negotiations (Art. 7). The Convention requires annual reviews by the CoP of implementation efforts and its overall effectiveness at meeting the objective of the Convention. Most importantly, the CoP has the authority to adopt amendments and protocols as they deem necessary to further the goals of the Convention (Art. 15).[8] Given that adherence to the commitments of the Convention as drafted will have but a minimal impact on limiting temperature increases, it is likely that the CoP will spend considerable time attempting to strengthen the agreement. It is during this Phase II of negotiations that serious commitments are likely to occur.

One of the first issues that the CoP will have to resolve is the actual scope of the agreement. While several countries pushed for specific timetables for reducing individual gases, others, notably the U.S., initially rejected the notion of firm commitments for individual gases, arguing instead that any approach to climate change must be comprehensive in scope, covering all of the key greenhouse gases. Under such an approach, which would have to include a weighting mechanism for the gases, individual countries could choose which gases to target for emissions reductions. Theoretically, under this approach, it might be feasible to achieve a reduction in a country's total greenhouse gas contribution while actually increasing emissions of CO_2 (Chapter 10).

The Framework Convention did not resolve this ideological difference, requiring only that countries aim to return emissions to 1990 levels of "anthropogenic emissions of CO_2 and other greenhouse gases not controlled by the Montreal Protocol." And while it leaves open the possibility of adopting weights for the various gases (Art. 4.2.c), it does not rule out future protocols or amendments dedicated to individual gases.

VI. THE ROLE OF AGRICULTURE

As currently drafted, the Framework Convention's impact on emissions from the agricultural sector is unclear. In fulfilling their obligations under the Convention, some countries may decide to reduce agricultural emissions, while others may find it easier to reduce fossil fuel related emissions. Verification of agricultural emission reduction will be more complex and uncertain than those from fossil fuels (See Greene and Salt, Chapter 13). But, if further negotiations lead to broad acceptance of the comprehensive approach (described in section V), the agricultural sector would be more directly impacted.

The fundamental question is: How important a contributor to climate change is agriculture? An analysis by Duxbury and Mosier (Chapter 12) concludes that agriculture contributes between 26–31% of the total of anthropogenic contributions to climate change (Table 1-4). Of the remainder, energy contributes between 61–69%, while the other anthropogenic sources (landfills (CH_4), nylon production (N_2O), etc.) contribute between 5–8%. This analysis is in line with the estimates of Edmonds et al. (1992), but significantly higher than the estimates of others (U.S. EPA, 1989; Office of Technology Assessment, 1991). These differences are largely due to differing definitions of what exactly constitutes an agricultural source.

Table 1-4. Comparison of contributions of agriculture (Ag) and other (O) anthropogenic sources to global warming using the IPCC 20- and 100-year GWPs.

Gas	Anthropogenic emissions[a]			Share of GWP[b]			
				20 yr		100 yr	
	Total	Ag	O	Ag	O	Ag	O
	Tg	---------% of total anthropogenic------------------					
CO_2	28600	23	77	15	52	19	65
CH_4	360	46	54	14	16	5	6
N_2O	5	61	39	2	1	2	2
			Total	**31**	**69**	**26**	**73**

a From Watson et al. (1992)
b Using direct GWP only
Source: Adapted from Duxbury and Mosier, Chapter 12.

However, while the analysis by Duxbury and Mosier suggests an agricultural contribution of somewhere between one quarter and one third of total anthropogenic sources, this estimate probably overstates the importance of agricultural emissions for several reasons. First, this analysis includes CO_2 emissions from deforestation, currently estimated at approximately 20% of the total CO_2 emissions from all sources (IPCC, 1992). Clearly, some percentage of this annual amount is for purposes other than agriculture. Removing deforestation from the agricultural sector entirely, also an unfair assumption, reduces the contribution of agriculture to between 15–18% of the total.

Second, these source estimates do not differentiate between agricultural and industrial emissions in terms of the net emissions of these gases. Major sources of CH_4 include such diverse sources as leakages of natural gas, ruminants, and rice paddies. And yet their unit mass contribution to climate change may be very different. While a fossil fuel source results in the release of new carbon (in the form of CH_4) to the system, this is not so clear for rice and ruminants. What are now rice paddies, may have previously been wetlands or swamps, which are also major sources of CH_4 (IPCC, 1992b). In such cases, the cultivation of rice on that land may not increase the net flux of CH_4 to the atmosphere. And ruminants require careful evaluation also. During the feed growing process, CO_2 is removed from the atmosphere; and while most is returned as CO_2, some is recycled in the form of CH_4 (Drennen and Chapman, 1992). Ignoring this factor results in an overemphasis on the role of ruminant CH_4 by as much as 40% (Chapter 10).

These arguments are not meant to discount the importance of emissions from agriculture. But they do underscore the degree of uncertainty remaining in quantifying any individual activity's contribution to climate change and the complexity of issues confronting policy makers.

The agricultural sector will not, nor should it, avoid scrutiny in the search for ways to minimize greenhouse gas emissions. Duxbury and Mosier (Chapter 12) propose useful guidelines for determining whether individual agricultural emissions should be reduced, including:

- Does the activity contribute significantly to changing the climate?
- How essential is the activity?
- What is the cost-benefit of the proposed mitigation practice?

With these guidelines in mind, individual goals for minimizing emissions of carbon dioxide, methane, and nitrous oxide are discussed below.

A. Carbon Dioxide

The goal for CO_2 reductions from the agricultural sector should be to make sure that land is used as efficiently as possible, thereby maximizing the preservation of non-cultivated areas, such as forests. This goal is particularly important in light of predictions which suggest that the area of agricultural land could increase by 60% by 2025 (IPCC, 1992b). Such an increase could result in releases of soil carbon on the magnitude of 5–10 years of current fossil fuel emissions (IPCC, 1992b). Further, on cultivated lands, practices should aim to limit soil erosion, through minimization of tillage practices, and other conservation techniques. Another major goal should be to slow the current rate of deforestation and forest degradation.

B. Methane

Realistic options include the improvement of ruminant nutrition levels through feed additives. Waste management practices should strive to ensure aerobic conditions at all times to minimize potential CH_4 releases. Such practices include minimizing the lagooning of waste (unless a CH_4 collection system is present), prompt application of wastes as fertilizer on fields, and the use of wastes as a fuel source (common practice in many developing countries). Given the importance of rice in many cultures, it is unrealistic to expect countries such as China to reduce or curtail rice production. Instead, research needs to continue on minimizing CH_4 emissions from flooded paddies, possibly through the selection and development of cultivars and changes in the nutrient and water management practices.

C. Nitrous Oxide

Reducing N_2O emissions from the agricultural sector requires improved nitrogen management, with the goal of leaving as little residual N as possible in the soil during non-cropped periods of the year. Duxbury and Mosier (Chapter 12) recommend several practices for accomplishing this goal, which can be summarized as paying closer attention to the use and management of fertilizer techniques.

D. Agricultural versus Fossil-fuel Related Emissions

As discussed above, options do exist for reducing emissions from the agricultural sector. The actual atmospheric increase in CH_4 is 32 Tg per year; stabilization at current levels would require a reduction in annual emissions of 40–60 Tg (IPCC, 1992). Reducing emissions from agricultural sources by approximately 20% could achieve this goal. However, it may be easier, both administratively and financially, to focus first on non-agricultural sources. Large non-agricultural sources of CH_4 include emissions from coal mining (27–47 Tg per year), from natural gas systems (25–42 Tg per year), and from the oil industry (5–30 Tg per year) (IPCC, 1992). A background report for the IPCC (U.S./Japan Working Group on Methane, 1992) estimates potential reductions of 30–90% for coal mining and 20–80% for oil and natural gas systems. Achieving a 50% reduction from each of these sources would reduce total annual emissions by 30–50 Tg, enough to stabilize emissions. For coal mines, achieving this goal would require pre-mining degasification, and utilization of the CH_4 from ventilation air (U.S./Japan, 1992); most of the technology for accomplishing this has been demonstrated. For the oil and natural gas systems, achieving these emission reductions would require reduced venting and flaring during oil production and the detection and repair of leaky natural gas pipelines. The technologies to achieve these goals are also available (U.S./Japan, 1992).

Fairness is yet another reason for targeting non-agricultural sources (Shue, Chapter 11). Simply stated, the developed world's use of fossil fuels over time is the primary cause of global warming. Duxbury and Mosier (Chapter 12) calculated the relative importance of the emissions from the agricultural and energy sectors for three different countries, the U.S., India, and Brazil. Figure 1-3 shows their results for both a 20 and 100 year GWP. For the U.S., agricultural emissions are but a small component (5%) of the total contributions. This share is much larger for India (40–70%) and Brazil (70–80%), yet their average *per capita* emissions are extremely low compared to the U.S. From a standpoint of fairness, it is the contribution of the U.S. (and other industrialized countries) that should be reduced. This does not imply that developing countries have a right to unabated increases in emissions, but that

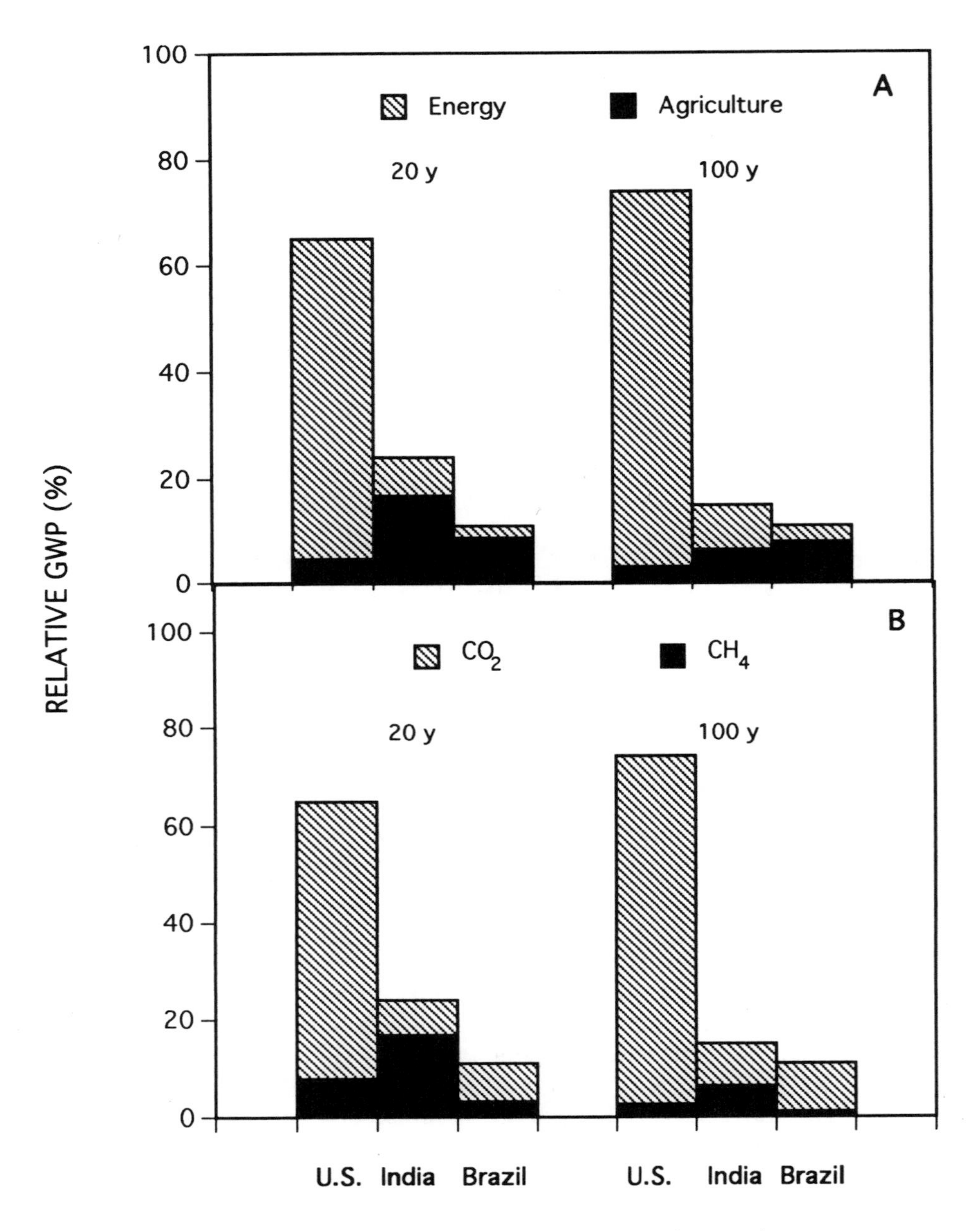

Figure 1-3. Global warming potentials created by (a) agriculture and energy sectors and (b) carbon dioxide and methane in the U.S., India, and Brazil.

the developed world must take the lead in reducing emissions. And it is quite likely that strong commitments by the developed world will lead to an explosion in new technologies that minimize greenhouse gas emissions; eventually, these technologies will diffuse to all countries of the world.

E. Verification and Compliance

Perhaps the thorniest problem with attempting to reduce agricultural emissions is verification of claimed emission reductions. This is due to the high level of uncertainty associated with current emission levels at the global level (see Chapter 13 for more complete discussion of this issue). A country could claim that ruminant emissions had been reduced by a certain percentage; verifying this claim would be difficult. But for ruminants, existing estimates of global emissions are based on estimates for just a handful of types of ruminants. Determining actual ruminant emissions from an individual country would require a detailed inventory of the size, quantity, and type of animals, along with detailed reports of climate variables and feeding patterns. Confirming a country's compliance with any pledge to reduce agricultural emissions could become a shell game.

Greene and Salt (Chapter 13) propose a reasonable solution to this problem. They suggest that, instead of attempting to negotiate commitments that include the option of specifically reducing agricultural emissions, future agreements should focus on agricultural management practices. Countries could agree on specific animal waste management practices which minimize CH_4 emissions, such as aerobic spreading of the wastes on fields or the installation of CH_4 collection systems on waste lagoons. The adoption of these management practices could then be directly monitored and would not require verifying actual emissions or detailed baseline estimates. This approach would be more acceptable to individual Parties concerned with invasive monitoring techniques and in general would be much simpler to negotiate. Further, this approach would have the additional benefit of promoting sound agricultural practices, a goal which few countries would find unacceptable.

VII. CONCLUSIONS AND RECOMMENDATIONS

In 1957, Revelle and Suess, referring to the likely impact of increasing CO_2 emissions on the future climate, noted that "human beings were now carrying out a large-scale geophysical experiment of a kind that could not have happened in the past, nor be repeated in the future." Thirty-six years later, this experiment continues. Jones (Chapter 2) reports that the earth has warmed 0.5°C since the mid 19th century. He is reluctant, however, to ascribe this increase to global warming. In the face of such continued uncertainty, policy makers find it difficult to draft legislation that would minimize the risks associated with climate change. This is especially true in this period of economic sluggishness and with pressures to solve more immediate problems.

The international community took the first step towards confronting climate change with the signing of the Framework Convention on Climate Change. But this first step is not enough; every day that passes will make effective action more difficult as the world pursues a development path based on the use of fossil fuels.

For agriculture, the risks of inaction are considerable. Increasing temperatures, shifting precipitation patterns, and increasing variability of severe climate events will challenge farmers. While adaptation strategies may minimize the impact for many, others, particularly those in the developing world, may not be able to adapt.

Policy makers cannot use uncertainty as an excuse for inaction; science will not be able to resolve much of the remaining uncertainty in the near term. Increased funding will help to resolve uncertainty, but it will not eliminate it. Decisions must be made now based on existing information. This action should be considered as an insurance policy against the threat of catastrophic damage. For the U.S., existing studies indicate that emissions could be reduced by 10–40% at zero cost or even at a net benefit to society (Cline, 1992; NAS, 1991; Rubin et al., 1992). These are actions that should be taken in any case.

What should be the international response to climate change? Initial emission reductions should come from the energy sector in the industrialized world. This makes sense from a standpoint of ability to pay, historical responsibility, the magnitude of opportunities available for achieving emission reductions in the energy sector, and the idea that many of the agricultural emissions are associated with subsistence level farming, providing a basic necessity. Further, these reductions must go beyond a return to 1990 emission levels, as currently suggested by the Framework Convention.

The agricultural sector can also make some contributions to emission reductions. However, rather than having a policy of allowing a country to meet its obligations through emission reductions in the agricultural sector, it might make more sense to target agricultural management practices as an additional step to fossil fuel derived emission reductions. This would involve setting standards for such practices as the handling of agricultural wastes or fertilizer use on rice paddies. This would significantly simplify the verification procedures and lead to sound agricultural practices. These are also actions that could be agreed upon in the short term.

Developing countries must also take responsibility for minimizing their growth in emissions. While it is clear that developed countries are responsible for the historical burden of greenhouse gas emissions, the sheer numbers of people living in the developing world make their cooperation imperative. A reasonable first step would be to eliminate subsidies for electricity and fossil fuel use (widespread in the developing world) (see Figure 3-2, World Bank, 1992), as artificially low prices result in inefficient use of these resources. Further actions will clearly require financial and technical assistance from the developed world. Such assistance should aim at improving the efficiency of existing fossil fuel uses, promoting renewable energy systems, minimizing deforestation, and controlling population growth.

There are several areas which do require additional research. First is improvement in the climate models, including improving resolution of the models (smaller grid sizes), making refinements in the coupling of ocean and atmospheric modules, and reducing problems relating to signal and noise uncertainties. Second, additional research is necessary to resolve the uncertainty associated with the CO_2 fertilizer effect. The studies in this book show how important assumptions about CO_2 fertilization are to the ability of agriculture to adapt to climate change. Until such uncertainty is resolved, studies which include a beneficial effect from CO_2 should also include other environmental factors, such as damage from increased ground-level ozone. It seems likely that the beneficial effect observed in controlled environments will be significantly less beneficial in the real world where other factors are included.

Finally, the responsibility for action is not limited to policy makers. Researchers have a clear responsibility to help bridge the existing communications gap between the policy and science communities. Bridging the gap requires that researchers share their results and concerns with their sponsors throughout the course of the research project. It also requires that researchers take their results to the decision makers, leaving the proverbial "ivory tower" and searching out those in power to make decisions. The IPCC has gone a long way towards bridging this gap, and further efforts on the part of both sides should help clarify the remaining obstacles to effectively confront the risks of climate change.

NOTES

[1] 1 Gt C = 1 Gigaton carbon = 1 billion metric tons = 10^{15}g carbon = 3.67 Gt CO_2.

[2] For a good review of modelling efforts, see IPCC (1990, pp. 75–91).

[3] Some greenhouse gases are more effective, on a unit basis, of affecting, or "forcing," the climate system. Indices for ranking this effectiveness, termed the Global Warming Potential (GWPs), have been proposed. Consensus values do not exist, due to uncertainties in the basic scientific mechanisms for the various gases, and disagreement about whether non-scientific concerns should be included. Chapters 10 and 12 cover this more completely.

[4] Higher temperatures cause the growing period to accelerate plant development (especially the grain filling stage) causing less grain yields.

[5] Vernalization is a minimum temperature requirement of some crops, e.g., winter wheat, that need cold winter temperatures to initiate the flowering process.

[6] The authors define world economic welfare as the change in consumer and producer surplus resulting from climate change as a percentage of the value of agricultural output in 1986. As a percentage of gross domestic product, these estimated welfare effects of global climate change are even smaller.

[7] Similar to Rosenzweig et al., the Adams et al. study used two climate change scenarios based on the results of doubled CO_2 experiments conducted with two general circulation models: (1) Goddard Institute of Space Studies (GISS) model, and (2) Geophysical Fluid Dynamics Laboratory (GFDL) model. The GISS model predicts that a doubled CO_2-induced climate change would cause an increase in average U.S. temperature of 4.6°C and an 8% increase in precipitation. The more extreme GFDL model predicts an temperature increase of 5.0°C and a decrease in precipitation of 0.8% for the U.S.

[8] The first meeting of the Conference of the Parties is tentatively scheduled for January 1995. Technically, this first meeting must take place within one year of the date that the Framework Convention enters into force, which requires ratification by 50 countries.

LITERATURE CITED

Adams, R.M. 1989. Global climate change and agriculture: An economic perspective. *Am. J. of Agri. Econ.* 71:1272–79.

Adams, R.M., B.A. McCarl, D.J. Dudek, and J.D. Glyer. 1988. Implications of global climate change for Western agriculture. *Western Journal of Agricultural Economics.* 13:348–56.

Adams, R.M., C. Rosenzweig, R.M. Peart, J.T. Ritchie, B.A. McCarl, J.D. Glyer, R.B. Curry, J.W. Jones, K.J. Boote, and L.H. Allen, Jr. 1990. Global climate change and U.S. agriculture. *Nature.* 345:219–24.

Aman, A.C., Jr. 1993. The Montreal Protocol and the future of global legislation. *Law and Policy.* Vol. 15. (In press).

Cheng, C.C. and B.A. McCarl. 1989. *The Agricultural Sector Model.* Texas A&M University, Department of Agricultural Economics.

Cline, W. 1992. *The Economics of Global Warming.* Institute for International Economics. Washington, D.C.

Clinton, W.J. 1993. Remarks by the President in Earth Day Speech. White House. Office of the Press Secretary. April 21.

Crosson, P. 1993. Impacts of climate change on the agriculture and economy of the Missouri, Iowa, Nebraska, and Kansas (MINK) region. In *Agricultural Dimensions of Global Climate Change.* eds. H.M. Kaiser and T.E. Drennen, Chapter 6. St. Lucie Press: Florida.

Drennen, T. 1993. After Rio: The status of climate change negotiations. In *Agricultural Dimensions of Global Climate Change.* eds. T. Drennen and H.M. Kaiser, Chapter 10. St. Lucie Press: Florida.

Drennen, T. and D. Chapman. 1992. Negotiating a response to climate change: The role of biological emissions. *Contemporary Policy Issues.* 10(3):49–58.

Duxbury, J. and A.R. Mosier. 1993. Status and issues regarding agricultural emissions of greenhouse gases. In *Agricultural Dimensions of Global Climate Change.* eds. T. Drennen and H.M. Kaiser, Chapter 12. St. Lucie Press: Florida.

Edmonds, J., J. Callaway, and D. Barns. Agriculture in a comprehensive trace-gas strategy. 1992. In *Economic Issues in Global Climate Change: Agriculture, Forestry, and Natural Resources.* eds. J. Reilly and M. Anderson. Westview Press: Boulder.

Framework Convention on Climate Change. 1992. UN, A/AC.237/18, May 15.

Greene, O. and J. Salt. 1993. Greenhouse gas emissions due to agriculture: Monitoring and verification issues. In *Agricultural Dimensions of Global Climate Change.* eds. T. Drennen and H.M. Kaiser, Chapter 13. St. Lucie Press: Florida.

Hoffert, M. 1992. Climate sensitivity, climate feedbacks and policy implications. In *Confronting Climate Change: Risks, Implications, and Responses.* ed. I. Mintzer. Cambridge University Press: NY.

International Environment Reporter (IEP). 1992. Nations agree to cuts in production of methyl bromide, faster CFC phase-out. Dec. 2. pp. 769–71.

Intergovernmental Panel on Climate Change. 1990. *Climate Change: The IPCC Impact Assessment.* Australian Government Publishing Service: Canberra, Australia.

Intergovernmental Panel on Climate Change. 1990b. *Climate Change: The IPCC Scientific Assessment.* J.T. Houghton et al., eds. Cambridge University Press.

IPCC. 1992. *Climate Change 1992. The Supplementary Report to the IPCC Scientific Assessment.* eds. J.T. Houghton, B.A. Callandar, and S.K. Varney. Cambridge University Press: Cambridge.

IPCC. 1992b. Agriculture, Forestry, and Other Human Activities. *IPCC 1992 Supplement: Working Group III Response Strategies.* Draft, January.

Kaiser, H.M., S.J. Riha, D.S. Wilks, and R. Sampath. 1993. Adaptation to climate change at the farm-level. In *Agricultural Dimensions of Global Climate Change.* eds. H.M. Kaiser and T.E. Drennen, Chapter 7. St. Lucie Press: Florida.

Karl, T.R., H. Diaz, and T. Barnett. 1990. *Climate Variations of the Past Century and the Greenhouse Effect* (a report based on the First Climate Trends Workshop). National Climate Program Office/NOAA: Rockville, MD.

Kettunen, L., J. Mukula, V. Pohjonen, O. Rantanen, and U. Varjo. 1988. The effects of climatic variations on agriculture in Finland. In *The Impact of Climatic Variations on Agriculture, Vol. 1: Assessments in Cool Temperate, and Cold Regions.* eds. M. Parry, T. Carter, and N. Konijn. Kluwer Academic Publishers: Dordrecht, The Netherlands.

Montreal Protocol on Substances that Deplete the Ozone Layer. 1987. Reprinted in *Ozone Diplomacy: New Directions in Safeguarding the Planet.* 1991. R. Benedick. Harvard University Press: Cambridge, MA.

National Academy of Sciences (NAS). 1991. *Policy Implications of Greenhouse Warming.* National Academy Press: Washington, D.C.

Office of Technology Assessment (OTA). 1991. *Changing by Degrees: Steps to Reduce Greenhouse Gases.* U.S. Congress: Washington, D.C.

Pittock, A. 1989. The greenhouse effect, regional climate change, and Australian agriculture. Paper presented at the Fifth Agronomy Conference of the Australian Society of Agronomy. Perth, Australia.

Pittock, A. and H. Nix. 1986. The effects of changing climate on Australian biomass production: A preliminary study. *Clim. Change.* 8:243–55.

Ravishankara, A.R., S. Solomon, A.A. Turnipseed, and R.F. Warrent. 1993. Atmospheric lifetimes of long-lived halogenated species. *Science.* 259:194–99.

Revelle, R. and H. Suess. 1957. Carbon dioxide exchange between atmosphere and ocean and the question of an increase of atmospheric CO_2 during the past decades. *Tellus.* 9: 18–28.

Rosenzweig, C. 1989. Global climate change: Predictions and observations. *American Journal of Agricultural Economics*. 71:1265–71.

Rosenzweig, C. and M.L. Parry. 1993. Potential impacts of climate change on world food supply: A summary of a recent international study. In *Agricultural Dimensions of Global Climate Change*. eds. H.M. Kaiser and T.E. Drennen, Chapter 5. St. Lucie Press: Florida.

Rubin, E., R. Cooper, R. Frosch, T. Lee, G. Marland, A. Rosenfeld, and D. Stine. 1992. Realistic mitigation options for global warming. *Science*. 257: 148–66.

Santer, B. 1985. The use of general circulation models in climate impact analyses: A preliminary study of the impacts of a CO2 induced climate change on West European agriculture. *Clim. Change*. 7:71–93.

Smit, B. 1989. Climate warming and Canada's comparative position in agriculture. *Clim. Change Dig*. 87(2):1–9.

Shue, H. 1993. A framework for analysis of global climate and international justice. In *Agricultural Dimensions of Global Climate Change*. eds. T. Drennen and H.M. Kaiser, Chapter 11. St. Lucie Press: Florida.

Tobey, J., J. Reilley, and S. Kane. 1992. Economic implications of global climate change for world agriculture. *J. of Agri. and Res. Econ*. 17:195–204.

United States Environmental Protection Agency (U.S. EPA). 1989. *Policy Options for Stabilizing Global Climate*. Draft Report to Congress. U.S. EPA: Washington, D.C.

U.S. Environmental Protection Agency. 1990. *The Potential Effects of Global Climate Change on the United States, Vol. 1: Regional Studies*. Report to Congress. Washington, D.C.

United States/Japan Working Group on Methane. 1992. *Technological Options for Reducing Methane Emissions: Background Document of the Response Strategies Working Group*. Draft. IPCC Response Strategies Working Group, January.

Walker, B., M. Young, J. Parslow, K. Crocks, P. Fleming, C. Margules, and J. Landsberg. 1989. *Global Climate Change and Australia: Effects on Renewable Natural Resources*. CSIRO Division of Wildlife and Ecology: Canberra, Australia.

Williams, G., R. Fautley, D. Jones, R. Stewart, and E. Wheaton. 1988. Estimating effects of climatic change on agriculture in Saskatchewan, Canada. In *The Impact of Climatic Variations on Agriculture, Vol. 1: Assessments in Cool Temperate, and Cold Regions*. eds. M. Parry, T. Carter, and N. Konijn. Kluwer Academic Publishers: Dordrecht, The Netherlands.

Wolfe, D.W. and J.D. Erickson. 1993. Carbon dioxide effect on plants: Uncertainties and implications for modeling crop response to climate change. In *Agricultural Dimensions of Global Climate Change*. eds. H.M. Kaiser and T.E. Drennen, Chapter 8. St. Lucie Press: Florida

2

IS CLIMATE CHANGE OCCURRING?

EVIDENCE from the INSTRUMENTAL RECORD

P.D. Jones*
Climatic Research Unit, University of East Anglia

I. INTRODUCTION

Temperature data from land and marine areas form the basis for many studies of climatic variations on local, regional and hemispheric scales, and the global mean temperature is a fundamental measure of the state of the climate system. In this chapter, it is shown that the surface temperature of the globe has warmed by about 0.5°C since the mid-nineteenth century. This is an important part of the evidence in the "global warming" debate (Folland et al., 1990, 1992). How certain are we about the magnitude of the warming? Where has it been greatest? How much of the warming is due to the increased concentration of greenhouse gases in the atmosphere? This chapter addresses these and related issues.

II. DATA

A. Land Regions

Knowledge of the past climate of the globe is most detailed since 1850. The Vienna Meteorological Congress of 1873 led to the establishment of many national meteorological agencies and to an expansion of instrumental recording. Prior to this time, data was confined to Europe, eastern North America, and limited coastal parts of Africa, South America, and Asia. By the 1920s, the only major land areas without meteorological instrumentation were some interior parts of Africa, South America, Asia, Arctic coasts and the whole of Antarctica.

Using the most comprehensive compilation of instrumental climate data, several investigators (Bradley et al., 1985; Jones et al., 1986a, c; Jones, 1988) have produced a grid-point data set of surface air temperature anomalies for each month from January 1851. To overcome the irregular distribution of the station network, the basic station data have been interpolated onto a 5° latitude by 10° longitude grid. The temperature data are expressed as anomalies (departures) from a common reference period (1951–70) to enable interpolation to be easily achieved.

Absolute mean temperatures recorded at neighboring stations vary because of factors such as site elevation and techniques used to calculate mean monthly temperature. Interpolation of data in absolute values is also affected both by these problems and by the varying numbers of stations. The use of anomaly values from a common reference period overcomes these problems. As a consequence, average hemispheric temperatures are calculated only in relative and not in absolute terms.

A number of problems have to be considered with regard to the raw station data before interpolation can begin. Each problem may affect regional and hemispheric average time series derived from the grid-point data. The problems may be grouped into three areas:

(1) Changes in station locations, observing schedules and practices, and in thermometer exposure;

(2) Increasing urbanization around many stations;

(3) Changes in the spatial coverage.

In order to assess homogeneity, we (Jones et al., 1986a, c) have checked each of the 3300 station records making visual comparisons of the differences between neighboring station time series. As a result of this exercise, some station time series were corrected while a number of records were discarded. Details of the station homogeneity exercise and the corrections/deletions applied are given elsewhere (Jones et al., 1985, 1986b).

Perhaps the most serious problem affecting the homogeneity of some temperature monitoring sites is increasing urbanization. This can lead to systematic biases (warming trends) in the records of affected sites. Although some of these sites will have been identified in the station homogeneity comparisons mentioned earlier, some only marginally affected are likely to pass unnoticed. The degree to which urbanization has affected hemispheric-scale temperature changes has been assessed by comparing regional-average series with data from specially developed rural station networks for certain areas of the world. Detailed analyses of the urbanization influence have been undertaken (Jones et al., 1989, 1990) using rural networks for the contiguous United States (Karl et al., 1988; Wigley and Jones, 1988; Karl and Jones, 1989), for the western part of the former Soviet Union, eastern Australia, and eastern China. Comparisons with the rural networks show that the greatest "residual" warming (grid-rural) in the Jones et al. (1986a, c) data set is apparent over the contiguous United States (about 0.15°C over the 1901–

84 period). The additional warming in the other industrial areas is barely noticeable. As other less-industrialized regions of the world are likely to be less affected than those studied, Jones et al. (1990) conclude that the residual urban warming in the Jones et al. (1986a, c) data set is at most 0.05°C over the last 100 years.

The problem of the gradually increasing spatial coverage has been assessed by Jones et al. (1986a, c) using a "frozen grid" technique. Hemispheric averages were calculated, for the 1941–80 period, with subsets of data that simulate the coverage during times of poorer coverage in the late nineteenth and early twentieth century. These averages were then compared with the record based on all available data. The results of the analyses suggest that changing spatial coverage has had little effect on hemispheric land temperature anomalies since 1900. Prior to this, there was additional uncertainty, though annual averages appear reliable since 1875 in the Northern Hemisphere and since 1890 in the Southern Hemisphere. More recent studies confirm these results (Madden et al., 1993).

B. Marine Regions

Land represents only 29% of the Earth's surface. Until recently it was assumed that changes in air temperature over the ocean were similar to those over the land. Recently-produced compilations of marine data taken by "ships of opportunity" now enable this assumption to be tested. Since the 1850s, ships have been obliged to take weather observations and measure the temperature of the sea surface. In the last twenty years, major international efforts have been made to put all of this log-book information into computer data banks. One such compilation is the Comprehensive Ocean-Atmosphere Data Set (COADS) produced by National Oceanic and Atmospheric Administration (NOAA) workers at Boulder, Colorado (Woodruff et al., 1987). Another is located at the U.K. Meteorological Office (Bottomley et al., 1990).

Unfortunately, just as with the land data, the marine data are subject to inhomogeneities. In most analyses (Bottomley et al., 1990; Folland et al., 1990; Jones and Wigley, 1990; Jones et al., 1991) temperatures over the ocean can be estimated using sea-surface temperature (SST) anomalies. This is done because the number of SST measurements are greater than simultaneously-taken marine air temperatures and their within-month variability is less (Trenberth et al., 1992). Although the difference between air and sea temperatures can be up to 6–8°C in absolute terms, during certain seasons at certain locations, in anomaly terms, the two marine data sets are in excellent agreement (Folland et al., 1984; Jones et al., 1986d; Bottomley et al., 1990). Sea-surface temperature anomalies are, therefore, an excellent surrogate for marine air temperature anomalies.

Problems with the homogeneity of sea-surface temperature (SST) data arise due to differences in the method of sampling the sea water. Before World War II, the sea water was collected in an uninsulated canvas bucket. There was a delay of a few minutes between sampling and measuring the temperature. During this time the water in the bucket generally cooled slightly by evaporative means. Since World War II, most readings have been made in the intake pipes through which sea water is taken on board ships to cool the engines. Comparative studies of the two methods indicate that bucket temperatures are cooler by 0.3–0.7°C (James and Fox, 1972). Correcting the SST data for this measurement change may, at first, seem like an intractable problem. Folland and Parker (1990, 1991) of the U.K. Meteorological Office, however, have developed a method for correcting the canvas bucket measurements based on physical principles related to the causes of the cooling. The cooling depends on the prevailing meteorological conditions, and so varies according to the time of year and location.

The coverage of SST measurements is largely determined by merchant shipping routes. Coverage in the nineteenth century was mainly confined to the Indian and Atlantic Oceans north of 40°S. Coverage in the Pacific improved during the twentieth century but was poor in the southeastern Pacific and everywhere south of 45°S, except near South America. Frozen grid analyses with the marine data (Bottomley et al., 1990; Jones et al., 1991) indicate that hemispheric estimates can be well-estimated from the sparse data availability in the late nineteenth and early twentieth centuries (Trenberth et al., 1992; Madden et al., 1993).

III. TEMPORAL VARIATION IN HEMISPHERIC MEAN TEMPERATURE

A. Land Regions

Time series of mean hemispheric seasonal and annual temperature anomalies are shown for both the Northern (Figure 2-1) and Southern Hemisphere (Figure 2-2). The features exhibited by the two sets of curves have been discussed elsewhere (Wigley et al., 1985, 1886; Jones et al., 1986a, c; Folland et al., 1990, 1992). In the annual series, both hemispheres show a warming of the order of 0.5°C since the late nineteenth century. The warming is considerably more erratic in the Northern Hemisphere (Figure 2-1) with a clear cooling of about 0.2°C between 1940 and 1970 in some seasons. In the Southern Hemisphere (Figure 2-2), the warming is more monotonic and there is no visual evidence of cooling after 1945.

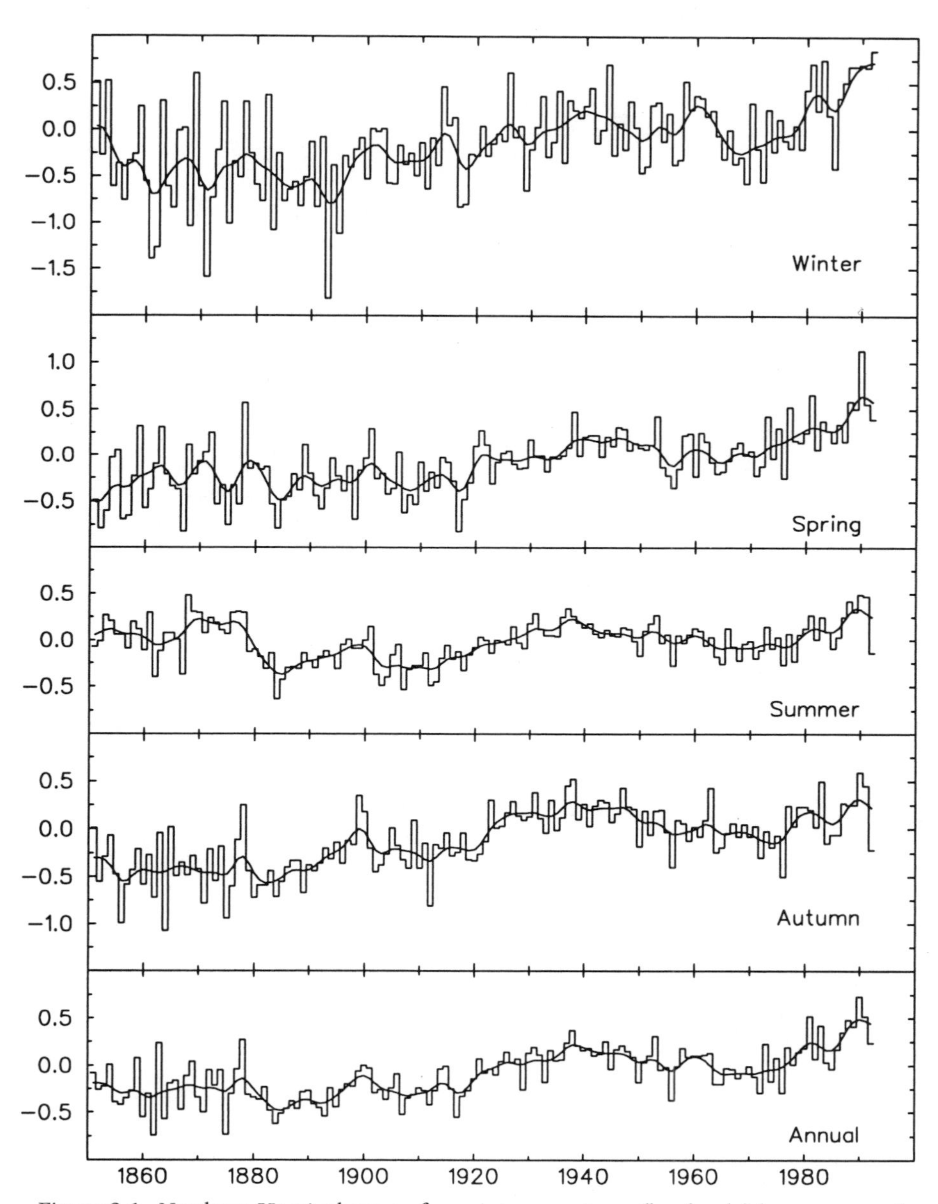

Figure 2-1. Northern Hemisphere surface air temperatures (land only) by season and year, 1851–1992. Data are expressed as anomalies from 1951–1970. Northern Hemisphere winters in this and subsequent figures are dated by the year in which the January occurs. Smooth curves in this and subsequent plots were obtained by using a 10–year Gaussian filter.

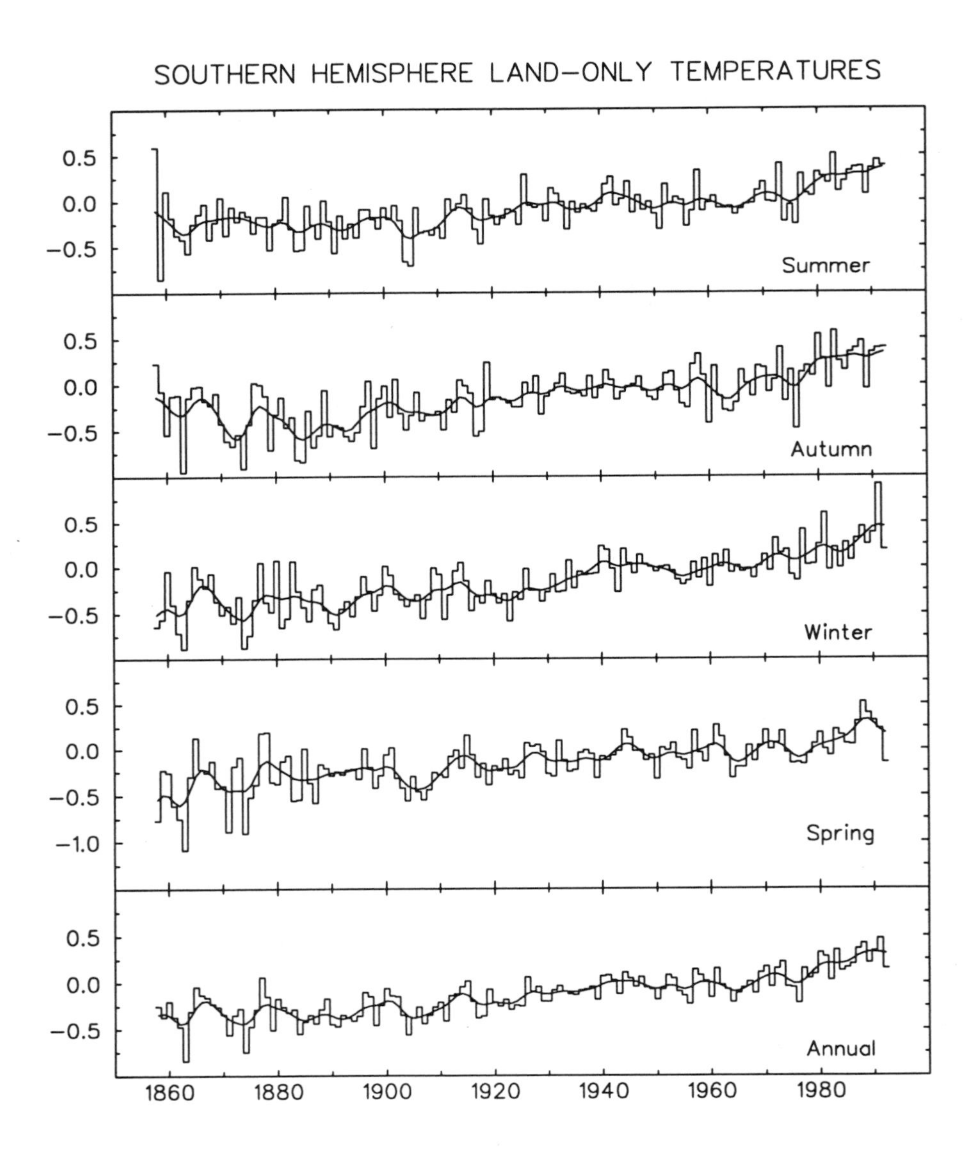

Figure 2-2. Southern Hemisphere surface air temperatures (land only) by season and year, 1858–1992. Southern Hemisphere summers in this and subsequent figures are dated by the year in which the January occurs. Data are expressed as anomalies from 1951–1970.

The various seasonal curves in Figures 2-1 and 2-2 show considerable variation in periods of warming and cooling. The greater seasonal differences occur over the Northern Hemisphere. In summer and autumn, the 1980s are hardly warmer than the temperature levels of the 1930s and 1940s. In winter and spring, the 1980s are clearly the warmest decade. All seasons except summer show the long term warming evident in the annual data. In summer, the 1850s to 1870s were as warm as the two most recent decades. The cooling from the 1870s to the 1880s is the most pronounced feature of the summer data. Over the Southern Hemisphere, there is more agreement between the seasonal series.

In both hemispheres, there is greater year-to-year variability during the nineteenth century. Most of this increase is due to the sparser spatial coverage at that time. Although the individual years may be less reliable than twentieth century values, the frozen-grid analyses undertaken by Jones et al. (1986a, c) indicate that the decadal-scale temperature fluctuations are reliably reproduced by the available data. Thus, the relative warmth of the 1850s to 1870s compared to the 1880s in some seasons, in both hemispheres, is probably real. This has important implications for a number of other hemispheric temperature analyses (e.g., Hansen and Lebedeff, 1987, 1988; Vinnikov et al., 1990) which begin their analyses around 1880. Warming rates measured over 1881 to 1990 are slightly greater than rates calculated over 1861 to 1990.

Longer temperature records exist for parts of Europe and eastern North America (Jones and Bradley, 1992). These longer site records confirm the warming from the late nineteenth century, as do most paleoclimatic records (Bradley and Jones, 1992). The European records show that the 1880s were the coldest decade, at least since 1700. Therefore, part of the 0.5°C warming since the 1880s may reflect this unusually low starting point.

Most of the focus of past and future temperature change has been concerned with mean temperatures. Maximum and minimum temperature data are not routinely exchanged between countries. Few General Circulation Model (GCM) experiments have considered changes in maximum and minimum temperatures. For estimating future changes, the diurnal cycle (range of daily maximum minus minimum temperature) in GCM control runs is often poorly simulated. The issue is, however, particularly important with regard to climate change impact studies. Projected climate impacts will vary widely for a scenario assuming equal temperature increases for both maximum and minimum temperatures (what has been assumed until now) compared to a scenario which assumes an increase in daily minimum temperatures only.

Recently developed data sets (Karl et al., 1993) have enabled analysis of maximum and minimum temperatures to be made for 37% of the global land mass. These indicate that over the 1951–90 period, minimum temperatures have risen at a rate three-times that of maximum temperatures. The reduction in the diurnal temperature range is approximately equal to the temperature increase. The changes in the temperature range are detectable in all of the regions studied for all seasons.

B. Land and Marine Regions

In Jones and Wigley (1990) and Jones et al. (1991), the grid-point land and grid-box marine data are combined into a 5° x 5° grid-box data set. Where land and marine data overlap, the resulting value in a grid box is the average of the land and marine components. Time series of mean hemispheric annual and seasonal temperature anomalies based on the combined (land-plus-marine) data set are shown in Figure 2-3 (Northern Hemisphere) and Figure 2-4 (Southern Hemisphere). In the combined data set, temperatures are expressed as departures from the 1950–79 period. Most of the features exhibited by the two sets of curves have been discussed elsewhere (Folland et al., 1990, 1992; Jones and Wigley, 1990). The two hemispheric annual curves and their average (the global series), Figure 2-5, represent the major observational evidence in the "global warming" debate.

In many respects the year-to-year and decadal-scale variations in Figures 2-3 and 2-4 are dampened versions of those seen in Figures 2-1 and 2-2. There are, however, subtle differences. Variability in the nineteenth century in the combined data set is not greater than during the twentieth century. The cooling evident in some of the Northern Hemisphere seasons for "land only" data is now hardly apparent. Variability in the combined data set is now similar between the two hemispheres. In the "land only" series the Northern Hemisphere variability was visually greater.

IV. THE SPATIAL PATTERN OF WARMTH DURING THE 1981–1990 DECADE

Seasonal temperature anomalies for the Northern and Southern Hemispheres, relative to the 1950–1979 reference period have been plotted in Figures 2-6 (NH) and 2-7 (SH). Annual temperature anomalies for the two hemispheres are shown in Figure 2-8.

Most areas of the globe experienced above-normal (1950–1979) average temperatures for the 1981–1990 decade; however, there was great variability from season to season in the Northern Hemisphere (Figure 2-6). Much of the decadal warmth was apparent during the December–January–February (DJF) and March–April–May (MAM) seasons, with the warmth located over northern and central Asia and over northwestern North America. During June–July–August (JJA) anomalies were smaller in magnitude over North America and

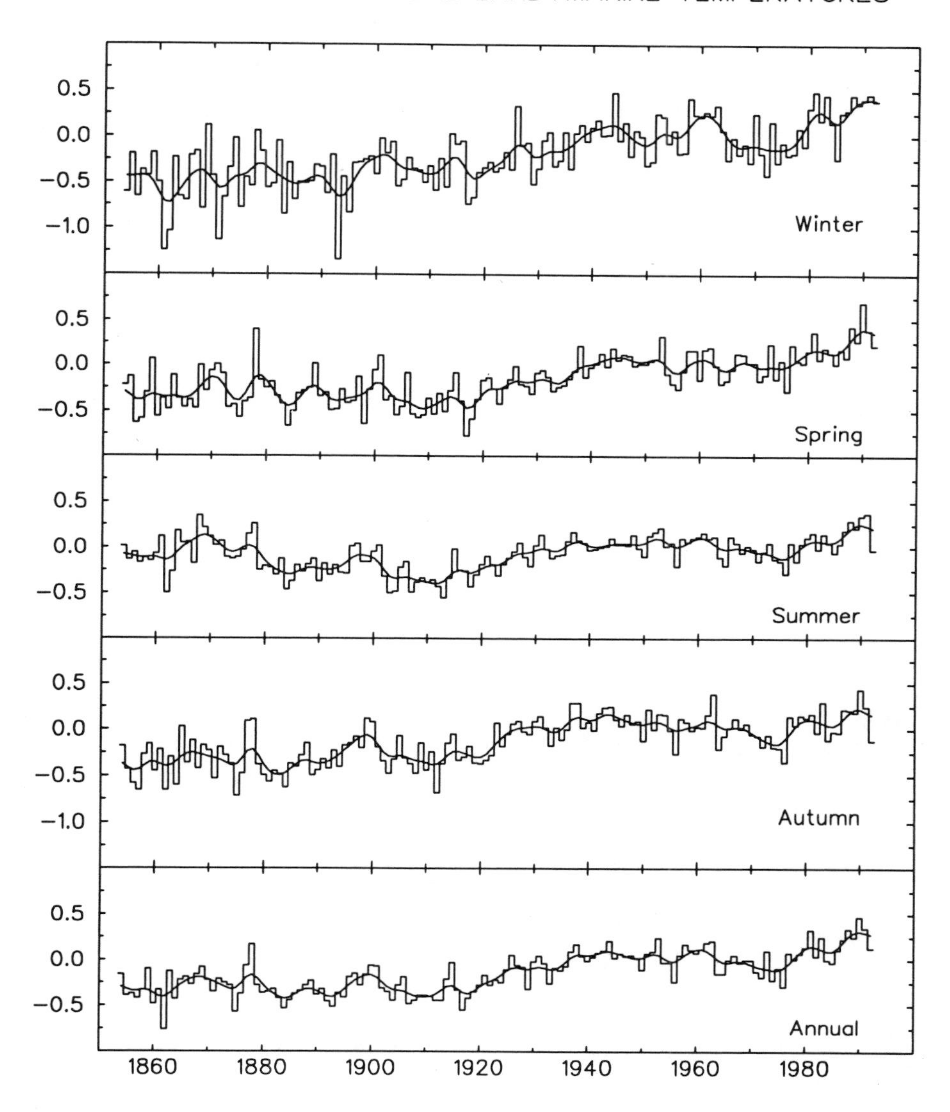

Figure 2-3. Northern Hemisphere surface air temperatures (land-plus-marine data) by season, 1854–1992. Data are expressed as anomalies from 1950–1979.

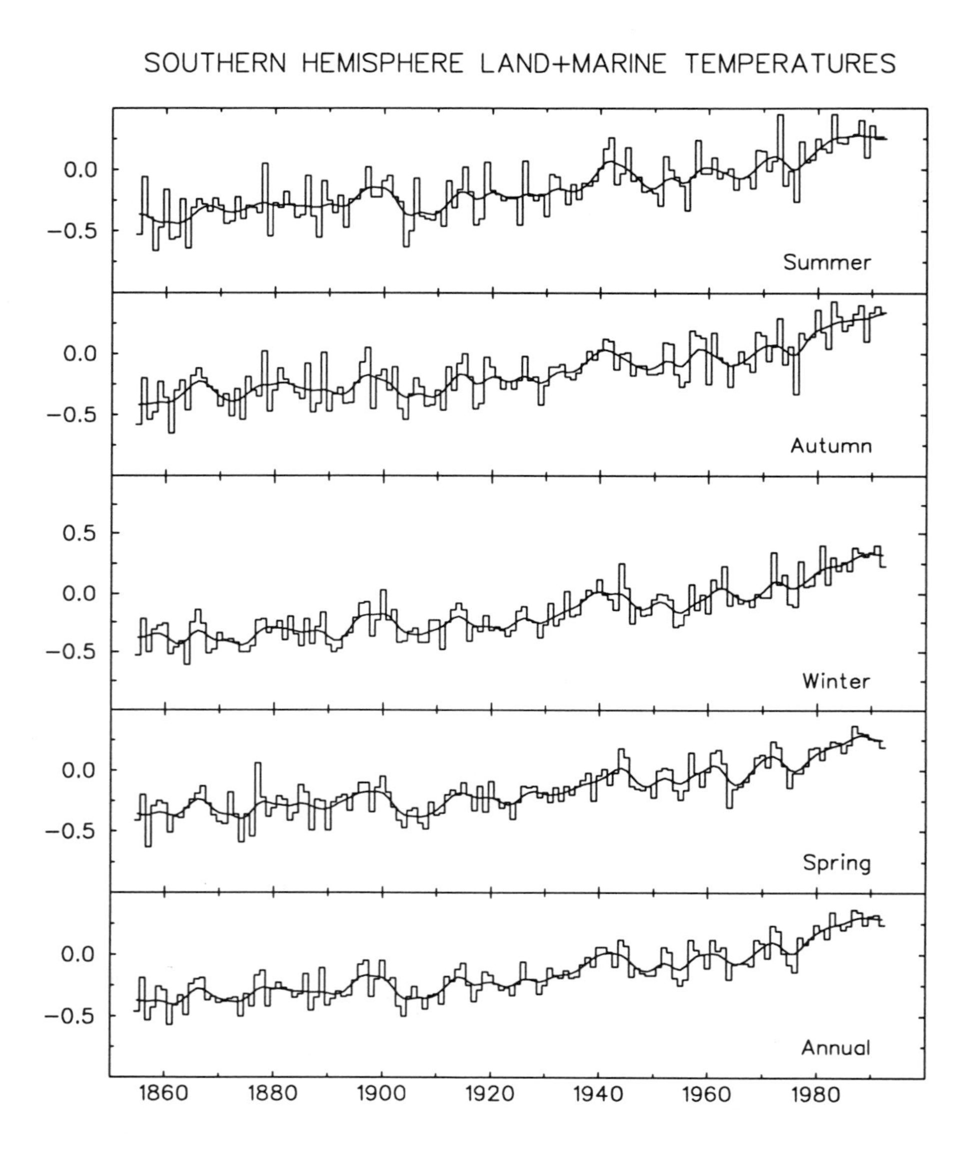

Figure 2-4. Southern Hemisphere surface air temperatures (land-plus-marine data) by season, 1854–1992. Data are expressed as anomalies from 1950–1979.

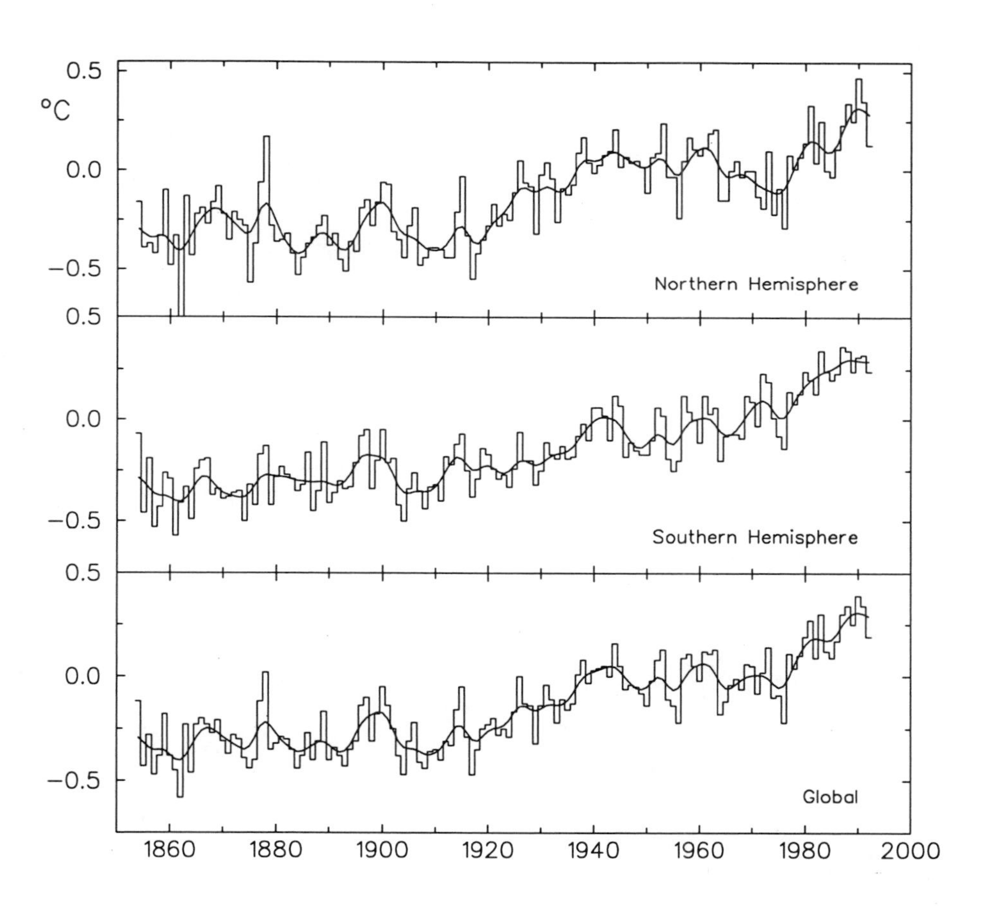

Figure 2-5. Northern, Southern Hemisphere and Global annual surface air temperatures (land-plus-marine data), 1854–1991. Data are expressed as anomalies from 1950–1979.

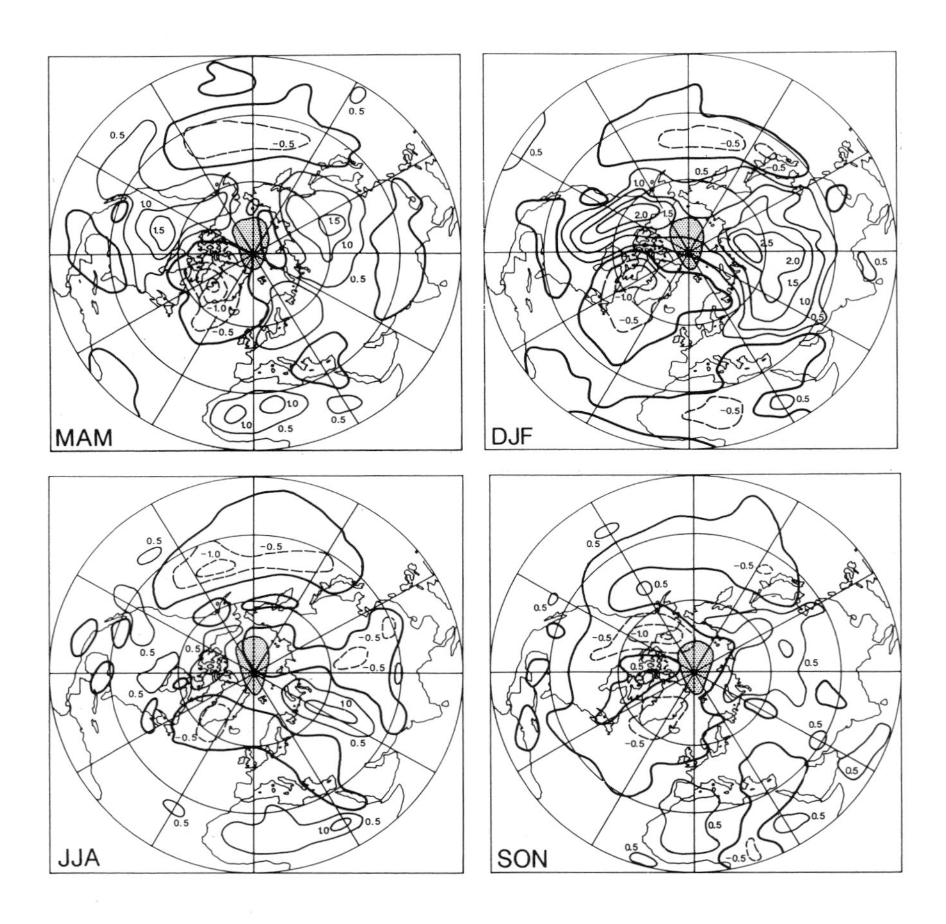

Figure 2-6. Seasonal temperature anomalies for the Northern Hemisphere for 1981–1990. Anomalies are based on the 1950–1979 period and the contour interval is 0.5°C, with negative anomalies dashed. Shaded areas indicate regions with insufficient data.

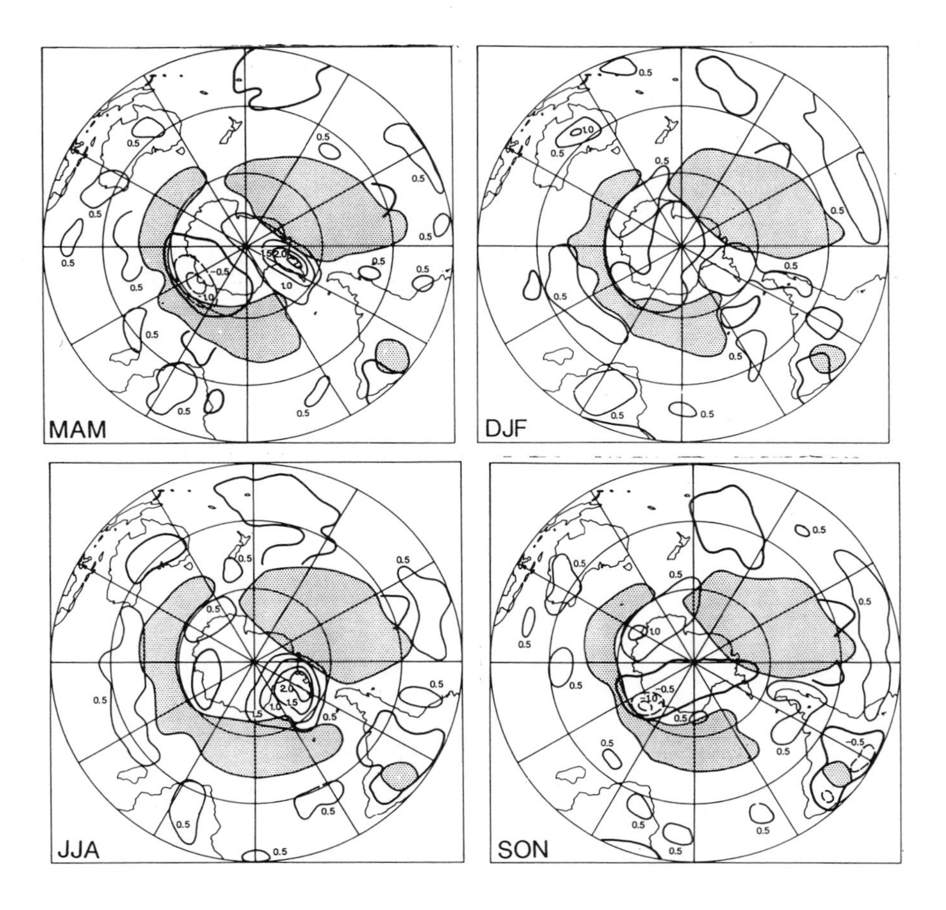

Figure 2-7. Seasonal temperature anomalies for the Southern Hemisphere for 1981–1990. Anomalies are based on the 1950 – 1979 period and the contour interval is 0.5°C, with negative anomalies dashed. Shaded areas indicate regions with insufficient data.

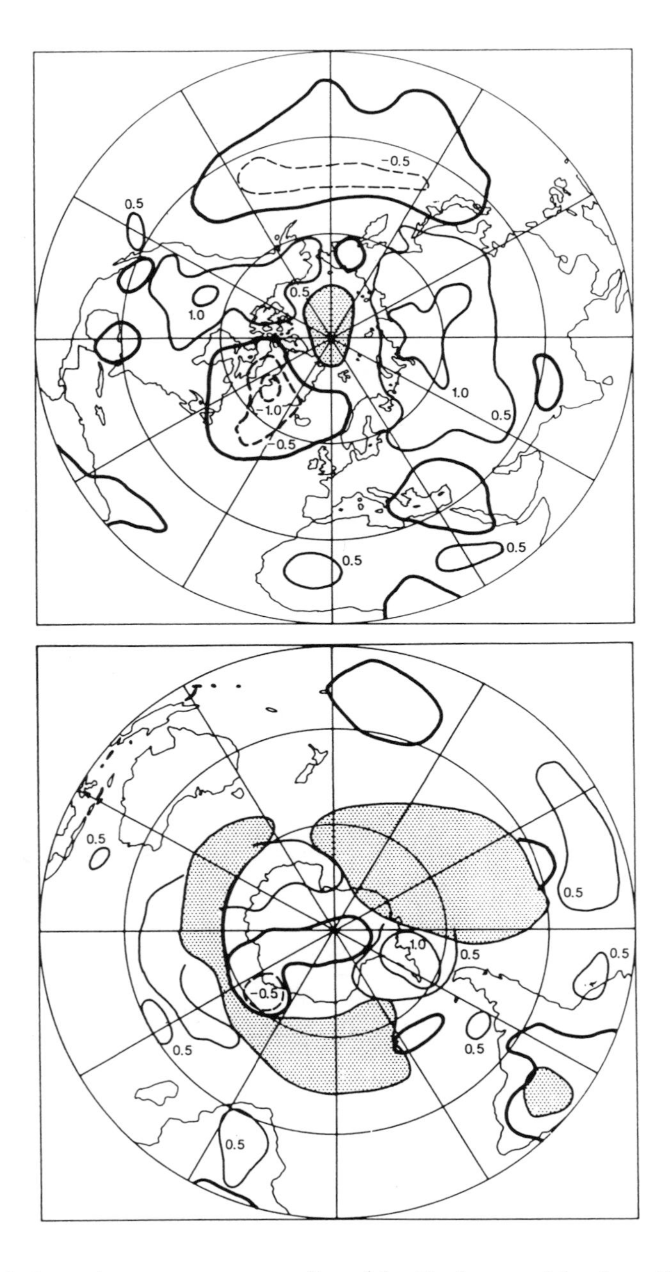

Figure 2-8. Annual temperature anomalies of the Northern and Southern Hemisphere for 1981–1990. Anomalies are based on the 1950–1979 period and the contour interval is 0.5°C with negative anomalies dashed. Shaded areas indicate regions with insufficient data.

negative over parts of Asia. For September–October–November (SON) temperatures were below-normal over western Canada and Alaska but slightly above-normal over Siberia. Over northern Africa temperatures were above normal during MAM, JJA, and SON but below-normal in DJF. Major areas of relatively consistent below-normal temperatures included the central North Pacific Ocean and the northern North Atlantic region encompassing Iceland and Greenland.

Over the Southern Hemisphere (Figure 2-7) anomalies were generally smaller in magnitude, except over parts of Antarctica. Almost all of the hemisphere was warmer than normal. The major exception to this was part of eastern Antarctica during MAM and SON. In tropical and temperate latitudes, Australia, the Indian Ocean, and the eastern equatorial Pacific were warmer than normal. The regions with the greatest warming over Antarctica were the Australian sector in JJA, SON, and DJF and the Antarctic Peninsula in MAM and JJA.

The annual anomaly maps for the two hemispheres (Figure 2-8) obviously encompass the average features of the four seasons. Over the Northern Hemisphere relative warmth prevailed over most of Eurasia, especially Siberia, the western-half of North America, and western Africa. Temperatures were cooler during the decade over the northern Pacific Ocean and the northern Atlantic region surrounding Greenland and Iceland. Most regions of the Southern Hemisphere have experienced warmth during the decade except for parts of eastern Antarctica, the Amazon Basin, and part of the southwestern Pacific. Warmth during the decade was greatest over the Indian Ocean, southern Africa, the eastern equatorial Pacific Ocean, the Antarctic Peninsula, and the Australian sector of Antarctica. An in-depth climate assessment of the 1981–1990 decade has been produced by Ropelewski and Halpert (1991).

V. CONCLUSIONS

The world has warmed by about 0.5°C since the late nineteenth century. Careful analysis of the constituent land and marine data assures us that we can be fairly confident of this result. Inhomogeneities in land-based station time series have been corrected and all marine (SST) data prior to 1941 have been adjusted for the change from bucket to intake measurements. Analysis using rural data sets allows us to conclude that any remaining urbanization influences in global land averages are slight, with an order of magnitude smaller than the 0.5°C warming. The greatest remaining uncertainty is associated with the adjustments that must be applied to the nineteenth century SST temperatures. Although this uncertainty is small (0.1–0.2°C) (Bottomley et al., 1990; Jones et al., 1991), its magnitude is sufficiently large to allow a wide range of possible warming since the mid-nineteenth century (0.3–0.6°C) (Folland et al., 1990).

The warming since the last century has not been equal everywhere, nor equal throughout the seasons. Although most areas have warmed since the turn of the century, several small areas have cooled. The 1980s are clearly the warmest decade recorded. Some of the greatest warming has occurred over the continental areas of the Northern Hemisphere. Despite this, the warming is more spatially coherent and significant over mid- to lower-latitude regions of both hemispheres, particularly tropical oceans. Temperature change has also varied seasonally. In the Northern Hemisphere, for example, the warming is greater and more significant in spring and least in autumn. Recent analyses of maximum and minimum temperature trends over some land areas for 1950–90 indicate that minimum temperatures have risen at three-times the rate of maximum temperatures (Karl et al., 1993).

The time series of hemispheric and global temperature averages exhibit variability on both high- and low-frequency time scales. At present it is not possible to ascribe with any degree of confidence a cause for the long term warming of 0.5°C, although the greenhouse effect does fit some of the facts. The recent warming, for example, is a case where the evidence is somewhat equivocal. Some of the warming may simply be due to the trend calculation beginning in late nineteenth century which was clearly one of the coldest periods of the last 500 years. While the equilibrium results from General Circulation Models (GCMs) suggest that the world should be warming, they also indicate that the warming should be greatest in polar regions and probably strongest in the Northern Hemisphere. Recent warming has been stronger in middle latitudes, particularly in the Southern Hemisphere. The most detailed discussion of the global warming issue and the detection of the greenhouse effect is given by Wigley and Barnett (1990).

* This work was supported by the Atmospheric and Climate Change Research Division of the U.S. Department of Energy under Grant No. DE-FG02-86-ER60397.

LITERATURE CITED

Bottomley, M., C.K. Folland, J. Hsiung, R.E. Newell, and D.E. Parker. 1990. *Global Ocean Surface Temperature Atlas (GOSTA).* H.M.S.O: London.

Bradley, R.S., P.M. Kelly, P.D. Jones, C.M. Goodess, and H.F. Diaz. 1985. *A Climatic Data Bank for Northern Hemisphere Land Areas, 1851–1980.* U.S. Dept. of Energy, Carbon Dioxide Research Division, Technical Report TRO17.

Bradley, R.S. and P.D. Jones, eds. 1992. *Climate Since A.D. 1500.* Routledge: London.

Folland, C.K., T.R. Karl, N. Nicholls, B.S. Nyenzi, D.E. Parker, and K.Ya. Vinnikov. 1992. Observed climate variability and climate. In *Climate Change 1992. The Supplementary Report to the IPCC Scientific Assessment,* eds. J.T. Houghton, B.A. Callander, and S.K. Varney, 135–70. Cambridge University Press: New York.

Folland, C.K., T.R. Karl, and K.Ya. Vinnikov. 1990. Observed climate variations and change. In *Climate Change: The IPCC Scientific Assessment,* eds. J.T. Houghton, G.J. Jenkins, and J.J. Ephraums, 195–238. New York: Cambridge University Press.

Folland, C.K. and D.E. Parker. 1990. Observed variations of sea surface temperature. In *Climate-Ocean Interaction,* ed. M.E. Schlesinger, 31–52. NATO Workshop, Oxford. Kluwer Academic Press: Hingham, MA.

Folland, C.K. and D.E. Parker. 1991. Worldwide surface temperature trends since the mid 19th century. In *Greenhouse-gas-induced Climatic Change: A Critical Appraisal of Simulations and Observations,* ed. M.E. Schlesinger, 173–94. Elsevier Scientific Publishers: New York.

Folland, C.K., D.E. Parker, and F.E. Kates. 1984. Worldwide marine fluctuations, 1856-1981. *Nature* 310:670–73.

Hansen, J.E. and S. Lebedeff. 1987. Global trends of measured surface air temperature. *J. Geophys. Res.* 92:13345–72.

Hansen, J.E. and S. Lebedeff. 1988. Global surface temperatures: update through 1987. *Geophys. Res. Letters* 15:323–26.

James R.W. and P.T. Fox. 1972. Comparative sea-surface temperature measurements. *Marine Science Affairs Report* No.5, WMO 336.

Jones, P.D. 1988. Hemispheric surface air temperature variations: Recent trends and an update to 1987. *Journal of Climate* 1:654–60.

Jones, P.D. and R.S. Bradley. 1992. Climatic variations in the longest instrumental records. In *Climate Since A.D. 1500,* eds. R.S. Bradley and P.D. Jones, 246–68. Routledge: London.

Jones, P.D., P.Ya. Groisman, M. Coughlan, N. Plummer, W.C. Wang, and T.R. Karl. 1990. Assessment of urbanization effects in time series of surface air temperature over land. *Nature* 347:169–72.

Jones, P.D., P.M. Kelly, C.M. Goodess, and T.R. Karl. 1989. The effect of urban warming on the Northern Hemisphere temperature average. *Journal of Climate* 2:285–90.

Jones, P.D., S.C.B. Raper, R.S. Bradley, H.F. Diaz, P.M. Kelly, and T.M.L. Wigley. 1986a. Northern Hemisphere surface air temperature variations: 1851–1984. *Journal of Climate and Applied Meteorology* 25:161–79.

Jones, P.D., S.C.B. Raper, B.S.G. Cherry, C.M. Goodess, and T.M.L. Wigley. 1986b. *A Grid Point Surface Air Temperature Data Set for the Southern Hemisphere, 1851–1984.* U.S. Dept. of Energy, Carbon Dioxide Research Division, Technical Report TRO27.

Jones, P.D., S.C.B. Raper, B.D. Santer, B.S.G. Cherry, C.M. Goodess, P.M. Kelly, T.M.L. Wigley, R.S. Bradley, and H.F. Diaz. 1985. *A Grid Point Surface Air Temperature Data Set for the Northern Hemisphere.* U.S. Dept. of Energy, Carbon Dioxide Research Division, Technical Report TRO22.

Jones, P.D., S.C.B. Raper, and T.M.L. Wigley. 1986c. Southern Hemisphere surface air temperature variations: 1851-1984. *Journal of Climate and Applied Meteorology* 25:1213–30.

Jones, P.D. and T.M.L. Wigley. 1990. Global warming trends. *Scientific American* 263:84–91.

Jones, P.D., T.M.L. Wigley, and G. Farmer. 1991. Marine and land temperature data sets: A comparison and a look at recent trends. In *Greenhouse-gas-induced Climatic Change: A Critical Appraisal of Simulations and Observations,* ed. M.E. Schlesinger, 153–72. Elsevier Scientific Publishers: New York.

Jones, P.D., T.M.L. Wigley, and P.B. Wright. 1986d. Global temperature variations, 1861-1984. *Nature* 322:430–4.

Karl, T.R., H.F. Diaz, and G. Kukla. 1988. Urbanization: Its detection and effect in the United States climate record. *Journal of Climate* 1:1099–123.

Karl, T.R. and P.D. Jones. 1989. Urban bias in area-averaged surface air temperature trends. *Bulletin of the American Meteorological Society* 70:265–70.

Karl, T.R., P.D. Jones, R.W. Knight, G. Kukla, N. Plummer, K.P. Gallo, and J. Lindseay. 1993. Asymmetric trends of daily maximum and minimum temperature: Empirical evidence and possible causes. *Bulletin of the American Meteorological Society.* In press.

Madden, R.A., D.J. Shea, G.W. Branstator, J.J. Tribbia, and R. Weber. 1993. The effects of imperfect spatial and temporal sampling on estimates of the global mean temperature: Experiments with model and satellite data. *J. Climate.* In press.

Ropelewski, C.F. and M.S. Halpert, eds. 1991. *Climate Assessment: A Decadal Review 1981–1990.* Climate Analysis Center, NOAA, U.S. Dept. of Commerce.

Trenberth, K.E., J.R. Christy, and J.W. Hurrell. 1992. Monitoring global monthly mean surface temperature. *J. Climate* 5:1405–23.

Vinnikov, K. Ya., P.Ya. Groisman, and K.M. Lugina. 1990. The empirical data on modern global climate changes (temperature and precipitation). *J. Climate* 3:662–77.

Wigley, T.M.L., J.K. Angell, and P.D. Jones. 1985. Analysis of the temperature record. In *Detecting the Climatic Effects of Increasing Carbon Dioxide*, eds. M.C. MacCracken and F.M. Luther, 55–90. U.S. Dept. of Energy, Carbon Dioxide Research Division.

Wigley, T.M.L. and T.P. Barnett. 1990. Detection of the Greenhouse effect in the observations. In *Climatic Change: The IPCC Scientific Assessment*, eds. J.T. Houghton, G.J. Jenkins, and J.J. Ephraums, 239–65. Cambridge University Press: New York.

Wigley, T.M.L. and P.D. Jones. 1988. Do large–area–average temperature series have an urban-warming bias? *Climatic Change* 12:313–19.

Wigley, T.M.L., P.D. Jones, and P.M. Kelly. 1986. Empirical climate studies: warm world scenarios and the detection of climatic change induced by radiatively active gases. In *The Greenhouse Effect, Climatic Change, and Ecosystems*, eds. B. Bolin, B.R. Döös, J. Jäger, and R.A. Warrick, 271–323. Wiley, SCOPE series: New York.

Woodruff, S.D., R.J. Slutz, R.J. Jenne, and P.M. Steurer. 1987. A Comprehensive Ocean-Atmosphere Data Set. *Bulletin American Meteorological Society* 68:1239–50.

3

SOME ISSUES in DETECTING CLIMATE CHANGE INDUCED by GREENHOUSE GASES USING GENERAL CIRCULATION MODELS

Benjamin D. Santer
Program for Climate Model Diagnosis and Intercomparison
Lawrence Livermore National Laboratory

I. INTRODUCTION

In the last five years, a number of papers have been published in the scientific literature which have had as their focus the detection of greenhouse-gas-induced climate change in observed climate records. These studies can be divided into two general types—those dealing with the detection of a model-predicted signal in the observed data, and those concerned solely with the analysis of model data.

Investigations of the first type often have as their starting point the greenhouse gas (GHG) signal predicted by a climate model, and then attempt to find this signal in observed records of surface temperature (Barnett, 1986; Barnett and Schlesinger, 1987; Santer et al., 1991, 1993a) or in records of temperature change in the lower stratosphere and troposphere (Karoly, 1987, 1989). The results of such comparisons of model and observed data have been inconclusive (Wigley and Barnett, 1990). While they have failed to provide convincing statistical evidence for the existence of a GHG signal in the observations, they have also pointed out that there are many possible (and plausible) explanations for such failure. Studies of this type have focused attention on problems of methodology (which statistical tools should we use in comparing model and observed data?), and on the difficulties involved in establishing a unique cause-and-effect link between changes in the climate of the last century and changes in greenhouse gases. They have also indicated that there are a number of areas in which the observed instrumental records of climate change and the model-predicted GHG signals show qualitative agreement (Houghton et al., 1990).

The second type of study focuses on model data only, and has had two purposes. The first purpose is to learn something about the natural variability[1] of the climate system. This provides us with information about the background "noise" of the climate system in the absence of any change in greenhouse gases caused by human activities. The second purpose is to identify climate variables which may be sensitive and highly specific indicators of GHG-induced climate change—in other words, variables which respond to changes in greenhouse gases in a unique way that cannot be confused with the natural variability of climate, and that is also very different from the response to changes in other external factors.

The aim of this chapter is not to provide a comprehensive overview of previous detection studies and their principal findings. Instead, we will examine why detection of GHG-induced climate change is difficult, and address some of the problems associated with the use of model data in detection studies. The main issues addressed are:

- Model signal uncertainties
- Natural variability uncertainties
- The attribution problem

Before discussing these issues, the chapter begins with a brief introduction to climate models and an explanation of why models are essential tools in GHG detection studies. Some historical background to the different types of greenhouse warming experiments which have been performed is also presented.

II. CLIMATE MODELS AND GREENHOUSE WARMING EXPERIMENTS

There is no direct historical or paleoclimatic analogue for the rapid change in atmospheric CO_2 which has taken place over the last century (Crowley, 1991). This means that we cannot use paleoclimatic data[2] or instrumental records in order to predict the regional and seasonal patterns and rate of climate change over the next century. We must therefore rely on numerical models of the Earth's climate system in order to make such predictions.

A large number of different numerical models have been used to study the effects of greenhouse gases on climate. The simplest of these consider the radiation budget at a single point on the Earth's surface. The most complex models attempt to simulate the full three-dimensional circulation of the atmosphere and ocean. For example, a typical fully-coupled ocean-atmosphere general circulation model (O/AGCM) generally divides the atmosphere and ocean into a number of discrete layers (extending from the bottom of the ocean to the top of the atmosphere), with each layer consisting of a two-dimensional grid of thousands of points. The model then solves equations for the transport of heat, momentum, moisture (in the atmosphere), and salinity (in the ocean) on this three-dimensional grid. A typical horizontal resolution in current O/AGCMs is 4° latitude x 5° longitude. Physical processes which occur on spatial scales smaller than the mesh of this grid (such as cloud formation) are parameterized—that is, their properties depend on the values of climate variables which are averaged over the 4° x 5° grid-cell. The bottom topography of the ocean and the land surface orography are represented in a realistic way, but are smoothed to correspond to the resolution of the model.

In O/AGCMs, as in the real world, the atmosphere and ocean communicate with each other, exchanging heat and momentum. The time scales of most atmospheric phenomena, such as frontal systems (with time scales of several days) and high pressure blocks (time scales of weeks) are much faster than typical ocean time scales (of the order of centuries for the deep ocean circulation). The interaction between these fast and slow time scales can lead to a rich and complex spectrum of climate variability. It is essential to incorporate the coupling between the fast and slow components of the climate system in order to model the natural variability and to understand how climate might change in response to gradually increasing concentrations of greenhouse gases (Hasselmann, 1988). Without an O/AGCM in which the ocean model is capable of realistically absorbing and redistributing heat from the atmosphere,[3] we will not have confidence in projections of the time evolution of the climatic response to GHG-forcing.

It is important to realize that the development of sophisticated O/AGCMs is a dynamic process. Such models evolve as computational speed and storage evolve, and as our understanding of the physics of the climate system improves. Until very recently, for example, most of our information concerning the possible climate response to GHG increases came from so-called equilibrium response experiments (see Schlesinger and Mitchell, 1987, and Mitchell et al., 1990). Such experiments generally used a relatively sophisticated atmospheric GCM, coupled to a much simpler model of the top layer of the ocean (the mixed-layer, usually the uppermost 50–100 m). The experimental set-up involved instantaneously doubling the atmospheric CO_2 concentration (e.g., from 330 to 660 ppm), and then simulating the climate response over a period of 20-50 years. Because the mixed-layer of the ocean has a rapid response time (10–15 years), and since the longer time scales of the intermediate and deep ocean were neglected, such experiments allowed the climate system to reach a new equilibrium state within a relatively short time (10–20 years). The investigator then compared a sample of the model's new equilibrium climate with a climate sample from a control run without doubling of atmospheric CO_2 in order to learn something about the physics of the response.

In the real world, of course, CO_2 is increasing gradually and does not instantaneously double its atmospheric concentration. The more relevant question is how the climate system—including the deep ocean, with its longer time scales—will respond to slowly increasing GHG concentrations. It is only within the last few years that scientists have been able to address this question by performing transient response experiments with the O/AGCMs described above (e.g., Stouffer et al., 1989; Washington and Meehl, 1989; Cubasch et al., 1992). In a typical transient response experiment, an O/AGCM is forced by some scenario of how CO_2 and other greenhouse gases might change in the future. Scenarios which have been used in these experiments range from a simple linear increase in CO_2 (by 1% per year; Washington and Meehl, 1989) to the scenarios developed by the Intergovernmental Panel on Climate Change (IPCC; Houghton et al., 1990), which cover a range of assumptions about how world energy use and emissions might evolve over the next century.

Transient experiments add a new dimension to the detection problem. In addition to supplying information about the spatial pattern and seasonal features of the climate response, these experiments also tell us something about the time evolution of the response on scales of decades to centuries. The time evolution is now of direct interest: the problem is to determine whether the trend which describes the climate system's response to GHG forcing (the signal) is much larger than the decadal-to-century time scale trends which occur through natural variability alone (the noise of the climate system).

III. MODEL SIGNAL UNCERTAINTIES

As we have seen in the previous section, we are forced to rely on numerical models of the climate system in order to obtain information about the possible climate response to GHG changes. The aim of this section is to consider the major uncertainties involved in these projections of GHG-induced climate change. These uncertainties have been divided into six categories, which cover the uncertainties associated with:

- Errors in simulating current climate in uncoupled models
- Errors in simulating current climate in coupled models
- Inclusion of all relevant feedbacks
- Non-uniqueness of the signal
- Future concentrations of greenhouse gases
- The cold start problem

Each of these categories will be discussed briefly.

A. Errors in Simulating Current Climate in Uncoupled Models

Our confidence in the predictive capability of O/AGCMs when used in greenhouse warming experiments is diminished by the knowledge that their individual atmospheric and oceanic components, when tested separately[4] to see how well they represent the current climate, still show large systematic errors. A number of recent studies documenting the performance of atmospheric GCMs (Gates et al., 1990, 1992; Boer et al., 1992) have shown that, although model performance has generally improved over the last decade, all atmospheric models still have systematic errors in their simulation of the current climate (Gates, 1992). The ocean GCMs presently in use also have systematic errors (e.g., Maier-Reimer et al., 1993), although their validation is more problematic due to the difficulty of obtaining sufficient observed data, particularly for the intermediate and deep ocean.

B. Errors in Simulating Current Climate in Coupled Models

Even if an atmospheric GCM and an oceanic GCM, when tested separately, performed perfectly in simulating the present climate, there would be no guarantee that they would be equally successful when coupled together. In fact, experience shows that the interactive coupling of atmosphere and ocean GCMs generally leads to a phenomenon known as climate drift—that is, the tendency of the climate system to drift into a new and unrealistic mean state (Gates et al., 1984; Washington and Meehl, 1989).[5]

We probably would not have much confidence in the predictive skill of the model if this new mean state were used as the starting point for a greenhouse warming experiment. In order to circumvent this problem, modelers usually use a technique known as flux correction. This is a way of ensuring that the coupled model maintains a realistic mean state (Sausen et al., 1988). In a typical coupled model, the surface fluxes from the atmosphere into the ocean (e.g., of heat, wind stress, and precipitation minus evaporation) and from the ocean into the atmosphere (e.g., of sea-surface temperature) are corrected, both spatially and over the seasonal cycle. Intuitively, one can think of these corrections as anomaly fields which are added to the computed fluxes, enabling the atmosphere and ocean to receive the fluxes that they need (rather than the uncorrected, erroneous fluxes that they get) in order to maintain a stable climate.[6]

At the present state of development of coupled models, the flux corrections which must be made are sometimes large relative to the flux changes predicted in greenhouse warming experiments, particularly in areas of strongly non-linear dynamics (sea-ice margins). Thus flux correction introduces an additional area of uncertainty in greenhouse warming experiments. Scientists are currently working to obtain a better understanding of the physics of atmosphere-ocean coupling, in order to reduce the magnitude of these corrections, and eventually to remove the need for an engineering solution to a scientific problem.

C. Inclusion of all Relevant Feedbacks

Let us assume that we have an O/AGCM which realistically simulates the present climate without relying on any form of flux correction. Would this be a guarantee that the model would successfully predict the climate response to increasing GHG concentrations? The answer is, "probably not." Successful simulation of the present climate is probably a necessary, but not a sufficient, condition to ensure successful simulation of future climate. To be confident that our model has predictive skill on time scales of decades or longer, we would have to be sure that it incorporates all of the physics and feedback mechanisms that are likely to be important as GHG concentrations increase.

There are a number of reasons why it is difficult to feel confident that we have not forgotten anything important. We know, for example, that the feedbacks between clouds and the surface radiation budget are poorly understood. Cloud-radiation feedbacks involve such factors as the height, thickness, percentage coverage, and optical properties[7] of clouds. Recent studies have shown that different schemes for modelling cloud formation processes can lead to substantially different results in greenhouse warming experiments (Mitchell et al., 1989).

Numerous other examples are possible. Thus, we know that O/AGCMs lack an interactive biosphere and treat surface hydrology in a relatively crude way. Most models do not explicitly consider the radiative effects of aerosols, or of greenhouse gases other than CO_2 and water vapor. They do not incorporate an interactive carbon cycle model, so it is difficult to determine whether a CO_2-induced change in climate could influence the uptake of atmospheric CO_2 by the deep ocean, and hence feedback on the climate change. In summary, therefore, we hope that current O/AGCMs incorporate all of the important physics and feedback mechanisms necessary to model the effect of increasing GHG concentrations on climate—but we cannot guarantee this.

D. Non-Uniqueness of Model GHG Signal

In any transient experiment with a fully-coupled O/AGCM, the model's own internally-generated natural variability will be superimposed on the true time-dependent GHG signal (Santer et al., 1993a). In the presence of substantial natural climate variability, the GHG signal is not uniquely defined. This is illustrated in Figure 3-1, which shows the time evolution of zonally-averaged annual mean surface air temperature changes[8] in a 100-year greenhouse warming experiment recently performed with a coupled O/AGCM (Cubasch et al., 1992). In this experiment, the model was forced by time-varying GHG concentrations specified in the IPCC Scenario A ("Business-as-Usual"; Houghton et al., 1990).

The two panels of Figure 3-1 show different definitions of the signal. In the upper panel, the changes in surface air temperature in the experiment have been defined by subtracting the average pattern of surface air temperature in the first 10 years of the control run. In the lower panel, the changes are defined relative to each individual year of the control run.[9] Although both definitions show the same qualitative picture of a slowly-emerging greenhouse warming signal, with the largest temperature changes at high latitudes in both hemispheres, the precise details differ. The two definitions yield substantially different temperature changes in the region 45°S–90°S (which in this region translate into differences of two to three decades in the onset of warming). These differences are due to the natural variability which the model demonstrates in the control run (see Figure 3-2).

A further complication is the so-called "initial condition" problem. The experiment shown in Figure 3-1 commenced with a GHG concentration equivalent to that which existed in the atmosphere in 1985. Even in 1985 we had only limited observations about the climatic state[10] of the ocean, particularly the intermediate and deep ocean. This has the consequence that the experiment started from an ocean state which did not exactly correspond to the "true" ocean state which existed in 1985 but was imperfectly observed. Obviously, as one goes further back in time, sparser observations make it increasingly difficult to reconstruct a three-dimensional picture of the ocean's temperature and salinity.

It is only within the last few years that we have started to realize the implications of our imperfect knowledge of such initial conditions. While the pioneering work of Lorenz (1984) illustrated that the results of simple, "three equation" climate models are highly sensitive to initial conditions, such ideas have only recently been tested in the context of experiments with O/AGCMs. Recently, Cubasch et al. (1993) performed a suite of three greenhouse warming experiments with the Hamburg O/AGCM. The three transient experiments "forced" the coupled model with identical increases in green-

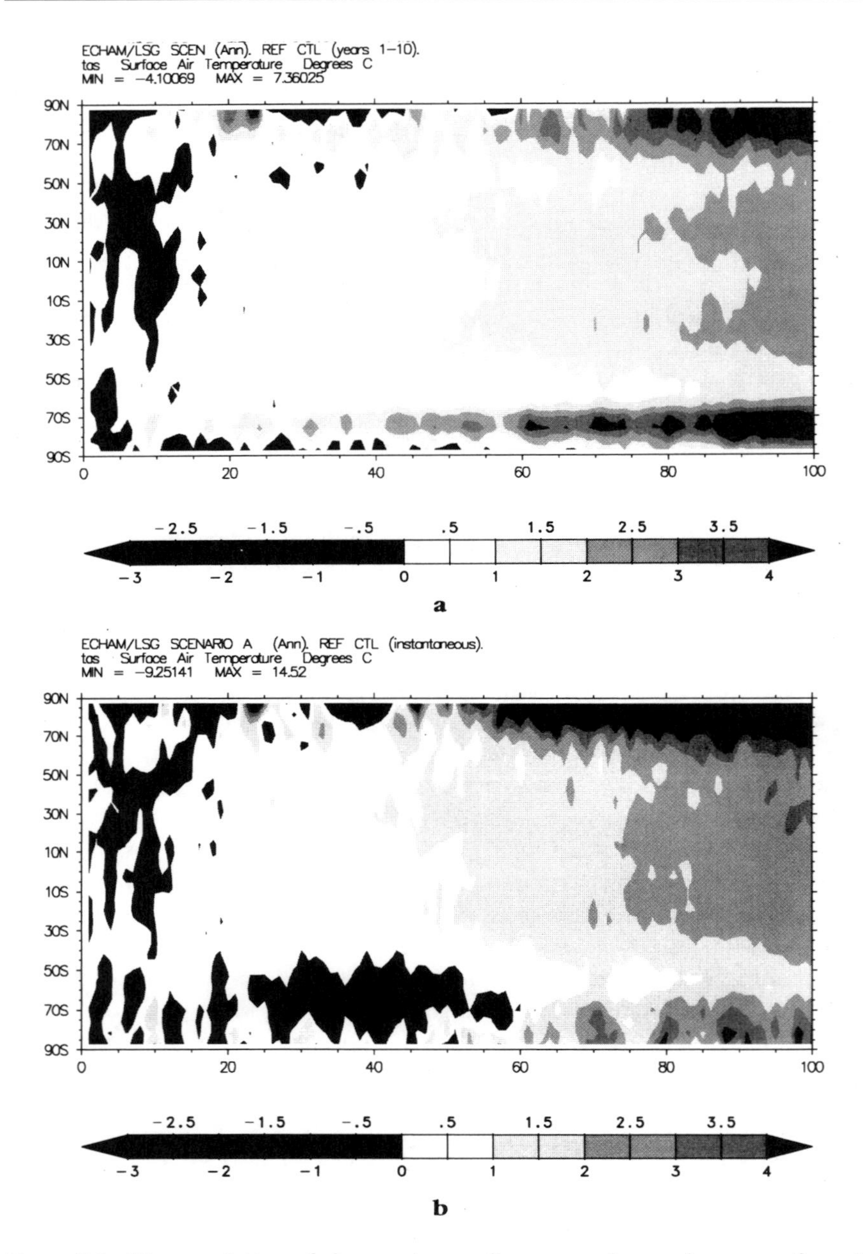

Figure 3-1. Time evolution of changes in zonally-averaged annual mean surface air temperature in the 100-year Scenario A experiment performed by Cubasch et al. (1992). Changes are expressed relative to the average of the first 10 years of the control run (Definition 1; panel 3-1a) or the instantaneous state of the control run (Definition 2; panel 3-1b). The space-time evolution of the signal differs for the two definitions.

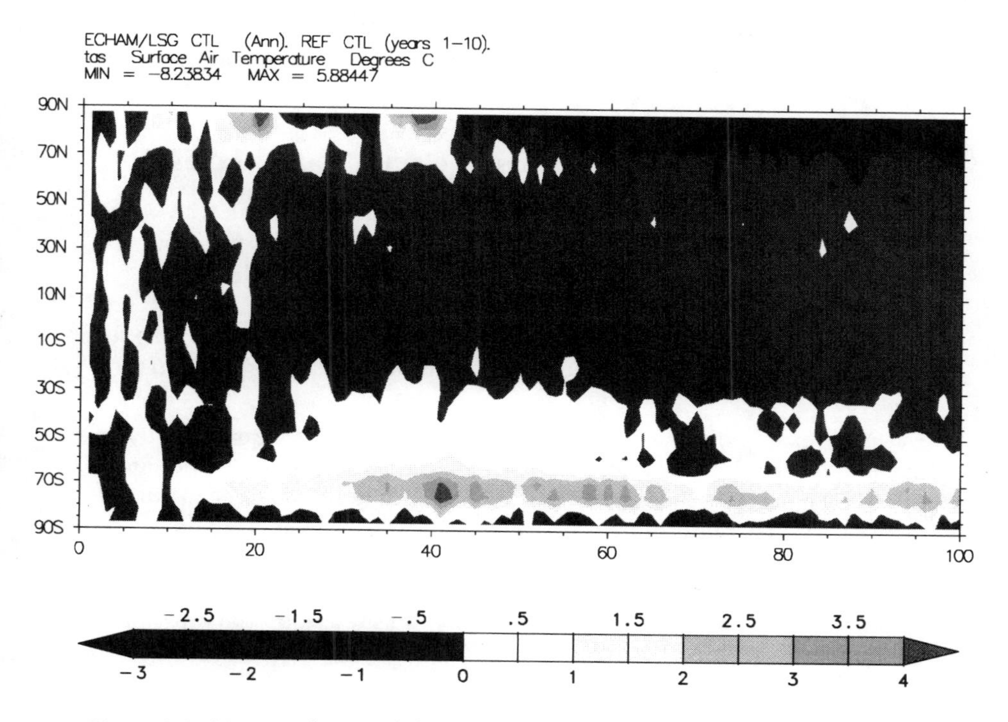

Figure 3-2. Time evolution of changes in zonally-averaged annual mean surface air temperature in the 100-year control integration performed by Cubasch et al. (1992). Changes are expressed relative to the average of the first 10 years of the control run. Note the large variability at high latitudes in both hemispheres.

house gases (the equivalent CO_2 concentrations from 1985-2035 specified in the IPCC Scenario A), but each experiment started from different initial conditions. The initial conditions were three different "snapshots" of the Cubasch et al. (1992) 100-year control run (taken at years 30, 60 and 90). Together with the original 100-year Scenario A experiment performed by Cubasch et al. (1992), this suite of four experiments provides some insights into the sensitivity of the greenhouse warming signal to the initial conditions of the climate system.

If we assumed that the four greenhouse warming experiments were not sensitive to the initial conditions, we would expect them to evolve in a similar way over space and time. We found, however, some notable differences in the space-time structure of the surface temperature signal in the four experiments. This suggests that (even for a single O/AGCM) we might have to perform a large number of transient greenhouse warming experiments in order to obtain a good idea of the statistical properties of the climate system's response to GHG increases. The same argument obviously applies to the natural variability properties simulated in a control run.

E. Future Concentrations of Greenhouse Gases

A further uncertainty is that we have no convenient crystal ball with which we can peer into the future and see how atmospheric GHG concentrations will change over the next 50-100 years. This uncertainty regarding the forcing has at least two aspects—the difficulty of predicting future GHG emissions, and uncertainties regarding the global carbon cycle, which will determine how the emitted CO_2 is partitioned and cycled between the atmosphere, ocean, and biosphere.

A simple example shows how such forcing uncertainties are translated into uncertainties regarding the magnitude and space-time evolution of the signal. Recall that Figure 3-1a illustrated the zonally-averaged annual mean surface air temperature changes for the Cubasch et al. (1992) Scenario A experiment performed with the Hamburg O/AGCM. Figure 3-3 shows the corresponding temperature changes from an experiment with a lower level of GHG forcing (Scenario D). In Scenario D, some relatively optimistic assumptions are made regarding the reduction of GHG emissions after the year 2000 (see Houghton et al., 1990).[11] A comparison of these two figures shows that there are differences both in absolute magnitude[12] and in the space-time evolution of the surface temperature signal in the two experiments.

F. Cold Start Problem

The final area of uncertainty in defining a transient GHG signal with an O/AGCM is the so-called "cold start" problem. This problem is related to the experimental set-up. Experiments such as those performed by Cubasch et al. (1992) were started with the atmosphere and ocean at equilibrium with respect to an equivalent CO_2 concentration of 360 ppm, which approximately corresponds to 1985 concentrations. In the real world, of course, there have been substantial changes in GHG concentrations before 1985. The neglect of this previous history of the GHG forcing means that we are neglecting any GHG-induced warming of the ocean which has taken place prior to 1985. This is the "cold start" error—the ocean has not been "warmed up" before the start of the experiment. Using simple linear models, Hasselmann et al. (1993) estimated that this error may be as large as 0.4°C after 50 years of the Cubasch et al. (1992) Scenario A experiment.

Obviously, it would be more realistic to start a transient greenhouse warming experiment with atmospheric GHG concentrations appropriate to 1900 or even earlier. A rigorous investigation of the cold start error will require

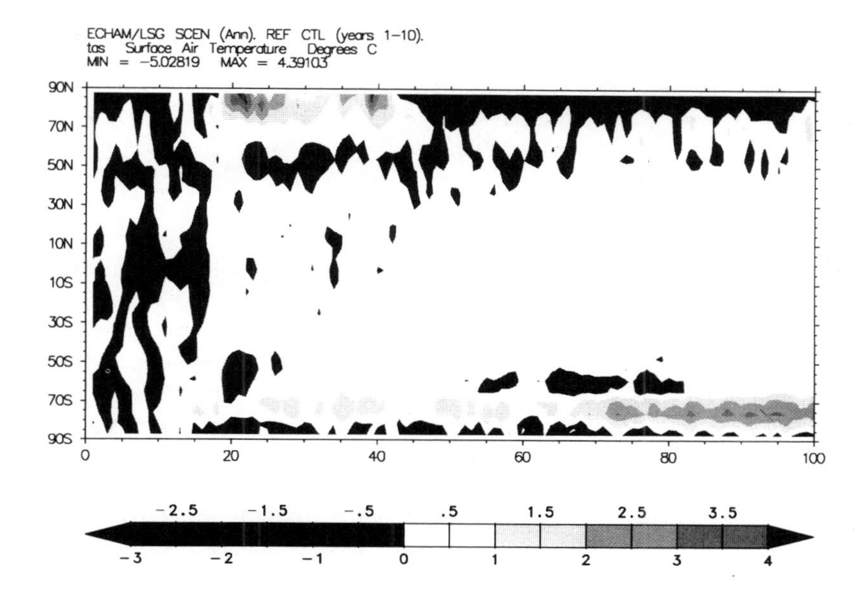

Figure 3-3. Time evolution of changes in zonally-averaged annual mean surface air temperature in the 100-year Scenario D experiment performed by Cubasch et al. (1992). Changes are expressed relative to the average of the first 10 years of the control run. Note the differences in absolute magnitude and space-time evolution of the signal relative to Scenario A (Figure 3-1a).

such an experiment. At present, however, even a single 100-year O/AGCM integration on a state-of-the-art supercomputer could consume many months of computer time. It is also worth noting that many such experiments would be required. Due to the twin problems of model-generated natural variability and sensitivity to initial conditions (Section III-4), a single experiment could not provide a single, definitive estimate of the magnitude of the cold start error.

IV. NATURAL VARIABILITY UNCERTAINTIES

The previous section discussed the uncertainties associated with the use of models to define the climate change signal likely to result from future changes in atmospheric GHG concentrations. But the climate change signal is only one part of the detection problem. In order to say something meaningful about when (or even whether) we could expect to detect a GHG signal, we need to have good estimates of the natural variability of the climate system on time scales of decades to centuries. Information on natural climate variability can be derived from three sources: instrumental records, paleoclimate records, and numerical models. This section considers some of the uncertainties associated with each of these sources.

A. Instrumental Records

In an ideal world, we would have observed surface air temperature over the last 100-150 years using an observing strategy and a network of observing stations well-suited to the detection of GHG-induced climate change. Our "ideal strategy" would have specified that there should be no changes in the instrumentation used to record temperature,[13] that observing times, frequencies, or practices should not change, that the location and elevation of observing stations should not change, and that stations should not be located in or near rapidly-growing urban areas. Our "ideal network" would have had enough stations to obtain a reasonable record of temperature variations over the entire Earth's surface (at least on spatial scales of several hundred kilometers), with no little or no change in the number of observing stations as a function of time.

Unfortunately, neither the ideal observing strategy nor the ideal observing network exists. (See Chapter 2 for further discussion of this problem.) All of the problems alluded to above make it difficult to reconstruct a homogeneous, spatially-complete picture of surface temperature changes over the last century (e.g., see Folland et al., 1990; Jones et al., 1991). We can attempt to correct some of these errors, such as the effects of urban warming on temperature records (Jones et al., 1989, 1990). Other problems, such as the deterioration in the spatial coverage of the observing network as one goes further back in time, are essentially insolvable.

Recently, it has been suggested that satellite records of near-surface temperature may provide the solution to some or all of these problems (Spencer and Christy, 1990). While satellite-derived data are spatially complete, they measure temperature in the lower troposphere and not at the Earth's surface. A more serious failing is the short length of available record (a decade or less). Satellite data cannot provide us with information about the natural variability of the climate system on time scales of decades.

B. Paleoclimate Records

Changes in climate affect a wide range of biological, chemical, and geological processes. As a result, climatic information is naturally recorded in tree rings, ice cores, coral reefs, laminated sediments, etc. (e.g., Crowley and North, 1991; Bradley and Jones, 1992; Briffa et al., 1992). If we can understand the recording mechanism—for example, the process by which climate imprints itself on tree growth and annual ring formation—then we have the potential to unlock a wealth of climate information stored in paleoclimate records.

Unfortunately, unravelling the history of climatic variability contained in such records is not a simple task. For example, many types of trees are more sensitive to moisture stress than to temperature, or may respond to non-climatic factors (e.g., changes in management practices). This makes it difficult to extract a temperature signal from the noise introduced by the variations in other factors which affect tree ring width. More importantly, spatial coverage is poor for paleoclimate data which can resolve annual temperature variability, and it is difficult to date and cross-check the climate information extracted from different locations (e.g., land and ocean) or from different proxy sources. For these reasons and many others, scientists have been unable to use paleoclimate data in order to reconstruct a satisfactory, spatially-complete picture of climate variability over the past 1,000 years.

C. Numerical Models

Numerical models provide another means of investigating the decadal- to century-time scale variability of the climate system. Some of the first model-based studies of natural variability used simple energy balance models (EBMs; e.g., Hasselmann, 1976). Such models generally solve equations for the heat balance of a highly-idealized representation of the Earth's atmosphere and ocean (Lemke, 1977). By forcing an EBM with white noise—for example, by heat flux anomalies which are essentially random in time,[14] and thus can be thought of as characteristic of day-to-day weather noise—it was possible to investigate the relationship between daily weather noise and the model's internally-generated variability on time scales of years to centuries. These early studies, together with more recent EBM studies by other groups (e.g., Wigley and Raper, 1990, 1991; Kim and North, 1991), have demonstrated that even simple EBMs can generate decadal- to century-time scale surface temperature fluctuations as an integrated response to daily-time scale random weather fluctuations.

While EBMs successfully reproduce many details of observed surface temperature variability on the annual- to decadal-time scales (Kim and North, 1991), they are not as useful on longer time scales, since they cannot explicitly simulate the horizontal and vertical transport of heat, salt, and momentum necessary for an accurate representation of the ocean circulation. It is therefore necessary to use more sophisticated models in order to obtain information on century-time scale natural variability.

Ideally, it would be desirable to study the century-time scale natural variability of the climate system by performing an ensemble of long (>1000 years) control runs with a fully-coupled O/AGCM. Due to computational restrictions, however, state-of-the-art O/AGCMs have generally been integrated for 100-200 years only (Stouffer et al., 1989; Cubasch et al., 1992). This is too short to obtain reliable information about the statistical properties of climate on the century time scale.[15]

This does not mean that we have to wait several years for the next generation of supercomputers before performing experiments which supply useful information about century-time scale natural variability. One possible answer is to extend the philosophy of noise-forced EBMs to its logical conclusion, and to force an uncoupled OGCM by white noise. This is computationally efficient (since the main computational burden in O/AGCM experiments is the atmosphere), which means that it is relatively inexpensive to integrate an uncoupled OGCM for several thousand years. The assumption underlying this type of experiment is that the ocean (with its very long time scales) is the most important player in determining the climate system's century-time scale natural variability, and that the atmosphere is a more or less passive "slave" whose behavior is relatively unimportant in terms of long time scale climate variability. The motivation is to see whether the ocean has preferred patterns and time scales of variability when forced by atmospheric weather noise.

This approach has recently been tested by Mikolajewicz and Maier-Reimer (1990) in a 3,800-year experiment. They forced the Hamburg OGCM with freshwater flux anomalies[16] which were white in time but had an amplitude and spatial structure characteristic of observed freshwater flux anomalies. The response of the ocean was extremely complex. Using advanced statistical techniques, it was possible to isolate different ocean "modes" of natural variability, each with its own characteristic time scale and spatial pattern. The dominant mode was found to have a time scale of roughly 320 years, and to describe the movement of large-scale salinity anomalies through the Atlantic via the model's conveyor belt circulation.[17] This mode shows up clearly in spectra of ice volume, mass transport, heat fluxes, and many other ocean variables (see Figure 3-4). It is interesting to note that temperature records reconstructed from Greenland ice cores also have a dominant mode of variability with a time scale just in excess of 300 years (Mikolajewicz and Maier-Reimer, 1991)—a tantalizing correspondence between the model and the real world, which deserves further investigation.

One drawback with using noise-forced OGCMs for studying century-time scale natural variability is that we do not know whether they are truly representative of the long time scale variability likely to occur in a fully-coupled O/AGCM. Although the important dynamics of the ocean are reproduced, and

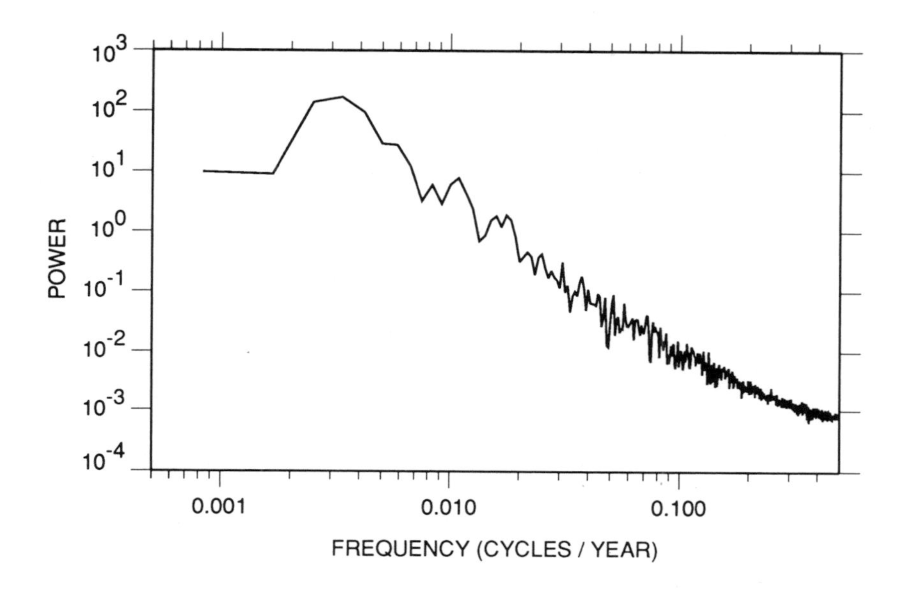

Figure 3-4. Spectrum of mass transport through the Drake Passage from the 3,800-year "ocean only" experiment performed by Mikolajewicz and Maier-Reimer (1990). Results are for a chunk length of 1,200 years, and are averaged over three non-overlapping chunks. Input time series were normalized. Note that the spectrum has maximum power at a period of approximately 320 years. The mode of variability associated with this spectral peak describes the movement of large-scale salinity anomalies through the Atlantic.

the effective forcing by the atmosphere is probably adequately represented by white noise, the model does not include any feedback with the atmosphere. In the Mikolajewicz and Maier-Reimer experiment (1990), for example, wind stress, freshwater fluxes and surface temperature were fixed at their climatological monthly mean values,[18] so that the atmosphere was unable to respond to any change in oceanic circulation, and in turn modify the pattern or time scale of the dominant ocean variability modes. The fact that surface temperature was held fixed at its climatological values in this experiment also has the consequence that the information we are probably most interested in for GHG detection purposes—the century time scale variability of surface temperature—is unavailable.

D. Implications of Natural Variability Uncertainties for Future Research

Without reliable estimates of decadal- to century-time scale natural variability, we will not be able to say anything meaningful about how long it will take to detect a GHG signal, or even whether a GHG signal can be detected at all. It is important to make a concerted effort to reduce the uncertainties in our knowledge of long time scale natural variability. This can be done in a number of different ways:

- By attempting to validate the variability data from noise-forced "ocean only" experiments and fully-coupled O/AGCM control runs. Rigorous validation will involve making comparisons with appropriate paleoclimate data. The development of a paleoclimate data set suitable for validation purposes will require a major international effort to date and cross-check the information from different geographical locations and different proxy sources.
- By exploring the sensitivity of results from noise-forced OGCMs to the horizontal and vertical resolution of the model and to the precise details of the forcing (e.g., the amplitude, correlation time, and correlation scale of the forcing, whether the forcing is applied to the fresh water fluxes, wind stress, etc.).
- By performing many 100–200-year control runs with a single O/AGCM, with each experiment starting from different (but plausible) initial conditions of the climate system (see Section III-D), or a much smaller number of long integrations (≥1,000 years). This will provide some insight into the statistical properties of the climate system on the century time scale.
- By trying to determine how much of the variability we see in control runs with fully-coupled O/AGCMs is *bona fide* natural variability of the coupled system, and how much is residual climate drift due to inadequacies in the flux correction scheme.
- By studying the model-dependence of natural variability results—for example, whether the modes of variability simulated by two totally different OGCMs are at all comparable.

The validation of model variability data will be a difficult task. While the century time scale variability in paleoclimate records reflects the response of the climate system to a complex mixture of external forcing factors (solar variability, volcanic aerosols, etc.) and internally-generated variability, the variability simulated by an O/AGCM represents *only* the natural variability of the coupled ocean-atmosphere system. In order to validate model variability in a more meaningful way, it may be necessary to perform experiments in which a coupled model is forced by the past forcing histories for volcanic aerosols and changes in solar luminosity.

We will always have to live with model-based uncertainties in defining the regional and seasonal details of a GHG signal. The fidelity with which models simulate decadal—to century-time scale natural variability should be testable, however, only if we can be clever enough to extract information from the silent biological, chemical, and geological witnesses to climatic change.

V. THE ATTRIBUTION PROBLEM: ESTABLISHING THE CAUSAL LINK

Let us assume that we have actually managed to detect the GHG signal predicted by a model in the observed record of surface temperature changes. This means that we have used some statistical technique to compare the model signal with the observed data. The correspondence between the two is so striking that we conclude (on the basis of some statistical test at a prescribed level of significance) that our result could not be due to chance alone.

This result does *not* mean that we have established a clear causal link between changes in GHG concentrations and changes in surface temperature. In order to attribute the change in climate to the change in greenhouse gases, we would have to rule out all other possible explanations for the climatic change. We would have to demonstrate in a convincing way that changes in solar luminosity, volcanic aerosols, sulfate aerosols, or other external forcing factors could not have resulted in the observed surface temperature changes. We would also have to demonstrate that the internally-generated variability of the climate system on time scales of decades to centuries could not be confused with a slowly-evolving GHG signal. Finally, we would have to show that no combination of these external forcing changes or internal natural variability could explain the observed changes.

Given the uncertainties in our understanding of natural variability (see Section IV) and in our knowledge of the history of solar and volcanic forcing (and other forcing mechanisms), it is easy to see why attribution is a much more difficult task than detection. For example, evidence from experiments investigating the model response to changes in the solar constant suggests that the pattern of surface temperature change may be similar to the response pattern obtained in greenhouse warming experiments (Wigley and Jones, 1981). This result may be due to the fact that the surface temperature changes are at least partly due to feedback mechanisms (such as ice-albedo feedback) which respond to *both* GHG and solar forcing.

Another example is the vertical profile of temperature change (stratospheric cooling and tropospheric warming), a common feature of greenhouse warming experiments. Recent work by Santer et al. (1993b) with the Hamburg O/AGCM suggests that (at least in this particular model) the simulated natural variability pattern can look similar to the profile of vertical temperature change which the model predicts in greenhouse warming experiments. This means that natural variability could mimic a GHG signal.

Clearly, it will be difficult to solve the attribution problem if we use a single variable only, such as temperature changes at the Earth's surface. By considering a number of climate variables simultaneously, we will probably have a better chance of defining a climatic fingerprint which is unique to changes in greenhouse gases (Madden and Ramanathan, 1980; MacCracken and Moses, 1982). The key to any "fingerprint" strategy is that we cannot use a hundred different climate variables simultaneously. We should focus on those variables for which suitable observed data exist, and which have high signal-to-noise ratios in the model data (in other words, variables which provide much more information about a GHG signal than about the model's own natural variability).

VI. CONCLUSIONS

State-of-the-art O/AGCMs are the product of large research teams and represent an enormous investment in terms of time, money, and scientific effort. Running these models for long greenhouse warming experiments may require months of CPU time, even on the fastest supercomputers. In view of this investment and the complexity of the models themselves, there is a tendency to regard the results of greenhouse warming experiments conducted with O/AGCMs as being "engraved in stone." The reasons that we have given above suggest that this would be a mistake. State-of-the-art coupled models can give us internally-consistent pictures of a possible future climate, and can teach us about the physical mechanisms which are likely to be important in changing climate. We should regard them as instruments for making intelligent guesses about future climate, rather than as instruments for making definitive predictions.

We have seen that model uncertainties affect both our predictions of GHG-induced climate change and our estimates of decadal- to century-time scale natural variability. The policy implications of these uncertainties are profound. Over the next 10 years it is possible, and perhaps even probable, that detection studies will provide some evidence in support of a link between changes in climate and changes in greenhouse gases. However, it is highly unlikely that such evidence will be sufficient to rule out natural variability and non-GHG forcing factors as explanations for the observed changes. It is also unlikely (particularly in the light of results from the Cubasch et al. (1993) Monte Carlo experiments) that research advances over the next decade will enable us to attach a high confidence level to our predictions of the timing and regional details of GHG-induced climate change. Decisions relating to energy and emissions policies, therefore, must be made against a background of probabilities rather than a background of certainties.

For scientists, the course of action is clear—to pursue research designed to reduce both signal and noise uncertainties. This is the only way that we will be able to detect a climate change signal and attribute this convincingly to changes in CO_2 and other greenhouse gases. For policy makers, the course of action is less clear, but scientific uncertainty does not relieve them of the responsibility to develop energy and emissions policies which address environmental problems with a time scale significantly longer than four years.

Acknowledgments

The author would like to thank Larry Gates and Karl Taylor for valuable comments and suggestions, and Uli Cubasch and Uwe Mikolajewicz for supplying model data.

NOTES

[1] Natural variability can be defined as that portion of the total variability of climate which has nothing to do with changes in GHG concentrations, or with changes in other external factors which influence climate (e.g., the sun's radiation output or volcanically-ejected dust), and is solely due to the internal dynamics of the atmosphere and ocean.

[2] By paleoclimatic data, we mean climatic data that can be inferred from such sources as tree rings, ice cores, layered lake sediments, etc. (see, for example, Bradley and Jones, 1992), providing information about climate variability on time scales of centuries to thousands of years.

[3] This redistribution occurs both horizontally, via currents such as the Gulf Stream and Kuroshio, and vertically, in regions where water denser than underlying water masses can sink (e.g., in areas of the North Atlantic or the Antarctic Circumpolar Current) or less-dense water upwells.

[4] By "testing separately," or "running in uncoupled mode," we mean that the atmospheric GCM is driven by the observed record of monthly-mean sea-surface temperature (SST) and sea-ice distribution, for example over the period 1979-1988, rather than by the SST and sea-ice fields predicted by an ocean model (which will have their own sources of error). This enables us to isolate errors which are due to the atmospheric model only. A similar procedure is used to evaluate errors in the ocean model component.

[5] For example, ocean temperatures which are much colder than those observed on average, or an unrealistic distribution of sea-ice.

[6] Note that once the flux corrections have been calculated, they remain invariant from year-to-year in a control run or experiment performed with the coupled model.

[7] For example, the size distribution of water droplets or ice particles within the cloud, or the number of cloud condensation nuclei.

[8] The zonal average is the mean change along each band of latitude.

[9] In other words, year 1 of the experiment minus year 1 of the control, etc.

[10] For example, in terms of the three-dimensional structure of temperature and salinity.

[11] The IPCC Scenario D assumes a 50% reduction in CO_2 emissions from 1985 levels by the

middle of the next century, largely from a shift to renewable and nuclear energy (Houghton et al., 1990).

[12] The globally-averaged annual temperature change after 100 years is over four times larger in Scenario A than in Scenario D (2.6° versus 0.6°, respectively).

[13] Or that there should be sufficient overlap between an old instrument and its replacement in order to calibrate the new instrument, and prevent discontinuities in the record.

[14] But which have an amplitude (and possibly also a spatial structure) typical of observed heat fluxes.

[15] Even if computational restrictions did not exist, we have the additional problem that at least some of the variability exhibited by a fully-coupled O/AGCM may be spurious climate drift attributable to inadequacies in the flux correction scheme (see Section III-B). Distinguishing between residual drift and real natural variability of the coupled system is a difficult task (see Santer et al., 1993b).

[16] Note that the freshwater fluxes from the atmosphere into the ocean are determined by the net balance between the processes of precipitation, evaporation, and surface runoff.

[17] The term "conveyor belt circulation" refers to the horizontal and vertical movement of large water masses within and between ocean basins.

[18] The spatially-coherent, temporally-white freshwater flux anomalies used to force the model were superimposed on the climatological mean fresh water fluxes.

LITERATURE CITED

Barnett, T.P. 1986. Detection of changes in global tropospheric temperature field induced by greenhouse gases. *J. Gèophys. Res.* 91:6659–67.

Barnett, T.P. 1991. An attempt to detect the greenhouse-gas signal in a transient GCM simulation. In *Greenhouse-Gas-Induced Climatic Change: A Critical Appraisal of Simulations and Observations* ed. M.E. Schlesinger, 559–68. Elsevier: Amsterdam.

Barnett, T.P. and M.E. Schlesinger. 1987. Detecting changes in global climate induced by greenhouse gases. J. *Geophys. Res.* 92:14772–80.

Boer, G.J., K. Arpe, M. Blackburn, M. Déqué, W.L. Gates, T.L. Hart, H. le Treut, E. Roeckner, D.A. Sheinin, I. Simmonds, R.N.B. Smith, T. Tokioka, R.T. Wetherald, and D. Williamson. 1992. Some results from an intercomparison of the climates simulated by 14 atmospheric general circulation models. J. *Geophys. Res.* 97: 12771–86.

Bradley, R.S. and P.D. Jones. 1992. *Climate Since A.D. 1500.* Routledge: London. 679 pp.

Briffa K.R., P.D. Jones, T.S. Bartholin, D. Eckstein, F.H. Schweingruber, W. Karlén, P. Zetterberg, and M. Eronen. 1992. Fennoscandian summers from A.D. 500: Temperature changes on short and long timescales. *Climate Dynamics.* 7:111-19.

Cubasch, U., K. Hasselmann, H. Höck, E. Maier-Reimer, U. Mikolajewicz, B.D. Santer, and R. Sausen. 1992. Time-dependent greenhouse warming computations with a coupled ocean-atmosphere model. *Climate Dynamics.* 8:55–69.

Cubasch, U., B.D. Santer, A. Hellbach, G. Hegerl, H. Höck, E. Maier-Reimer, U. Mikolajewicz, A. Stössel, and R. Voss. 1993. Monte Carlo climate change forecasts with a global coupled ocean-atmosphere model. *Climate Dynamics.* In press.

Crowley, T.J. 1991. Utilization of paleoclimate results to validate projections of a future greenhouse warming. In *Greenhouse-Gas-Induced Climatic Change: A Critical Appraisal of Simulations and Observations.* ed. M.E. Schlesinger, 35–45. Elsevier: Amsterdam.

Crowley, T.J. and G.R. North. 1991. *Paleoclimatology.* Oxford University Press: New York, 339 pp.

Folland, C.K., T. Karl, and K. Ya Vinnikov. 1990. Observed climate variations and change. In *Climate Change. The IPCC Scientific Assessment.* eds. J.T. Houghton, G.J. Jenkins, and J.J. Ephraums, 195–238. Cambridge University Press: Cambridge.

Gates, W.L. 1992. AMIP: The Atmospheric Model Intercomparison Project. *Bull. Am. Met. Soc.* 73:1962–70.

Gates, W.L., Y.J. Han and M.E. Schlesinger. 1984. The global climate simulated by a coupled atmosphere-ocean general circulation model: preliminary results. *Climatic Research Institute Report No. 57.* Corvallis: Oregon, 31 pp.

Gates, W.L., J.F.B. Mitchell, G.J. Boer, U. Cubasch, and V.P. Meleshko. 1992. Climate modelling, climate prediction and model validation. In *Climate Change 1992, the Supplementary Report to the IPCC Scientific Assessment.* eds. J.T. Houghton, B.A. Callander, and S.K. Varney, 97–134. Cambridge University Press: Cambridge.

Gates, W.L., P.R. Rowntree, and Q-C. Zeng. 1990. Validation of climate models. In *Climate Change 1992, the Supplementary Report to the IPCC Scientific Assessment.* eds. J.T. Houghton, B.A. Callander, and S.K. Varney, 93–130. Cambridge University Press: Cambridge.

Hasselmann, K. 1976. Stochastic climate models. Part I: Theory. *Tellus.* 28:473–85.

Hasselmann, K. 1988. Some problems in the numerical simulation of climate variability using high-resolution coupled models. In *Physically-Based Modelling and Simulation of Climate and Climatic Change.* ed. M.E. Schlesinger, 583–605. Kluwer: Dordrecht.

Hasselmann, K., R. Sausen, E. Maier-Reimer, and R. Voss. 1993. On the cold start problem in transient simulations with coupled ocean-atmosphere models. *Climate Dynamics.* In press.

Houghton, J.T., G.J. Jenkins, and J.J. Ephraums. 1990. *Climate Change. The IPCC Scientific Assessment.* Cambridge University Press: Cambridge, 365 pp.

Jones, P.D., P. Ya Groisman, M. Coughlan, N. Plummer, W.C. Wang, and T.R. Karl. 1990. Assessment of urbanization effects in time series of surface temperature over land. *Nature.* 347: 169–72.

Jones, P.D., P.M. Kelly, C.M. Goodess, and T.R. Karl. 1989. The effect of urban warming on the Northern Hemisphere temperature average. J. *Climate.* 2:285–90.

Jones, P.D., T.M.L. Wigley, and G. Farmer. 1990. Changes in hemispheric and regional temperatures 1851 to 1988 over land and marine areas. In *Greenhouse-Gas-Induced Climatic Change: A Critical Appraisal of Simulations and Observations.* ed. M.E. Schlesinger, 153–72. Elsevier: Amsterdam.

Karoly, D.J. 1987. Southern Hemisphere temperature trends: A possible greenhouse gas effect? *Geophys. Res. Lett.* 14: 1139–41.

Karoly, D.J. 1989. Northern Hemisphere temperature trends: A possible greenhouse gas effect? *Geophys. Res. Lett.* 16:465–8.

Kim K.Y. and G.R. North. 1991. Surface temperature fluctuations in a stochastic climate model. *J. Geophys. Res.* 96:18573–80.

Lemke, P. 1977. Stochastic climate models. Part 3: Application to zonally-averaged energy models. *Tellus.* 29:385–92.

Lorenz, E.N. 1984. Irregularity: a fundamental property of the atmosphere. *Tellus.* 36A: 98–110.

MacCracken, M.C. and H. Moses. 1982. The first detection of carbon dioxide effects. Workshop Summary, 8-10 June 1981, Harpers Ferry, West Virginia. *Bull. Amer. Met. Soc.* 63:1164–78.

Madden, R.A. and V. Ramanathan. 1980. Detecting climate change due to increasing carbon dioxide. *Science.* 209: 763–8.

Maier-Reimer E., U. Mikolajewicz, and K. Hasselmann. 1993. On the sensitivity of the global ocean circulation to changes in the surface heat flux forcing. *J. Phys. Oceanogr.* In press.

Manabe, S., R.J. Stouffer, M.J. Spelman, and K. Bryan. 1991. Transient responses of a coupled ocean-atmosphere model to gradual changes of atmospheric CO_2. Part I: Annual mean response. *J. Climate*. 4:785–818.

Mikolajewicz, U. and E. Maier-Reimer. 1990. Internal secular variability in an ocean general circulation model. *Climate Dynamics*. 4:145–56.

Mikolajewicz, U. and E. Maier-Reimer. 1991. One example of a natural mode of the ocean circulation in a stochastically forced ocean general circulation model. In *Strategies for Future Climate Research*. ed. M. Latif, 287–318. Max-Planck-Institut für Meteorologie: Hamburg.

Mikolajewicz, U., E. Maier-Reimer, and T.P. Barnett. 1992. Acoustic detection of greenhouse-induced climate changes in the presence of slow fluctuations of the thermohaline circulation. *J. Phys. Oceanogr.* In press.

Mitchell, J.F.B., S. Manabe, T. Tokioka, and V. Meleshko. 1990. Equilibrium climate change. In *Climate Change. The IPCC Scientific Assessment*. eds. J.T. Houghton, G.J. Jenkins, and J.J. Ephraums, 131–72. Cambridge University Press: Cambridge.

Mitchell, J.F.B., C.A. Senior, and W.J. Ingram. 1989. CO_2 and climate: a missing feedback? *Nature*. 341: 132–4.

Santer, B.D., T.M.L. Wigley, M.E. Schlesinger, and P.D. Jones. 1991. Multivariate methods for the detection of greenhouse-gas-induced climate change. In *Greenhouse-Gas-Induced Climatic Change: A Critical Appraisal of Simulations and Observations*. ed. M.E. Schlesinger, 511–36. Elsevier: Amsterdam.

Santer, B.D., T.M.L. Wigley, and P.D. Jones. 1993a. Correlation methods in fingerprint detection studies. *Climate Dynamics*. In press.

Santer, B.D., W. Brüggemann, U. Cubasch, K. Hasselmann, H. Höck, E. Maier-Reimer, and U. Mikolajewicz. 1993b. Signal-to-noise analysis of time-dependent greenhouse warming experiments. Part 1: Pattern Analysis. *Climate Dynamics*. In press.

Sausen, R., K. Barthel, and K. Hasselmann. 1988. Coupled ocean-atmosphere models with flux corrections. *Climate Dynamics*. 2:154–63.

Schlesinger, M.E. and J.F.B. Mitchell. 1987. Climate model simulations of the equilibrium climatic response to increased carbon dioxide. *Rev. of Geophys*. 25:760–98.

Spencer, R.W. and J.R. Christy. 1990. Precise monitoring of global temperature trends from satellites. *Science*. 247:1558–62.

Stouffer, R.J., S. Manabe, and K. Bryan. 1989. Interhemispheric asymmetry in climate response to a gradual increase of atmospheric CO_2. *Nature*. 342:660–2.

Washington, W.M. and G.A. Meehl. 1989. Climate sensitivity due to increased CO_2: Experiments with a coupled atmosphere and ocean general circulation model. *Climate Dynamics*. 4:1–38.

Wigley, T.M.L. and T.P. Barnett. 1990. Detection of the greenhouse effect in the observations. In *Climate Change. The IPCC Scientific Assessment*. eds. J.T. Houghton, G.J. Jenkins, and J.J. Ephraums, 239–56. Cambridge University Press: Cambridge.

Wigley, T.M.L. and P.D. Jones. 1981. Detecting CO_2-induced climatic change. *Nature*. 292:205–8.

Wigley, T.M.L. and S.C.B. Raper. 1990. Natural variability of the climate system and detection of the greenhouse effect. *Nature*. 344:324–7.

Wigley, T.M.L. and S.C.B. Raper. 1991. Internally-generated natural variability of global-mean temperatures. In *Greenhouse-Gas-Induced Climatic Change: A Critical Appraisal of Simulations and Observations*. ed. M. Schlesinger, 471–82. Elsevier: Amsterdam.

4

CURRENT CLIMATE and POPULATION CONSTRAINTS on WORLD AGRICULTURE

R. Gommes
Research and Technology Development Division, FAO, Rome

I. INTRODUCTION

World agriculture, whether in developed or developing countries, remains very dependent on climate resources. Even without projected changes in global temperature and rainfall patterns, several regions will experience difficulties in providing sufficient food to meet ever increasing demands due to population growth. This chapter examines the current climate and population constraints on world agriculture.

The first part of this chapter focuses on current climate constraints for agriculture for developing countries. For these countries, current fluctuations in agricultural production affect not only the income of the people but, in some cases, their very survival (Bach et al., 1981; IRRI, 1989; Riebsame, 1988). Perhaps the largest single driving force behind annual fluctuations of crop yields and food supply is variability in rainfall. Adequate supplies, at the crucial stages in crop development, can mean bounty or disaster for farmers. This does not mean that other factors, such as air moisture, fluxes in solar radiation, and temperatures are not important. Yet for any given season, these factors are probably less important than the timing and amount of rainfall.

The second section of this chapter focuses on population constraints on agriculture. With the world's population expected to continue growing at an exponential rate, several regions will face severe land constraints for planting crops. Efforts will continue to increase yields, but may not be possible in some areas of the world. In many recent instances of world hunger, there has clearly been a human component to the drought. The situation has been worsened by the cultivation of more marginal lands, which are more easily damaged by severe weather.

II. CLIMATE CONSTRAINTS

A. Yield Dependence on Rainfall

It is relatively difficult to express the degree of variability of crop yields or production with a simple statistic or index. Observed trends are due in most cases to changes in technology and management practices, but could also indirectly result from trends in weather or, even more directly, in population (trends in areas planted). One attempts to deal with these multiple effects by detrending the data, i.e., to express the variability relative to a linear or more complex trend. Even this, however, leads to difficulties linked to the high level of aggregation of national production statistics.

Consider the case of rice production in Italy and Bangladesh. As will be shown below, rice production in Bangladesh is exposed to major weather impacts and yet the detrended coefficient of variation (DCV[1]) is 10.8% in Italy and 7.1% in Bangladesh, indicating greater variability in Italy. This is explained by the fact that in Italy most rice is grown in a single and climatically rather homogeneous area. In Bangladesh, however, the crop is grown over a much larger area, following extremely different cultivation practices (from rainfed to floating), and with cropping intensities up to 300%.[2] The aggregation effect results in a lower *apparent* variability. Appreciable variability can be found in small countries or countries which grow only one crop a year (Sri Lanka 14%; Costa Rica 20.4%).

The discussion above aims at showing the importance of spatial variations of crop yields and production. The direct influence of weather is visible only at a very low level of aggregation, or where a dominant limiting factor affects relatively homogeneous areas, as is the case in the African Sahel.

Figure 4-1 shows crop yields in Africa from 1961 to 1990 (FAO, 1990). While the general continental trend is positive, several depressions relating to major crop failures in certain regions are evident. For example, during the peak of the 1984 drought, cereal production in Niger was disastrous (Figure 4-2). Note that there is a marked correlation between a crude rainfall index[3] (RI) and *per capita* cereal production (60% of their inter-annual variance is accounted for by RI). Interestingly, the area harvested *per capita* is relatively stable;[4] the coefficients of variation of yields (CV of Y), previously deflated for RI, are similarly stable, and surprisingly low.

B. Rainfall Trends

The trend illustrated above for Niger is a well known example of a recent and regional climatic fluctuation in the Sahel (Todorov, 1985; Glantz, 1987). Whether it is part of greenhouse gas induced warming, however, is still open

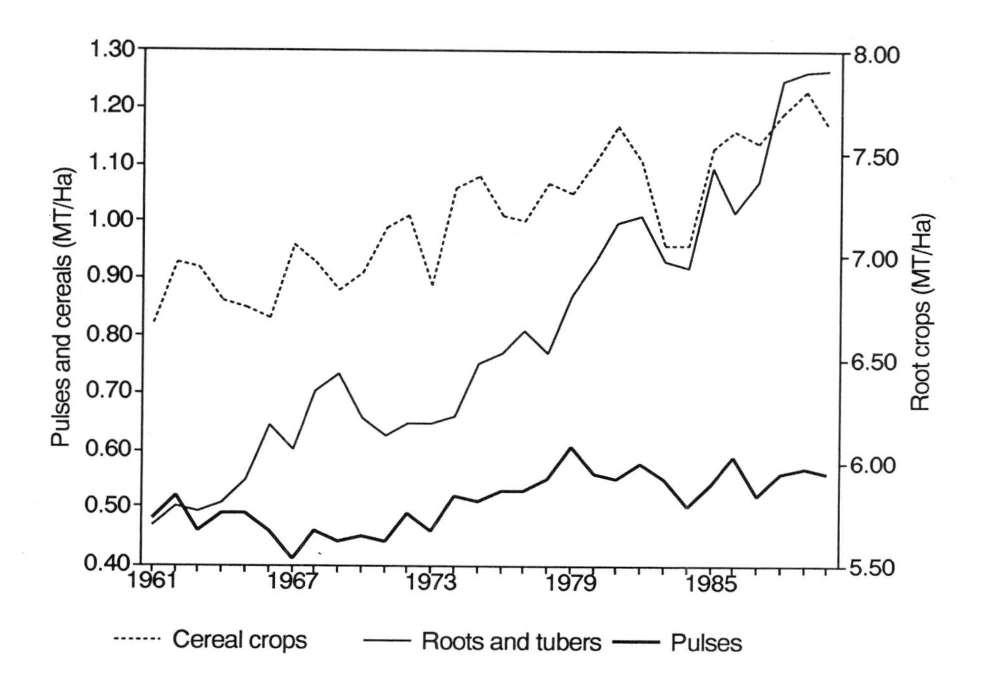

Figure 4-1. Yield trends of cereals, root-crops, and legumes in Africa between 1961 and 1990. Yields are in metric tons per hectare. Based on data from FAO (1990).

to debate. Changes in land use could provide an alternative explanation. Less well publicized are several exceptionally good years since 1988, which are often interpreted as a reversal of the trend to "desertification" by Sahelian farmers ("the rains have returned to the Sahel"). The natural variability in rainfall makes it difficult to state the causal relationships.

For some areas of the world, there are reported instances of upward trends in rainfall. Table 4-1 notes precipitation averages for four areas in Brazil. One notes that along the coast (Fortaleza and Rio de Janeiro), there has apparently been an increase in rainfall (compared with the previous normal period) accompanied by an increase in variability. Again, it is unlikely that this effect is due to a global climatological fluctuation; instead, it is probably ascribable to urbanization which increases convection and local rain. The effect is well documented in Brazil and other localities, even significantly smaller ones, in all tropical areas (de F. Monteiro, 1986; Oke, 1986).

Even in areas where rainfall amount does not increase, variability is affected. This results in more frequent thunderstorms and high intensity rainfall events with the all too well known negative consequences of violent

Table 4-1. Precipitation average annual totals (mm), and standard deviation in four Brazilian stations from contrasting climatic areas[a].

	1931-1960		1961-1990	
	Avg	Std. Dev.	Avg	Std. Dev.
Fortaleza	1260	385	1669	419
Rio de Janeiro	1088	211	1171	259
Manaus	2098	298	2238	301
Cuiabà	1370	233	1320	193

[a] Coastal: Fortaleza and Rio; Amazon region: Manaus; Mato Grosso: Cuiabà. The normals were recalculated to cover the rainy season (July–June). Refer to Table 4-2 for the coordinates of the stations.

Source: Data in the FAO agroclimatic database.

run-off, flooding, and deteriorating water quality in areas downwind from major cities.

C. Rainfall Variability

In technical terms, rainfall variability is difficult to define because the statistical distribution of rainfall has a positive skew, and because most series display a trend. However, defining variability is necessary for impact assessments as averages are usually insufficient, except for very broad planning. A comparison of four contrasting Brazilian stations, Figure 4-3, with the extremes observed in one of them, Figure 4-4, shows that the potential range of variation is such that Cuiabà (Mato Grosso), depending on the years, could be assigned to the much more humid Amazon basin climate.

Due to the skewed distribution of rainfall, the use of averages tends to overestimate actual expected rains. For instance, in May, amounts as high as the normal are recorded only in 36% of the years in Cuiabà; in the wetter Manaus, the frequency increases to 47%. This also means that the standard deviation and the coefficients of variation (CV) must be regarded with due caution.

As a consequence of the negative correlation between rainfall and Potential Evapotranspiration (PET), cropping season characteristics may be highly variable, particularly in transition climates (Rio de Janeiro). In these transition climates, where rainfall patterns are between unimodal and bimodal, farmers' decision making is even more difficult. The best strategy, in many cases, may be to plant at the beginning of a mid-season dry spell in order to benefit from the stored soil moisture. If the dry spell is not too long, well developed vegetative crops will be able to resume full growth with the next wet spell.

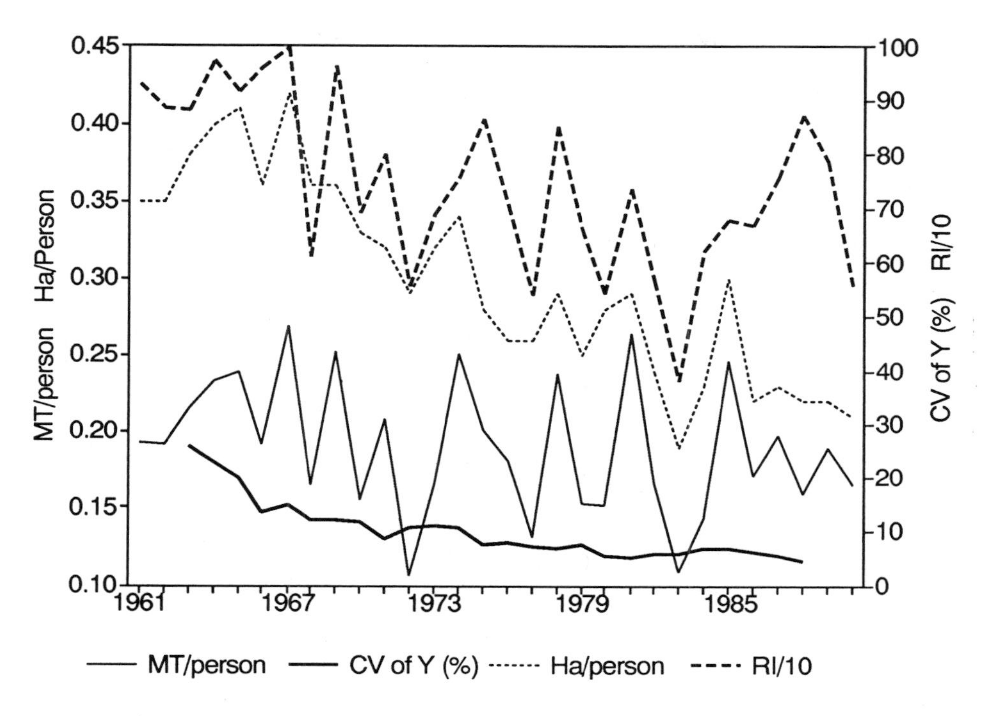

Figure 4-2. A comparison of *per capita* cereal production and area harvested in Niger (metric ton, urban population only) with an empirical rainfall index (RI). The coefficients of variation of yield (corrected for RI) are 5-year running averages. Based on data from FAO (1990) and from the IAO agroclimatic database.

Table 4-2 shows, for the same stations as in Figures 4-3 and 4-4, that the "average" season usually overestimates the length of actual seasons, and usually anticipates the actual average beginning by about up to two months (Rio de Janeiro). The standard deviations of both starting day and duration may also reach values close to three months (Manaus).

Clearly, a long season with high temperatures will allow a farmer to grow a crop with rather long cycles, but new problems of high pest and disease risk, combined with difficulty drying the harvested grain, and post harvest losses, frequently offset the benefits normally associated with long crop cycles.

There are many interesting situations where management techniques have been successfully used to reduce the problems caused by rainfall variability. One example occurs in the very low rainfall areas at the margin of the deserts (from Mauritania to the Sudan), which are characterized by very sparse vegetation and a short rainy season, normally not exceeding one month. These people use a water harvesting technique which consists of trapping

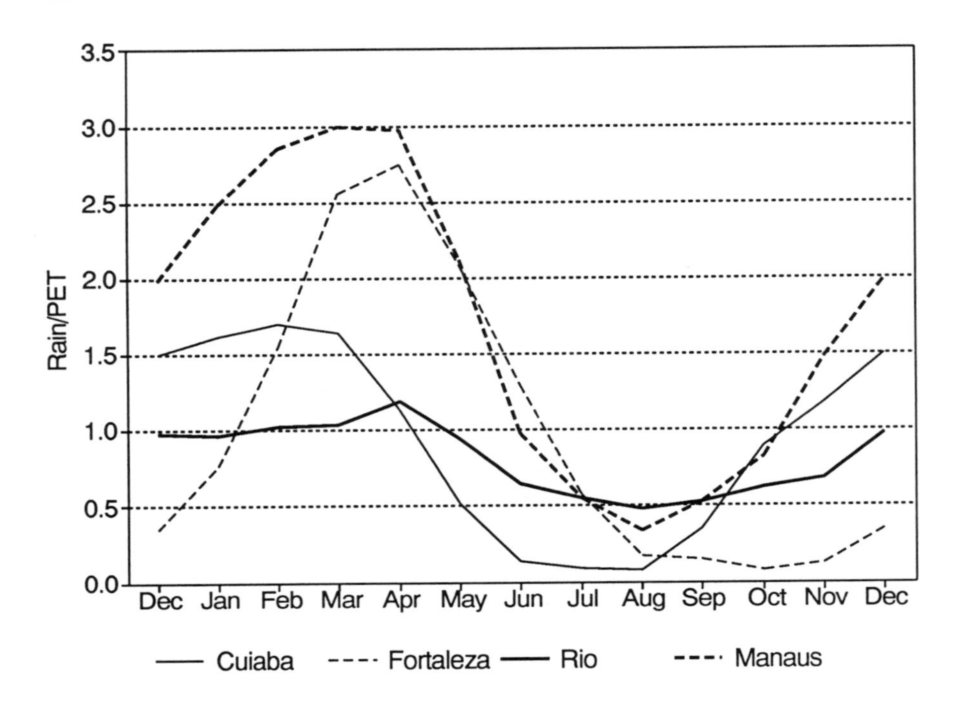

Figure 4-3. Average rainfall/PET (Potential evapotranspiration) ratios for 4 Brazilian stations. Based on data (1931–1990) from the FAO agroclimatic data base.

run-off and run-on water behind low (1 m) mud dams, and planting sorghum along the banks of the artificial pond, closer and closer to the dam as water infiltrates and evaporates. Crops grow under high radiation and low moisture, reducing the incidence of diseases. They mature over a longer period, further reducing risk, and are harvested when needed.

Finally, it is interesting to note that the risk may be higher in the transition zone where rainfall just supports rainfed crops than in both the dryer and wetter areas.

D. Persistence

Persistence is defined as the characteristic "inertia" of weather and climate which leads to the occurrence of dry or wet spells, and warm or cold periods. Persistence occurs at all time scales, from days to years. Persistence often appears in time series as pseudo-cycles, some of which (typically in the 2-15 year range) may be statistically significant. There is, however, no guarantee that the cycles will persist in the future. Their usefulness in agricultural

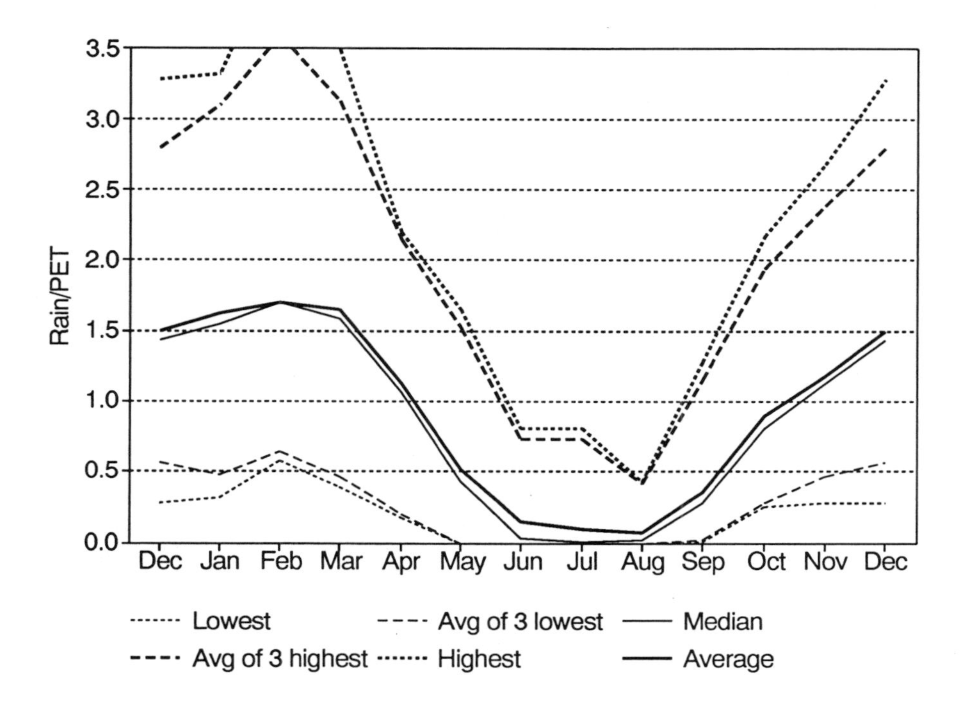

Figure 4-4. Range of variability of rainy seasons in Cuiabà (Rainfall/PET) for 1931–1990. For each month, the lowest and highest recorded values are shown next to the average of the three lowest records and the three highest records, and the average and the median. Based on data from the FAO agroclimatic database.

planning is very limited.

Regardless of the mechanism(s) behind persistence, they indirectly affect crop yields and production. Runs of either "good" or "bad" years are more likely than randomly distributed years. The implications for food supply in areas with limited or primitive storage facilities are serious. Table 4-3 illustrates the behavior of "good" (upper quartile) and "bad" (lower quartile) years in the area of southern Lake Victoria. The most likely sequence is "good after good" (0.404), followed by "bad after bad" (0.348). The probability of a "good" year following a "bad" year is only 0.206.

An example of an important consequence of persistence is the build-up of cattle stocks during runs of wet years. During subsequent dry years serious conflicts between nomads and farmers often result from herders moving into areas where crop farming economies predominate.

Table 4-2. Some characteristics of the rainy season in four Brazilian stations, 1931–1990.

Station	Cuiabà	Fortaleza	Rio	Manaus
Longitude	-56.12	-40.08	-43.17	-60.22
Latitude	-15.55	-7.52	-22.92	-3.13
Altitude	179.0	605.0	3.0	72.0
"Average season"[a]				
Starting day [b,c]	266	361	239	256
Duration	236	207	344	313
% years with one season only	73	78	18	68
% years with more than 2 seasons[d]	0	0	75	5
Years with 1 season				
Starting day [b,c]	267	8	292	248
Std. dev.	15	26	50	49
Duration	160	174	191	239
Std. dev.	73	42	70	96
Years with more than 1 season[e]				
Starting day [b,c]	273	23	309	290
Std. dev.	72	31	49	19
Duration	172	134	153	259
Std. dev.	49	48	45	18

[a]Long term averages
[b]Given as day numbers (1 = Jan. 1; 365 = Dec. 31)
[c]When rainfall amounts exceed half PET and are sufficient to sustain growth of germinating crops
[d]Occurs when a dry spell splits a long rainy season(s)
[e]Data corresponds to the longest normal season
Source: FAO agroclimatic database, FAO, 1978a, 1978b, 1980, 1981.

E. Extreme Weather

Weather elements can be subdivided into "standard" and "extreme" factors. The former are those which stay within the normal physiological amplitude for crops and may be used to forecast their behavior through crop modeling. The extreme factors involve much higher energies and can damage crops mechanically (like hail, extreme wind, or frost), but are difficult, if not impossible, to include in simulation models (Gommes and Nègre, 1992).

In so-called disaster-prone areas, like Bangladesh, Figure 4-5, where extreme events such as floods, droughts, and cyclones occur almost every year, the inter-annual variation of a risky crop production (like the Aman-season rice) may reach 1 million MT or more, values close to the increase brought about during the last 30 years by improved technology and new

Table 4-3. Transition matrix showing the probabilities that a maize yield in a given quartile class during year i will be followed by a yield in any other class during the following year i+1.

Year	i + 1				
	Quartile[a]		Q1	Q2	Q3
		0.348	0.174	0.272	0.206
	Q1				
i		0.305	0.158	0.196	0.341
	Q2				
		0.107	0.429	0.321	0.143
	Q3				
		0.242	0.192	0.162	0.404

[a]Q1, Q2, Q3 = lower, median and upper quartiles, respectively.
Source:Based on 1966-84 yield trials and 1932-84 weather data from Ukiriguru research station, Mwanza region (Gommes, 1985).

varieties.

In order to assess the impact of a weather disaster on crop production, one must link two fundamental aspects: first, the disaster proper, i.e., the destructive power of the event; secondly, the characteristics of the agricultural system which has been hit. To understand how a given agricultural system can be affected by a disaster, consider the case of rice in Bangladesh (Figure 4-5). Between 1970 and 1984, three major weather disasters occurred. In mid-November 1970, one of the worst cyclones in history caused between 200,000 and 400,000 deaths in Bangladesh alone; in 1978–1979 a severe drought befell the country, followed, in 1984, by extended heavy floods which lasted unusually long, from May to September. The result was, for each of the three extreme events, a sharp decrease in total rice production.

The 1970 cyclone struck the country in November. This coincided with the start of the Aman rice harvest, and a sizable fraction could not be harvested, leading to a drop in production of 1 million tons over the previous year. November is also the planting period for the irrigated dry season Boro rice crop. Following the cyclone, due to the disruption of economic activity, the destruction of infrastructure, and floods (including a tidal wave), the area cultivated with Boro rice decreased and planting or transplanting was delayed. This accounts for the sharp decrease in Boro production the following year (1971). The Aus rice crop was not directly hit, but it suffered as a result of the problems listed above, compounded by a war in 1971.

The role of cyclones was stressed in this section because they combine two very destructive factors: strong winds and intense rainfall. The damage caused (including long term effects) may be, in developing countries, of the same order of magnitude as GNP growth.

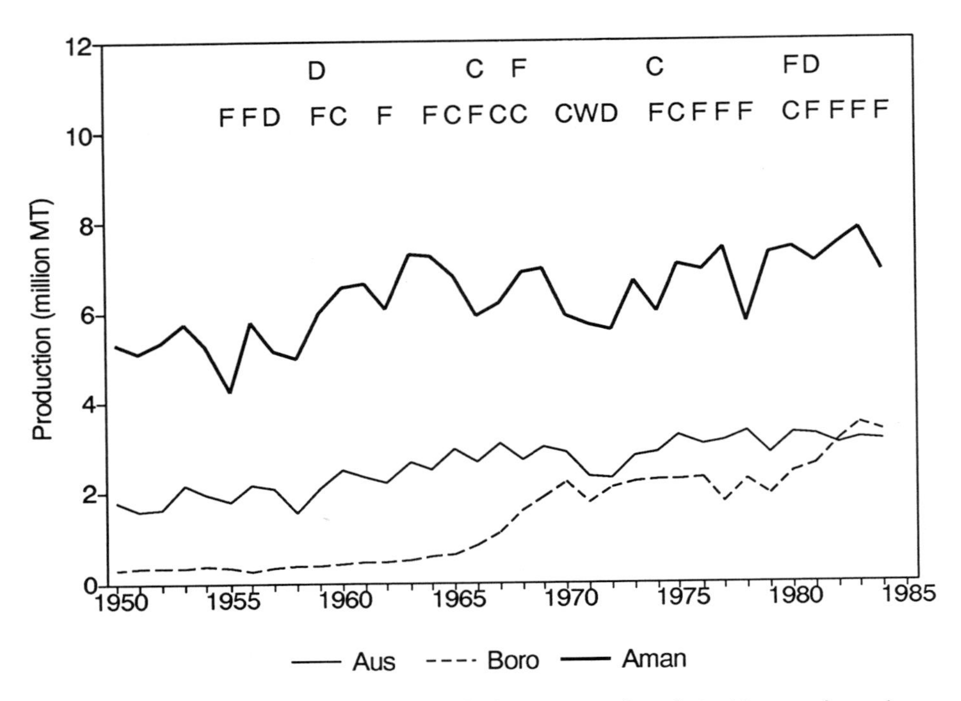

Figure 4-5. Rice production in Bangladesh, 1950–1984, with incidence of weather accidents given by D (drought), F (floods), C (cyclones), and W (war). The monsoon peaks around July; Boro is a dry-season winter (Rabi) irrigated rice (Nov–May), Aus is a pre-monsoon (Kharif) crop (Apr–Aug), while Aman rice (much of it floating) is a late monsoon crop (June–Dec). The indicated years correspond to the first year of the marketing season, which is conventionally reckoned from July through June. Graph based on data from Rahman (1985) and FAO (1990).

Other extreme events such as hail, frost, and fires appear to be marginal in comparison. Only the African droughts can provoke damage of a larger extent but, unlike tropical cyclones, which can be forecast statistically to some extent (Gray, 1990), droughts remain largely unpredictable. Well established connections between ENSO (El Niño/Southern Oscillations) and distant atmospheric features have been identified, covering such diverse effects as the Indian monsoon (Elliot and Angel, 1987), shrimp production in Louisiana (Childers et al., 1990), and wildland fires in the U.S. (Simard et al., 1985). In spite of promising research, little operational improvements in seasonal forecasts have yet been achieved.

F. Traditional Farm Management and Risk Reduction

Even under traditional farming conditions with little or no inputs, farmers face several management decisions: choice of crop mix and varieties (traditional varieties are sometimes extremely diversified), the location of the field (soil, slope), planting dates, and the timing and frequency of cultivation.

The overall goal for the farmer could be stated as one of "minimizing risk" or "maximizing or stabilizing income, production, and nutrition." In areas where farmers market their production, food and cash-crops frequently compete for the best land and the available labor. This is, in fact, one of the reasons why crops cannot always be planted at the optimum time, and timing can significantly affect the outcome of the farming season.

Figure 4-6 shows the average maize yields in the Eastern Mwanza region (southern Lake Victoria) as a function of planting date and soil type. About one third of the soil has water holding capacities between 35 and 50 mm, while only 2% stores between 90 and 115 mm. The best soils (Black Cotton Soils, 18%) can support short-cycled grains planted after the end of the rainy season. This figure provides a good example of how a farmer can plan his farming in a climate which is intermediate between unimodal and bimodal rainfall patterns. Clearly, the farmer should attempt to hit the optimal planting time, but if one is too late in planting, the results can be disastrous given the rapidly declining slope of Figure 4-6. On the other hand, the penalty for planting early is not as great, and on average, more stable production is achieved with the lower yields.

Assuming that land and labor are available, the proper combination of planting date and soil allows farmers to implement a wide array of possible *a priori* strategies.

III. HUMAN POPULATION CONSTRAINTS

A. Some Global Trends

According to the FAO (1990), the world's population (5.3 billion persons, with 4.0 in developing countries) is growing at an annual rate approaching 2% (Table 4-4). Rates in Africa are even higher, averaging 2.8%. Land resources of developing countries, however, are limited. FAO studies (FAO, 1989) indicate that in Asia, where population pressure is the highest, the limit of horizontal expansion of agriculture has been reached: the increase of arable land has virtually come to a stop (0.07% p.a.; FAO, 1990). A certain number of trends can be identified which result from population growth. At

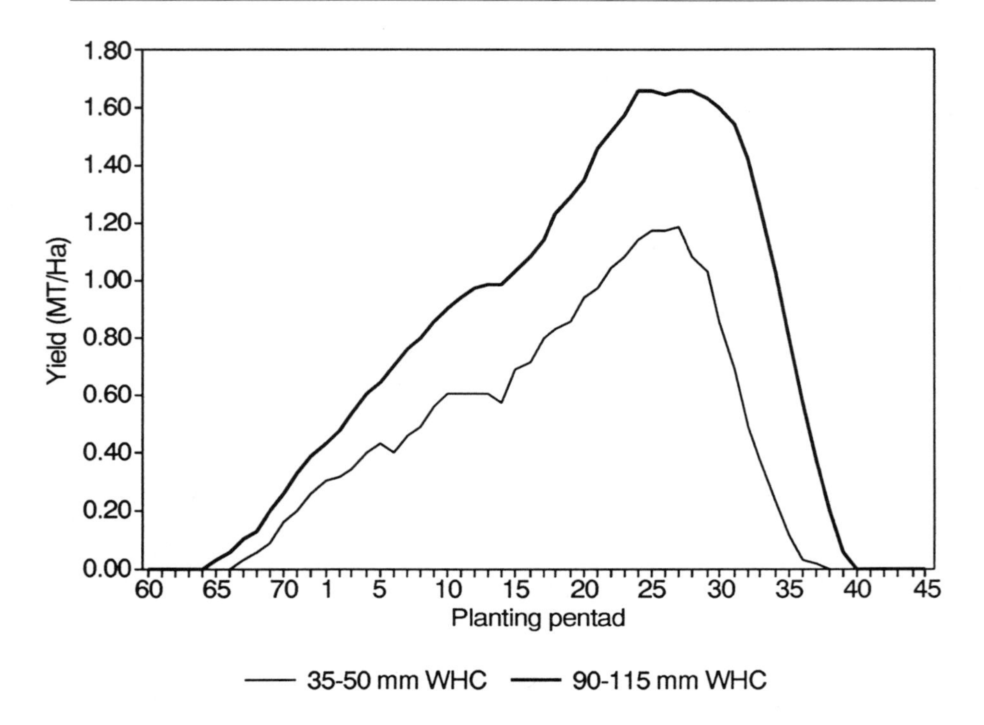

Figure 4-6. Average maize yield expected in East Mwanza region (Tanzania) as a function of planting date on two soil types with different Water Holding Capacities (WHC). The planting date is given in pentads (five-day periods): pentad 1 covers the days from January 1st to 5th, etc. Based on experimental data recorded between 1932 and 1984. After Gommes (1985).

a more local scale (national and below), vast disparities exist which may lead to surplus and deficit areas in the same country.

In Asia, a clear trend of increasing cereal yields (due to mechanization, better water management and inputs, new varieties, and higher cropping intensities[5]) has been sufficient to cover food requirements, although there are now indications that the trend is levelling off. However, more and more land is being diverted from cash crops to food crops, Figure 4-7, resulting in a loss of income for rural people.

In the Philippines, the area grown with food crops has stagnated since 1975. The cash crops shown in Figure 4-7 include sugarcane (about 75% of the total area). Areas under cash crops have been decreasing mainly due to unfavorable sugarcane markets. Even in developing countries, cash crops respond rather quickly to international market fluctuations. The pressure on land is frequently paralleled by an increasing variability in yields due to,

Table 4-4. Trends in world-wide population, food production, arable and irrigated land

	Average 1961-90 exponential growth rate (%)					
Continent	Population	Cereal[a]	Tuber[a]	Legume[a]	Arable Land[b]	Irrigated Land[c]
Africa	2.83	1.95	2.74	1.58	0.44	5.8
North & Central America	1.52	2.32	1.30	1.53	-0.59	9.6
South America	2.34	2.80	0.51	0.71	1.53	5.8
Asia[d]	2.11	3.30	1.95	0.26	0.07	31.0
Europe[d]	0.51	2.37	-1.27	1.57	-0.53	10.8
Oceania	1.72	3.14	1.62	13.76	1.23	4.1
World	1.88	2.66	0.78	0.83	0.12	15.2

[a]Refers to production
[b]All agricultural land, excluding perennial crops and permanent rangeland
[c]Percent of land under annual and permanent crops, average value from 1981-90
[d]Does not include the former USSR
Source: FAO (1990).

among other factors, the fact that high yielding varieties have higher water requirements which the variability of the water supply cannot always meet.

Singh (1989) indicates that increased variability also affects irrigated crops. For instance, in Thailand, an average of 12.6% of the rice land is not planted, and 20.8% not harvested, due to adverse weather. This results in only 66.6% of the potential rice area actually being productive in a given year.

Figure 4-8 illustrates another continental trend in Africa where cereals undergo a decline relative to the nutritionally poorer root crops, such as cassava. This is due to two main reasons: many root crops can grow on poor or degraded soils, and they are less weather sensitive, i.e., they yield more stable production than cereals.

Climatic factors affect population supporting capacities: land use and ownership systems are very dependent on population density. Even in the low rainfall areas of Africa, where so-called "contemplative" cattle herding (cattle numbers as a status symbol) predominates, there is now a trend towards more efficient meat and milk production.

In the wetter areas, more capital intensive farming is being developed. Where land becomes scarce because of increasing density, land ownership frequently changes from common to private, with parallel social changes. In the "mild" cases, this is a beneficial factor, as it favors medium and long term investments like land clearing, land improvement (fertilization), or perennial crops. In extreme cases (Bangladesh), many unskilled farm laborers lose their

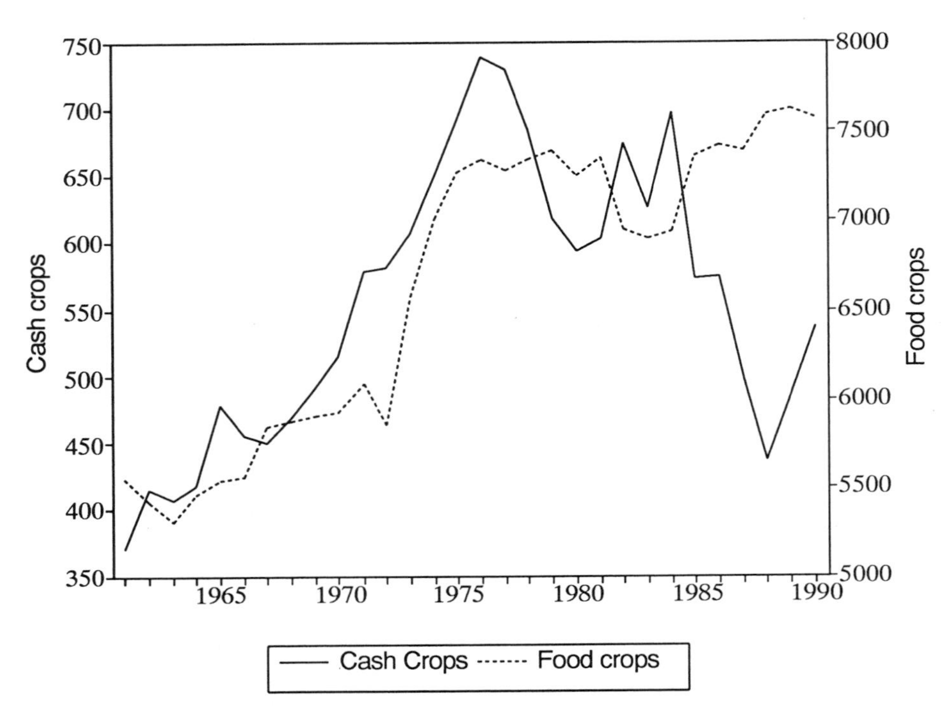

Figure 4-7. Areas under food and cash crop in the Philippines between 1961 and 1990 (thousand hectares). Based on data from FAO (1990).

income as a consequence of mechanization (diesel pumps for irrigation replacing hand pumps), creating situations of extreme rural poverty and massive migration to urban centers. In some countries, industrialization may offer an alternative source of income; but in many developing countries, due to population pressure, rural income and nutrition tend to decrease.

B. Interactions between Population Growth and Weather Risk

Even under stable climatic conditions, high population densities lead to less resilient agricultural production systems to the extent that certain systems amplify weather vagaries rather than buffering them. This has been referred to as the "human component of drought."

Land degradation, the cultivation of more marginal land (low natural fertility, low water storage capacity), and shorter fallow are three of the root causes. This is exemplified (Figure 4-9) based on a simulation of maize

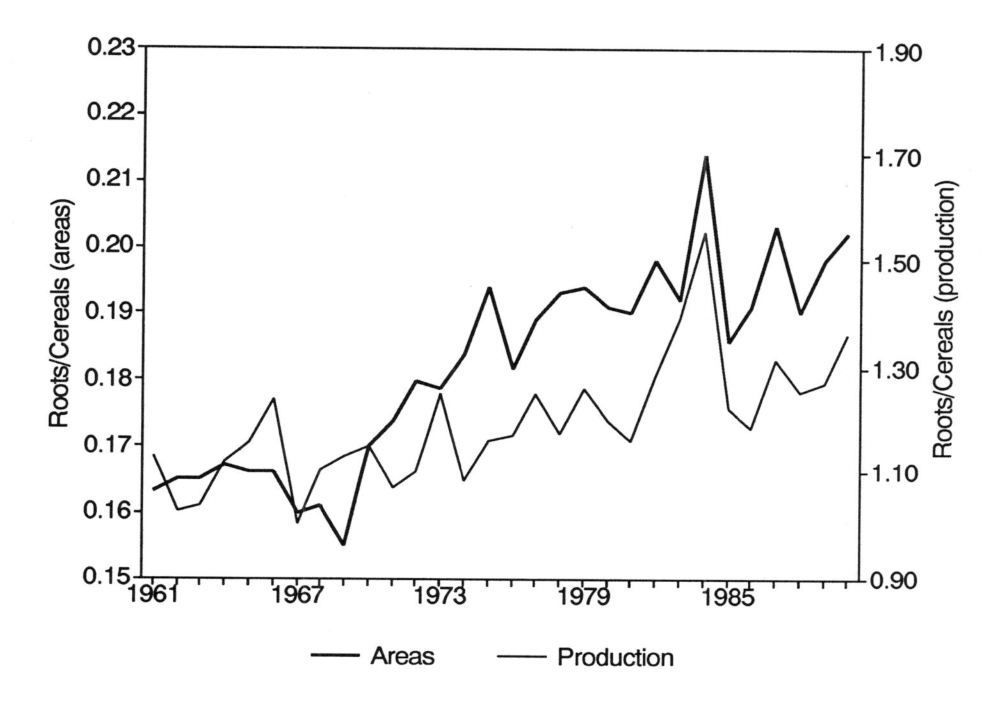

Figure 4-8. Relative importance of root crops and cereals in Africa (ratio of areas and ratio of productions), between 1961 and 1990. Based on data from FAO (1990).

production in the East Mwanza region (Lake Victoria, Tanzania), where 72% of the soils exhibit water holding capacities below 40 mm, and 23% store more than 130 mm, with very little intermediate soils available (De Pauw, 1984). The region experienced favorable rains during the 1960s, and maize gradually replaced the more resistant millet and sorghum. During the 1970s, the regions south of Lake Victoria, which were regarded as major producing areas, became very unstable rather abruptly.

Overstocking and overgrazing, unrealistic pricing policies, constant government efforts to impose the cultivation of cotton, shortage of labor (such as children attending school rather than herding cattle) are some of the additional factors which contributed to turning a food surplus area into a region with recurrent deficits culminating in a serious drought in 1983/84.

Paradoxically (but only apparently so) more "modern" (technological) farming may also lead to difficulties due, as hinted above, to high yielding varieties being less adaptable and more water demanding. Comparison of the recent evolution of cereal production in Niger (Figure 4-2) and Senegal (Figure

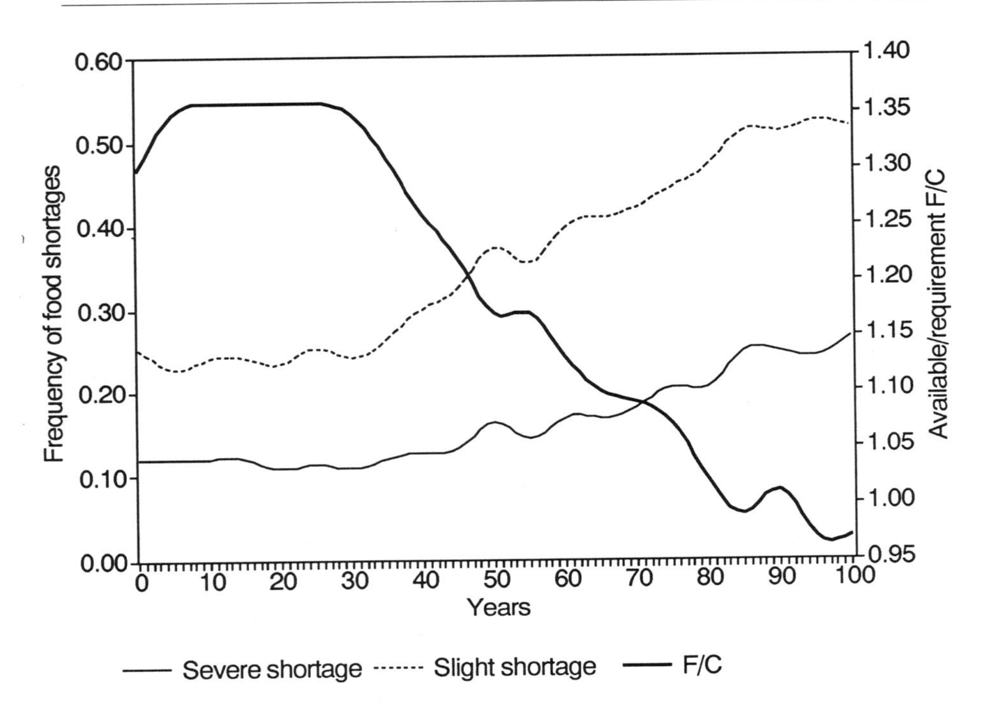

Figure 4-9. Simulated food security situation under the conditions of East Mwanza region (Tanzania). The population grows at a rate of 3% p.a. and the area cultivated by each person is assumed to be constant at 0.13 ha. Food requirements C are set at 182 kg maize/year. During surplus years, 50 kg can be carried over to the next season. Soils with high water storage capacity are used first. Actual weather characteristics corresponding to 1932/84 and a "safe" planting date were assumed (January, see figure 4-6). F is the available maize (current year's production + carry-over stocks). F/C is the average ratio of available maize over requirement. A slight shortage is defined as F<C; a severe shortage occurs whenever F<C/2. After Gommes (1985).

4-10) illustrates this fact. The two countries are comparable in many respects, for instance in their land use patterns, climate, and volume of food aid. The variability of rainfall, expressed by the rainfall index, follows roughly equivalent patterns, although rainfall is generally higher in Senegal.

This comparison also points out the consequences of two rather different strategies of coping with increased population. In Niger the area grown *per capita* remained roughly constant at 0.7 ha/person, and the production increase was achieved by expanding the traditional farming systems horizontally. In Senegal, on the other hand, the area grown (starting at 0.35 ha/person in 1961) has undergone a marked decrease, while yields have improved through the gradual adoption of more modern farming.

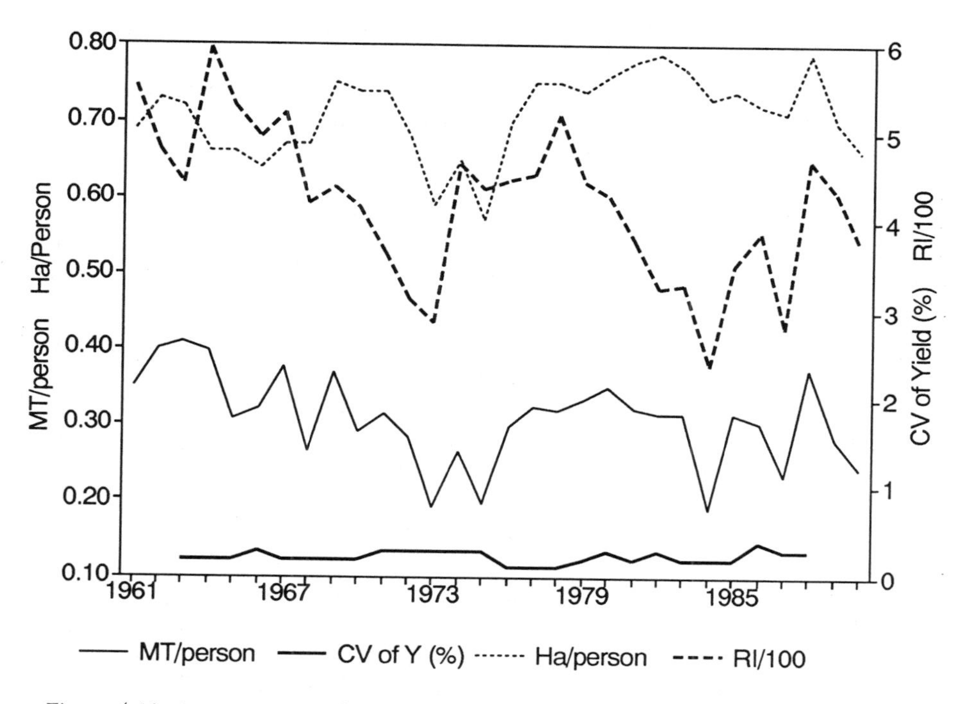

Figure 4-10. A comparison of *per capita* cereal production and area harvested (metric ton, rural population only) in Senegal with an empirical rainfall index (the RI is based on data from 18 stations, refer to note 3 for details). The coefficients of variation of yield (previously corrected for RI and detrended) are 5-year running averages. (Same sources of data as Figure 4-1).

In Senegal, contrary to Niger, the cereal yields have, as expected, a (statistically) significant trend component in addition to the RI influence. In absolute terms, the variability of yields (corrected for trend and RI) has been decreasing constantly, but it remains higher than in Niger. The *per capita* food production is more variable too, and the *per capita* cereal production is naturally lower, their rural population numbers being lower also (78% of rural population versus 88%).

It is obvious that a technically more advanced agricultural sector can be a solution to improving the nutritional status of increasing populations only if it is complemented by mechanisms that can absorb the potentially increased variability. Such mechanisms are more efficient markets, adequate storage for the marketable surpluses produced, crop diversification, and guidance on the proper use of all resources, including fertilizer and rainfall.

IV. CONCLUSIONS

So far, world food production has been able to keep pace with the increasing requirements of a human population growing at a rate close to 2% *per annum*, of which three quarters live in tropical developing countries. However, food production undergoes marked inter-annual and spatial variations which threaten food security. Such variations can be gradual when linked with changes in technology, largely predictable when deriving from market conditions, but almost predictable if they are brought about by weather vagaries.

Rainfall remains the dominant factor behind large scale irregularities in food supply. Several statistical characteristics of rainfall, in particular persistence, have a direct bearing on food supply: runs of consecutive "good" or "bad" years are more likely than average years. Further, more technically advanced agriculture, while being more productive, is also more dependent on climatic resources like rainfall. This can be corrected only through adequate storage and market infrastructures which are missing in many developing countries.

In addition, as a result of direct or indirect population pressure, many food production systems have become more fragile. Land resources are finite and, increasingly, cultivation intensification replaces horizontal expansion. Even under stable weather conditions, the variability of food production is increasing.

There are indications that, in many areas, the increase in production was possible only by losing income in rural areas and growing nutritionally less valuable crops.

For global climate simulations to realistically predict 21st century agriculture will require a considerable level of detail, including attention to the occurrence of extreme weather events, the actual statistical structure of weather, and present trends in world agriculture.

Significant qualitative changes may affect agriculture in the future, resulting from population pressure and adaptation to climate change. It is clear that new mechanisms will be needed if the world population is to develop sustainably and peacefully.

NOTES

[1] We define the "detrended coefficient of variation" as the standard deviation of departure from the linear trend divided by the average of the original series, in percent.

[2] Cropping intensity is the number of crops grown every year on the same land.

[3] For a given rainy season, the rainfall index is defined as the national average of total annual precipitation weighted, for each station, by its 1961–1990 normal. The 53 stations used were weighted by their normal rainfall, meaning low weight is given to

those areas which are less productive.

[4] It is the practice of the statistical services to report *harvested* areas, resulting in a "statistical yield" (Production/Area harvested). The *biologic* or *agronomic* yield would be the ratio between the production and the area sown and germinated. In poor years, only a fraction of the planted area may actually be harvested, resulting in reported yields being sometimes substantially higher than agronomic yields. This artificially reduces the actual variability of yields.

[5] Cropping intensities now reach 1.5 in Bangladesh, China, Indonesia, and Malaysia. In parts of China, three crops a year on the same plots are not uncommon (Singh, 1992).

LITERATURE CITED

Bach, W., J. Pankrath, and S. Schneider. eds. 1981. *Food-Climate interactions.* Proc. of an Internat. Workshop held in Berlin, 9–12 December 1980. D. Reidel Publishing Co.: Dordrecht, Boston, London.

Childers, D.L., J.W.Day, Jr., and R.A. Muller. 1990. Relating climatological forcing to coastal water levels in Louisiana estuaries and the potential importance of El Niño/Southern Oscillation events. *Clim. Res.* 1(1):31–42.

de F. Monteiro, C.A. 1986. Some aspects of the urban climates of tropical south America: The Brazilian contribution. In: *Urban Climatology and its applications, with special regard to tropical areas.* Proceedings of the technical WMO–WHO conference, Mexico D.F., 26–30 Nov. 1984. WMO No. 652., 166–98. WMO: Geneva.

De Pauw, E. 1984. *Soils, physiography and agroecological zones in Tanzania.* Crop Monitoring and Early Warning Project GCP/URT/047/NET. FAO/Kilimo: Dar Es Salaam, Tanzania.

Elliott, W.P. and J.K. Angel. 1987. The relations between Indian monsoon rainfall, the Southern Oscillation, and hemispheric air and sea temperature: 1884–1984. *J. Clim. Appl. Meteorol.* 26:943–48.

FAO. 1978a. *Report on the agro-ecological zones project. Vol. 1: Results for Africa.* World Soil Resources Report 48/1. FAO: Rome.

FAO. 1978b. *Report on the agro-ecological zones project. Vol. 2: Results for south-west Asia.* World Soil Resources Report 48/2. FAO: Rome.

FAO. 1980. *Report on the agro-ecological zones project. Vol. 4: Results for south-east Asia.* World Soil Resources Report 48/4. FAO: Rome.

FAO. 1981. *Report on the agro-ecological zones project. Vol. 3: Results for south and central America.* World Soil Resources Report 48/3. FAO: Rome.

FAO. 1989. *Rainfed agriculture in Asia and the Pacific region.* FAO Regional Office for Asia and the Pacific: Bangkok.

FAO. 1990. *1989 Production yearbook No. 43.* FAO Statistics series No.94. AGROSTAT-PC, the PC version (1990) of the FAO Production Yearbook was also used.

Glantz, M.H. 1987. Drought in Africa. *Sci. Am.* 256(6)34–40.

Gommes, R. 1985. Rainfed food-crop production in East Mwanza region, a semi-quantitative risk analysis. Proc. Workshop on Food Security and Nutrition Education, Mikumi, Tanzania. 27–31 May 1985. Jointly organized by IRA/UDSM/IUNS/SIDA/UNICEF/UTAFITI. Dar Es Salaam, Tanzania. 1–28.

Gommes, R. and Th. Nègre. 1992. *The role of agrometeorology in the alleviation of natural disasters.* Agrometeorology Series Working Papers No. 2. FAO: Rome.

Gray, W.M. 1990. *Forecast of Atlantic seasonal hurricane activity*. Department of Atmospheric Sciences, Colorado State Univ.: Fort Collins, CO.

IRRI. 1989. *Climate and Food Security*. Proc. of the Internat. Symposium on Climate Variability and Food Security in Developing Countries, 5–9 Feb. 1987, New Delhi, India. Organized by AAAS, INSA and IRRI. IRRI and AAAS: Manila and Washington.

Muchow, R.C. and J.A. Bellamy. eds. 1991. *Climate risk in crop production: Models and management for the semiarid tropics and subtropics*. CAB Int., Wallingford: Oxon, UK.

Oke, T.R. 1986. Urban Climatology and the tropical city. In *Urban Climatology and its applications, with special regard to tropical areas*. Proceedings of the technical WMO-WHO conference, Mexico D.F., 26–30 Nov. 1984. WMO No. 652., 1–25. WMO: Geneva.

Rahman, S.M. 1985. *Agroclimatic methods for Bangladesh*. Bangladesh Meteorol. Department: Dhaka.

Riebsame, W.E. 1988. *Assessing the social implications of climate fluctuations: A guide to climate impact studies*. UNEP: Nairobi, Kenya.

Simard, A.J., D.A. Haines, and W.A. Main. 1985. Relations between El Niño/Southern Oscillation anomalies and wildland fire activity in the United States. *Agr. For. Meteorol.* 36(2):93–104.

Singh, R.B. 1989. Contingency crop planning for rainfed areas. In FAO, 1989, 49-59.

Singh, R.B. 1992. Research and development strategies for increased and sustained production of rice in Asia and the Pacific region. RAPA 1992/17. FAO: Bangkok.

Todorov, A.V. 1985. Sahel: the changing rainfall and the "normals" used for its assessment. *J. Appl. Meteorol.* 24(2):97–107.

5

POTENTIAL IMPACTS of CLIMATE CHANGE on WORLD FOOD SUPPLY: A SUMMARY of a RECENT INTERNATIONAL STUDY[1]

Cynthia Rosenzweig
Columbia University and Goddard Institute for Space Studies

Martin L. Parry
Environmental Change Unit, Oxford University

I. INTRODUCTION

Recent research has focused on farm-level, regional, and national assessments of the potential effects of climate change on agriculture (see Chapter 6 by Crosson, and Chapter 7 by Kaiser et al.). These efforts have, for the most part, treated each region or nation in isolation, without relation to changes in production in other places. At the same time, there has been a growing emphasis on understanding the interactions of climatic, environmental, and social factors in a wider context (Parry, 1990), leading to more integrated assessments of potential impacts in national impact studies completed in the United States (Adams et al., 1990; Smith and Tirpak, 1989), Canada (Smit, 1989), Brazil (Magalhaes, 1992), and Indonesia, Malaysia, and Thailand (Parry et al., 1992). Regional studies have been conducted in high-latitude and semi-arid agricultural areas (Parry et al., 1988), and the U.S. Midwest (Rosenberg and Crosson, 1991). The results of these and other agricultural impact studies have been summarized in the Intergovernmental Panel on Climate Change (IPCC) Working Group II Report (IPCC, 1990b). Sensitivity studies of world agriculture to potential climate changes have indicated that the effect of moderate climate change on world and domestic economies may be small as reduced production in some areas is balanced by gains in others (Kane et al., 1991; Tobey et al., 1992).

The central aim of our recent study was to provide an integrated (i.e., combined biophysical and economic) global assessment of the potential effects of climate change on world food supply. Specifically, the objectives were to calculate quantitative estimates of climate change effects on the amount of food produced globally, world food prices, and the number of people at risk of hunger (defined as the population with an income insufficient to either produce or procure their food requirements) in developing countries.[2] The research involved estimating the responses of crop yields to greenhouse gas-induced climate change scenarios and then simulating the economic consequences of these potential changes in crop yields (Figure 5-1).

II. CROP YIELD METHODS

A. Estimation of Potential Changes in Crop Yield

Agricultural scientists in 18 countries (see Rosenzweig and Iglesias, 1993) estimated potential changes in national grain crop yields using compatible crop models and consistent climate change scenarios.[3] The crop models were those developed by the U.S. Agency for International Development's International Benchmark Sites Network for Agrotechnology Transfer (IBSNAT, 1989). The crops modeled were wheat, rice, maize, and soybeans. Wheat, rice, and maize account for approximately 85% of the world cereal exports; soybean accounts for about 67% of trade in protein cake equivalent. The crop models were run for current climate conditions and for climate conditions predicted by general circulation models (GCMs) for doubled atmospheric CO_2 levels at 112 sites (Figure 5-2). The direct effects of CO_2 on crop yields were also taken into account.

For the selected GCM climate change scenarios, site specific estimates of yield changes were aggregated to national levels for the modeled major crops. These national crop yield changes were then extrapolated to provide estimates of yield changes (for the three GCM scenarios) for the countries included in the Basic Linked System (BLS) world food trade model (discussed later in this chapter), and for the other crops included in the BLS.

B. Scenarios Based on General Circulation Model Results

The scenarios for this study were created by changing observed daily data from the current climate (1951-1980) according to doubled CO_2 simulations of three general circulation models available at the time of the initiation of the study in 1989. The GCMs used are those from the Goddard Institute for Space Studies (GISS), Geophysical Fluid Dynamics Laboratory (GFDL), and United

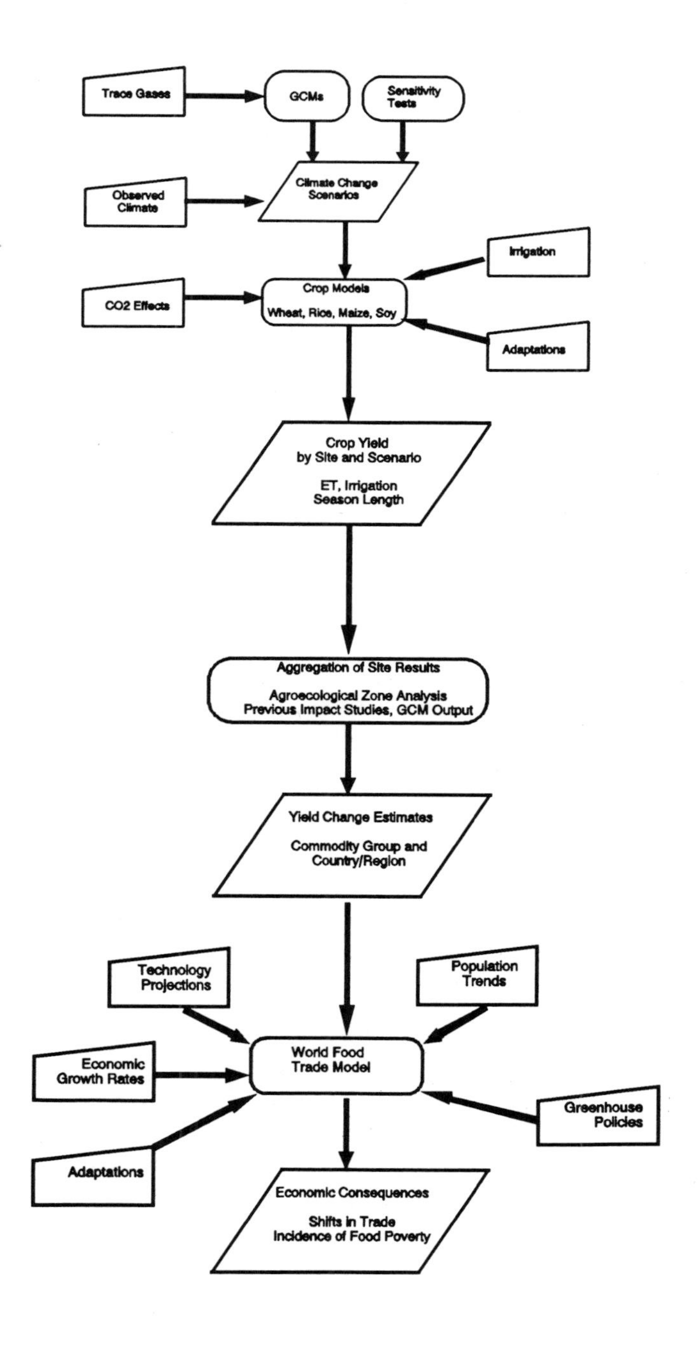

Figure 5-1. Key elements of the crop yield and world food study.

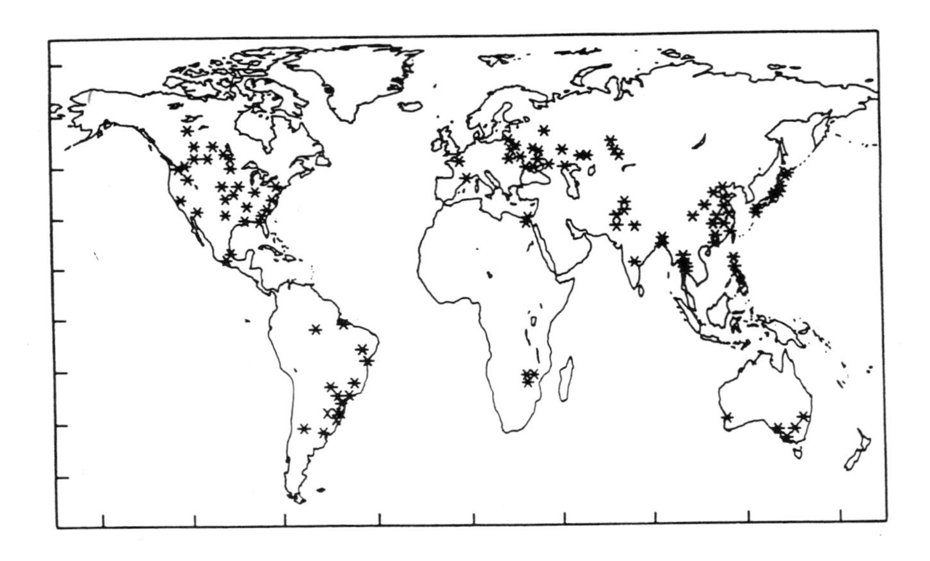

ARGENTINA	O.E. Sala and J.M. Paruelo
AUSTRALIA	B.D. Baer, W.S. Meyer and D. Erskine
BANGLADESH	Z. Karim, M. Ahmed, S.G. Hussain and Kh. B. Rashid
BRAZIL	O.J.F. de Siqueira, J.R.B. Farias and L.M.A. Sans
CANADA	M. Brklacich, R. Stewart, V. Kirkwood and R. Muma
CHINA	Z. Jin, D. Ge, H. Chen, J. Fang and X. Zheng
EGYPT	H.M. Eid
FRANCE	R. Delécolle, D. Ripoche, F. Ruget and G. Gosse
INDIA	D.G. Rao
JAPAN	H. Seino
MEXICO	D. Liverman, M. Dilley, K. O'Brien and L. Menchaca
PAKISTAN	A. Qureshi and A. Iglesias
PHILIPPINES	C.R. Escaño and L. Buendia
RUSSIA	G. Menzhulin, L. Koval and A. Badenko
THAILAND	M.L.C. Tongyai
U.S.	C. Rosenzweig, B. Curry, T.-Y. Chou, J. Ritchie, J. Jones and R. Peart
URUGUAY	W.E. Baethgen
ZIMBABWE	P. Muchena

Figure 5-2. Crop model sites and modelers.

Table 5-1. GCM climate change scenarios.

GCM	Year[a]	Resolution[b]	CO_2[c]	Change in Average Global Temperature	Change in Average Global Precipitation
GISS	1982	7.83° x 10°	630	4.2°C	11%
GFDL	1988	4.4° x 7.5°	600	4.0°C	8%
UKMO	1986	5.0° x 7.5°	640	5.2°C	15%

[a] When calculated.

[b] Latitude x longitude.

[c] Doubled CO_2 level (ppm)

Sources: GISS Hansen et al., 1983; GFDL Manabe and Wetherald, 1987; UKMO Wilson and Mitchell, 1987.

Kingdom Meteorological Office (UKMO) (Table 5-1). The temperature changes of these GCM scenarios (4.0–5.2°C) are at the upper end of the range (1.5 to 4.5°C) projected for CO_2-equivalent doubling by the IPCC (IPCC, 1990a, 1992). The GISS and GFDL scenarios, however, are near the mean temperature change (3.8°C) of recent doubled CO_2 experiments documented for atmospheric GCMs with a seasonal cycle and a mixed-layer ocean (IPCC, 1992). Mean monthly changes in temperature, precipitation, and solar radiation from the appropriate GCM grid box were applied to observed daily climate records to create climate change scenarios for each site.

C. Scientific Uncertainty of GCM Predictions

While GCMs currently provide the most advanced means of predicting the potential future climatic consequences of increasing amounts of radiatively active trace gases, their ability to simulate current climate varies considerably from region to region. They have been shown to simulate current surface air temperatures reasonably well, but do not reproduce current precipitation as accurately (IPCC, 1990a). Of special importance for agricultural climate change impacts, there is a notable lack of consensus among GCMs in prediction of regional soil moisture changes (Kellogg and Zhao, 1988). Furthermore, GCMs are not yet able to produce reliable projections of changes in climate variability, such as changes in the frequencies of drought and storms, even though these could affect crop yields significantly. (See Chapter 3 for further discussion of GCMs.)

D. CO_2 Level and Timing

For the crop modeling part of this study, climate changes from doubled CO_2 GCM simulations are utilized with an associated level of 555 ppm CO_2; the assumed timing for the BLS world food trade projections is that these conditions will occur in 2060.[4]

Because atmospheric concentrations of other greenhouse gases besides CO_2 (e.g., methane (CH_4), nitrous oxide (N_2O), and the chlorofluorocarbons (CFCs)) are also increasing, an "effective CO_2 doubling" has been defined as the combined radiative forcing[5] of all greenhouse gases having the same forcing as doubled CO_2 (usually defined as approximately 600 ppm; see Table 5-1 for doubling levels of the three GCMs used in this study). Level of CO_2 is important when estimating potential impacts on crops, because crop growth and water use have been shown to benefit from increased levels of CO_2 (Cure and Acock, 1986). A CO_2 level of 555 ppm was associated with the effective doubled CO_2 climate projections for use in the crop modeling simulations. This was based on the GISS GCM trace gas scenario A described in Hansen et al. (1988), in which the simulated climate had warmed to the effective doubled CO_2 level of about 4°C by 2060. This level assumes that non-CO_2 trace gases contribute approximately 15% of the change in radiative forcing from 300 to approximately 600 ppm.

E. Physiological Effects of CO_2

Most plants growing in experimental environments with increased levels of atmospheric CO_2 exhibit increased rates of net photosynthesis (i.e., total photosynthesis minus respiration) and reduced stomatal openings. (Experimental effects of CO_2 on crops have been reviewed by Acock and Allen (1985) and Cure (1985).) Partial stomatal closure leads to reduced transpiration per unit leaf area and, combined with enhanced photosynthesis, often improves water-use efficiency (the ratio of crop biomass accumulation or yield to the amount of water used in evapotranspiration). Thus, by itself, increased CO_2 can increase yield and reduce water use per unit biomass.

The crop models used in this study assume a beneficial effect due to increased atmospheric CO_2. Ratios were calculated between measured daily photosynthesis and evapotranspiration rates for a canopy exposed to high CO_2 values, based on published experimental results (Allen et al., 1987; Cure and Acock, 1986; Kimball, 1983), and the ratios were applied to the appropriate variables in the crop models on a daily basis. The photosynthesis ratios (555 ppm CO_2/330 ppm CO_2) for soybean, wheat and rice, and maize were 1.21, 1.17, and 1.06, respectively. Changes in stomatal resistance were based on experimental results by Rogers et al. (1983) (49.7/34.4 seconds/meter s m^{-1} for C3 crops, 87.4/55.8 s m^{-1} for C4 crops). Of course, there is always uncertainty regarding whether experimental results will be observed in the open field under conditions likely to be operative when farmers are managing crops. Plants growing in experimental settings are often subject to fewer environmental stresses and less competition from weeds and pests than are likely to be encountered in farmers' fields. (See Chapter 8 for a thorough discussion of the physiological effects of CO_2.)

F. Limitations of Crop Growth Models

The crop models embody a number of simplifications. For example, weeds, diseases, and insect pests are assumed to be controlled; there are no problem soil conditions (e.g., salinity or acidity); and there are no extreme weather events, such as tornados. The models are calibrated to experimental field data which often have yields higher than those currently typical under farming conditions. Thus, the effects of climatic change on yields in farmers' fields may be different from those simulated by the crop models.

The crop models simulate the current range of agricultural technologies available around the world, including the use of high-yielding varieties that are responsive to technological inputs, but by the year 2060 agricultural technology is likely to be very different. The models may be used to test the effects of some potential improvements in agricultural production, such as varieties with higher thermal requirements, and installation of irrigation systems, but do not include possible future improvements. (The BLS economic model used in the study does include future trends in yield improvement, but not technological developments induced by negative climate change impacts). Finally, models for crops such as millet and cassava were not yet sufficiently tested for use in this study. Potential yield changes of such crops, which may respond differently to both climate change and increases in CO_2, are needed for better assessment of climate change impacts in tropical and sub-tropical regions.

G. Farm-level Adaptations

In each country, the agricultural scientists used the crop models to test possible responses to the worst climate change scenario (this was usually, but not always, the UKMO scenario). These adaptations included change in planting date, change of cultivar, irrigation, fertilizer, and change of crop. Irrigation simulations in the crop models assumed automatic irrigation to field capacity when plant available water dropped to 50%, and 100% irrigation efficiency. All adaptation possibilities were not simulated at every site and country: choice of adaptations to be tested was made by the participating scientists, based on their knowledge of current agricultural systems.

For the economic analysis in the BLS, the crop model results reported by the participating scientists were then grouped into two levels of adaptation. Level 1 implies little change to existing agricultural systems reflecting farmer response to a changing climate. Level 2 implies more substantial change to agricultural systems, possibly requiring resources beyond the farmer's means.

Level 1 Adaptation includes:

1. Shifts in planting date (+/- 1 month) that do not imply major changes in crop calendar.
2. Additional application of irrigation water to crops already under irrigation.
3. Changes in crop variety to currently available varieties more adapted to the altered climate.

Level 2 Adaptation includes:

1. Large shifts in planting date (> 1 month).
2. Increased fertilizer application (included here because of implied costs for farmers in developing countries).
3. Installation of irrigation systems.
4. Development of new varieties (tested by manipulation of genetic coefficients in crop models).

Yield changes for both adaptation levels were based on crop model simulations where available. For the crops and regions not simulated, the negative impact was halved if adaptations were estimated to partially compensate for the negative effects of climate change; if compensation was estimated to be full, yield changes were set to 0. If yield changes were positive in response to climate change and the direct effects of CO_2, adaptation to produce even greater yield increases was not included, with the assumption that farmers would lack incentive to adapt further. The adaptation estimates were developed only for the scenarios which included the direct effects of CO_2 fertilization as these were judged to be most realistic.

H. Limitations of Adaptation Analysis

The adaptation simulations were not comprehensive because all possible combinations of farmer responses were not tested at every site. Spatial analysis of crop, climatic, and soil resources is needed to test fully the possibilities for crop substitution. Neither the availability of water supplies for irrigation nor the costs of adaptation were considered in this study; these are both critical needs for further research. (A related study on the Integrated Impacts of Climate Change on Egypt, which utilized the results of this work, does address future water availability for national agricultural production in that country (Strzepek et al., 1993).)

At the local level, there may be social or technical reasons why farmers are reluctant to implement adaptation measures, e.g., increased fertilizer application and improved seed stocks may be capital-intensive and not suited to indigenous agricultural strategies. Furthermore, such measures may not necessarily result in sustainable production increases (e.g., irrigation may

Table 5-2. Current world crop yield, area, production, and percent world production aggregated for countries participating in study.

Crop	Yield	Area	Production	Study Countries
	t/ha	ha x 1000	t x 1000	Percent
Wheat	2.1	230,839	481,811	73
Rice	3.0	143,603	431,585	48
Maize	3.5	127,393	449,364	71
Soybeans	1.8	51,357	91,887	76

Source: FAO, 1988.

eventually lead to soil salinization and lower crop yields). Thus, Level 2 Adaptation represents a fairly optimistic assessment of world agriculture's response to changed climate conditions as characterized by the GCMs tested in this study, possibly requiring substantial changes in current agricultural systems, investment in regional and national agricultural infrastructure, and policy changes. However, changes in regional, national, and international agricultural policies relating to farm-level adaptation were beyond the scope of the analysis.

I. Aggregation of Site Results

Table 5-2 shows the percentages of world production of wheat, rice, maize, and soybeans for the countries in which simulations were conducted. Simulations were carried out in regions representing 71–76% of the current world production of wheat, maize, and soybeans. Rice production was less well represented in the model simulations than the other crops, however, because India, Indonesia, and Vietnam have significant production areas which were not included in the study. Further research is needed in these key countries in order to improve the reliability of the projections of climate change impacts on rice production.

Crop model results for wheat, rice, maize, and soybean from the 112 sites in the 18 countries were aggregated by weighting regional yield changes (based on current production) to estimate changes in national yields. The aggregations were either calculated by the participating scientists or developed jointly with them (see Rosenzweig and Iglesias, 1993). The scientists in each country selected sites representative of major agricultural regions, described the regional agricultural practices, and provided production data for estimation of regional contributions to the national yield changes. Other production data sources included the United Nations Food and Agriculture Organization (FAO, 1988), the US Department of Agriculture (USDA) Crop Production Statistical Division, and the USDA International Service.

Table 5-3. Increment added to estimated yield change to account for direct effects of CO_2.

Crop	Percent[a]
Wheat	22
Rice	19
Soybeans	34
Coarse grains[b]	7
Other C3 crops	25
Other C4 crops	7

[a] Based on crop model simlulations
[b] Weighted by relative production of C3 and C4 crops constituting coarse grain production in the particular country or region.

J. Yield Change Estimates for Crops and Regions Not Simulated

Changes in national yields of other crops and commodity groups and regions not simulated were estimated based on three criteria: 1) similarities to modeled crops and growing conditions, 2) results from about 50 previously published and unpublished regional climate change impact studies (AIR Group, unpublished manuscript), and 3) projected temperature and precipitation changes (and hence soil moisture availability for crop growth) from the GCM climate change scenarios.

Estimates were made of yield changes with and without the direct effects of CO_2. Increments added to the estimated crop yield changes to account for direct CO_2 effects were based on average responses to CO_2 and climate change scenarios in the crop model simulations (Table 5-3). These increments differ from the photosynthesis ratios employed in the crop models because they incorporate the combined responses of the simulated crops to changes in photosynthesis, evapotranspiration, as well as climate. In the crop model simulations, the responses to CO_2 did not vary greatly across region and climate change scenario.

K. Limitations of Crop Yield Change Estimates

The primary source of uncertainty in the estimates lies in the sparseness of the crop modeling sites and the fact that they may not adequately represent the variability of agricultural regions within countries, the variability of agricultural systems within similar agro-ecological zones, or dissimilar agricultural regions. However, since the site results relate to regions that account for about 70 percent of world grain production, the conclusions concerning world totals of cereal production contained in this chapter are believed to be substantiated adequately.

Another source of uncertainty lies in the simulation of grain crops only, leading to estimation of yield changes for other commodities such as root crops and fruit, based primarily on previous estimates. The previous estimates tended to be less negative than the crop responses modeled in this study, and this introduced a bias in favor of these other crops in the world food trade model.

III. CROP YIELD RESULTS

A. Crop Yields without Adaptation

Table 5-4 shows wheat yield changes for doubled CO_2 climate change scenarios for the countries where crop model simulations were conducted (the yield changes include results from both rain-fed and irrigated simulations, weighted by current percentage of the respective practice). This indicates that changes of climate without the direct physiological effects of CO_2 cause decreases in simulated crop yields in all cases, while the direct effects of CO_2 mitigate the negative effects primarily in mid- and high-latitudes.

The magnitudes of the estimated yield changes vary by crop (Table 5-5). Global wheat yield changes weighted by national production are positive, while maize yield is most negatively affected, reflecting its greater production in low latitude areas where simulated yield decreases are greater. Simulated soybean yields are most reduced without the direct effects of CO_2, while maize production declines most with direct CO_2 effects, probably due to its lower response to the physiological effects of CO_2 on crop growth. Soybean is least affected in the less severe GISS and GFDL climate change scenarios when direct CO_2 effects are simulated, because soybean responds very positively to increased CO_2, but is the crop most affected by the high temperatures of the UKMO scenario.

The differences between countries in crop yield responses to climate change without the direct effects of CO_2 are primarily related to differences in current growing conditions. The crop modeling results showed that higher temperatures tend to shorten the growing period at all locations tested. At low latitudes, however, crops are currently grown at higher temperatures, produce lower yields, and are nearer the limits of temperature tolerances for heat and water stress. Warming at low latitudes thus results in accelerated growing periods for crops and greater heat and water stress, resulting in greater yield decreases than at higher latitudes. In many mid- and high-latitude areas, where current temperature regimes are cooler, increased temperatures do not significantly increase stress levels. At some sites near the high-latitude

Table 5-4. Current production and change in simulated wheat yield under GCM 2 x CO_2 climate change scenarios, with and without the direct effects of CO_2.

Country	Current Production			% Total	Change in Simulated Yields					
					without CO_2 Effects			with CO_2 Effects		
	Yield	Area	Production		GISS	GFDL	UKMO	GISS	GFDL	UKMO
	t/ha	ha x 1000	t x 1000		%	%	%	%	%	%
Australia	1.38	11,546	15,574	3.2	-18	-16	-14	8	11	9
Brazil	1.31	2,788	3,625	0.8	-51	-38	-53	-33	-17	-34
Canada	1.88	11,365	21,412	4.4	-12	-10	-38	27	27	-7
China	2.53	29,092	73,527	15.3	-5	-12	-17	16	8	0
Egypt	3.79	572	2,166	0.4	-36	-28	-54	-31	-26	-51
France	5.93	4,636	27,485	5.7	-12	-28	-23	4	-15	-9
India	1.74	22,876	39,703	8.2	-32	-38	-56	3	-9	-33
Japan	3.25	237	772	0.2	-18	-21	-40	-1	-5	-27
Pakistan	1.73	7,478	12,918	2.7	-57	-29	-73	-19	31	-55
Uruguay	2.15	91	195	0.0	-41	-48	-50	-23	-31	-35
Former USSR										
Winter	2.46	18,988	46,959	9.7	-3	-17	-22	29	9	0
Spring	1.14	36,647	41,959	8.7	-12	-25	-48	21	3	-25
USA	2.72	26,595	64,390	13.4	-21	-23	-33	-2	-2	-14
World[b]	2.09	231	482	72.7	-16	-22	-33	11	4	-13

[a] Results for each country represent the site results weighted according to regional production. The world estimates represent the country results weighted by national production.

[b] World area and production x 1,000,000.

Sources: Rosenzweig and Iglesias (1993).

Table 5-5. Changes in simulated wheat, rice, maize, and soybean yield.

	Change in Simulated Yields[a]					
	without CO_2 Effects			with CO_2 Effects		
Crop	GISS	GFDL	UKMO	GISS	GFDL	UKMO
	%	%	%	%	%	%
Wheat	-16	-22	-33	11	4	-13
Rice	-24	-25	-25	-2	-4	-5
Maize	-20	-26	-31	-15	-18	-24
Soybeans	-19	-25	-57	16	5	-33

[a] Crop yield changes were obtained by weighting site results first by regional production within countries and then by national contribution to total production simulated in the study.

Sources: Rosenzweig and Iglesias (1993).

boundaries of current agricultural production, increased temperatures can benefit crops otherwise limited by cold temperatures and short growing seasons, although the extent of soil suitable for expanded agricultural production in these regions was not studied explicitly. Potential for expansion of cultivated land is embedded in the BLS world food trade model (Section IV) and is reflected in shifts in production calculated by that model.

Simulated yield increases in the mid- and high-latitudes are caused primarily by:

1. Positive physiological effects of CO_2. At sites with cooler initial temperature regimes, increased photosynthesis more than compensated for the shortening of the growing period caused by warming.
2. Lengthened growing season and amelioration of cold temperature effects on growth. At some sites near the high latitude boundaries of current agricultural production, increased temperatures extended the frost-free growing season and provided regimes more conducive to greater crop productivity.

The primary causes of decreases in simulated yields are:

1. Shortening of the growing period. Higher temperatures during the growing season speed annual crops through their development (especially grain filling stage), allowing less grain to be produced. This occurred at all sites except those with the coolest growing season temperatures in Canada and the former USSR.
2. Decrease in water availability. This is due to a combination of increases in evapotranspiration rates in the warmer climate, enhanced losses of soil moisture and, in some cases, a projected decrease in precipitation in the climate change scenarios.
3. Poor vernalization. Vernalization is the requirement of some temperate cereal crops, e.g., winter wheat, for a period of low winter temperatures to initiate or

accelerate the flowering process. Low vernalization results in low flower bud initiation and ultimately reduced yields. Decreases in winter wheat yields at some sites in Canada and the former USSR were caused by lack of vernalization.

Figure 5-3 shows estimated potential changes in average national crop yields for the GISS, GFDL, and UKMO doubled CO_2 climate change scenarios allowing for the direct effects of CO_2 on plant growth. The maps are created from the nationally averaged yield changes for wheat, rice, coarse grains, protein feed, and other crops estimated for the BLS simulations for each country or group of countries. Yield changes for regions within countries are not represented. Latitudinal differences are apparent in all the scenarios. With direct CO_2 effects, high latitude changes are less negative or even positive in some cases, while lower latitude regions indicate more detrimental effects of climate change on agricultural yields.

The GISS and GFDL climate change scenarios produced yield changes ranging from +30 to -30%. Effects under the GISS scenario are, in general, more adverse than under the GFDL scenario to crop yields in parts of Asia and South America, while effects under the GFDL scenario result in more negative yields in the USA and Africa and less positive results in the former USSR. The UKMO climate change scenario, which has the greatest warming (5.2°C global surface air temperature increase), causes average national crop yields to decline almost everywhere (up to -50% in Pakistan).

B. Crop Yields with Adaptation

The adaptation studies conducted by the scientists participating in the project suggest that ease of adaptation to climate change is likely to vary with crop, site, and adaptation technique. For example, at present, many Mexican producers can only afford to use small doses of nitrogen fertilizer at planting; if more fertilizer becomes available to more farmers some of the yield reductions under the climate change scenarios might be offset. However, given the current economic and environmental constraints in Mexico, a future with unlimited water and nutrients is unlikely (Liverman et al., 1993). In contrast, switching from spring to winter wheat at the modeled sites in the former USSR produces a favorable response (Menzhulin et al., 1993), suggesting that agricultural productivity may be enhanced there, with the relatively easy shift to winter wheat varieties.

Yield estimates for the two levels of adaptation used in the BLS simulations for the UKMO scenario are shown in Figure 5-4. Level 1 Adaptation compensated for the climate change scenarios incompletely, particularly in the developing countries. For the GISS and GFDL scenarios, Level 2 Adaptation compensated almost fully for the negative climate change impacts. With the high level of global warming as projected by the UKMO climate change scenario, neither Level 1 nor Level 2 Adaptation fully overcomes the

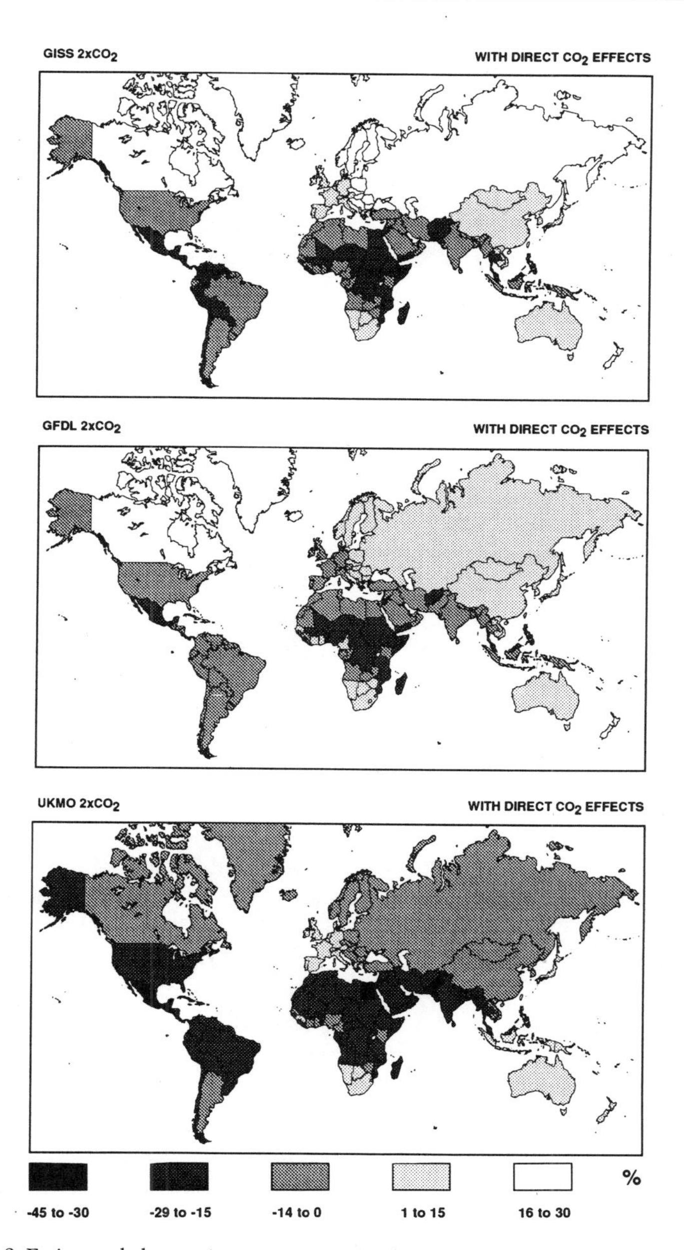

Figure 5-3. Estimated change in average national grain yield (wheat, rice, coarse grains, and protein feed) for the GISS, GFDL, and UKMO climate change scenarios with direct CO_2 effects. Results shown are averages for countries and groups of countries in the Basic Linked System world food trade model; regional variations within countries are not reflected.

negative climate change effects on crop yields in most countries, even when the direct CO_2 effects are taken into account.

IV. WORLD FOOD TRADE METHODS

A. The Basic Linked System

The Basic Linked System (BLS)[6] world food trade model consists of 16 linked, national (including the EC) models with a common structure, four models with country specific structure, and 14 regional group models. The 20 models in the first two groups cover about 80% of attributes of the world food system such as demand, land, and agricultural production. The remaining 20% is covered by 14 regional models for the countries which have broadly similar attributes (e.g., African oil exporting countries, Latin American high income exporting countries, Asian low income countries, etc.). The grouping is based on country characteristics such as geographical location, *per capita* income, and the country's position with regard to net food trade.

The BLS is a general equilibrium model system, with representation of all economic sectors, empirically estimated parameters, and no unaccounted supply sources or demand sinks. In the BLS, countries are linked through trade, world market prices, and financial flows (Figure 5-5). It is a recursively dynamic system: a first round of exports from all countries is calculated for an assumed set of world prices, and international market clearance is checked for each commodity. World prices are then revised, using an optimizing algorithm, and again transmitted to the national model. Next, these generate new domestic equilibria and adjust net exports. This process is repeated until the world markets for all commodities are cleared. At each stage of the reiteration, domestic markets are in equilibrium. This process yields international prices as influenced by governmental and inter-governmental agreements. The system is solved in annual increments, simultaneously for all countries. Summary indicators of the sensitivity of the world system used in this chapter include world cereal production, world cereal prices, and prevalence of population in developing countries at risk from hunger.

The BLS does not incorporate any climate relationships *per se*. Effects of changes in climate were introduced to the model as changes in the average national or regional yield per commodity. Ten commodities are included in the model: wheat, rice, coarse grains (feed grain), bovine and ovine meat, dairy products, other animal products, protein feeds, other food, non-food agriculture, and non-agriculture. Yield change estimates for coarse grains were based on the percentage of maize grown in the country or region; soybean crop model results were used to estimate the protein feed category; the estimates for the non-grain crops were based on the modeled grain crops

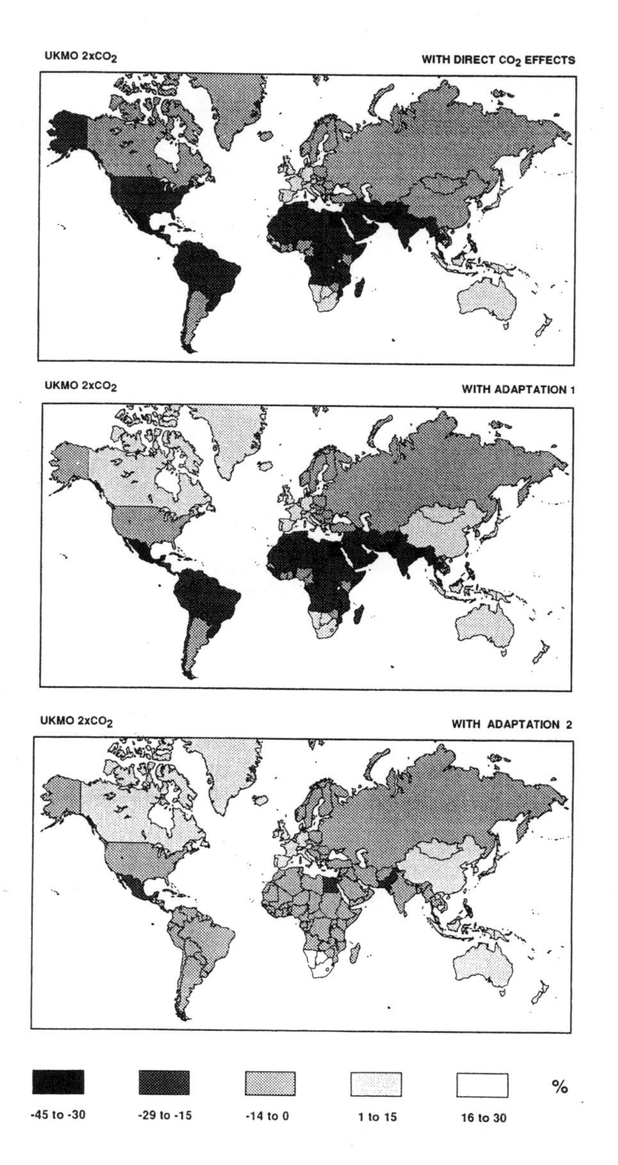

Figure 5-4. Estimated changes in average national grain yield (wheat, rice, coarse grains, and protein feed) under two levels of adaptation for the UKMO doubled CO_2 climate change scenario. Adaptation Level 1 implies minor changes to existing agricultural systems; Adaptation Level 2 implies major changes. Results shown are averages for countries and groups of countries in the Basic Linked System world food trade model; regional variations within countries are not reflected. Direct CO_2 effects on crop growth and water use are taken into account.

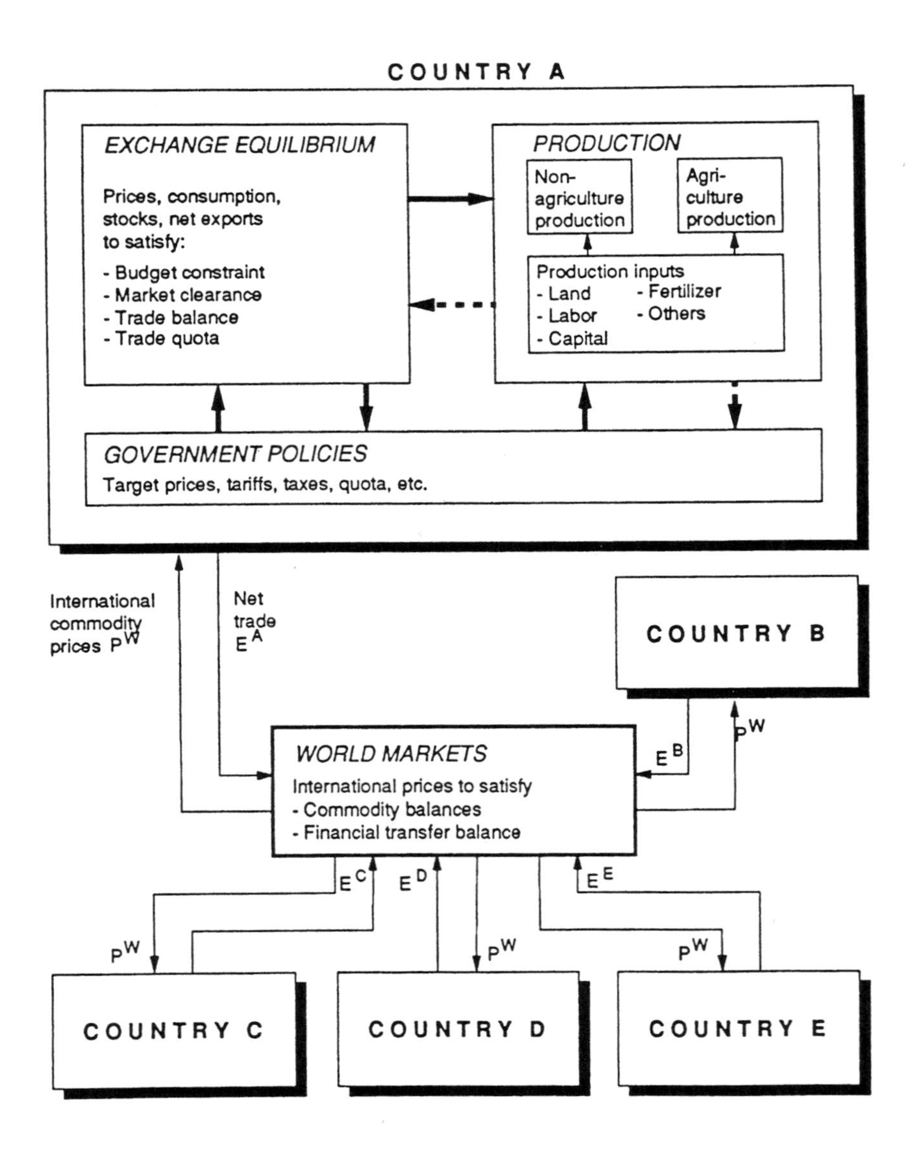

Figure 5-5. The Basic Linked System: Relationships between country components and world markets. Arrows to countries represent international commodity prices; arrows to world markets represent net trade.

and previous estimates of climate change impacts as described above. A positive bias toward non-grain crops was introduced by this procedure, since the previous estimates of yield changes of the non-grain crops were less negative than the modeled results from this study.

B. Economic Growth Rates

Economic growth rates are a product of several BLS functions. Non-agricultural production utilizes a Cobb-Douglas production function with labor and capital as production factors. Non-agricultural labor input depends primarily on population growth and somewhat on relative prices between agriculture and non-agriculture by means of a sector migration function. Capital accumulation depends on investment and depreciation, which in turn depend on saving and depreciation rates. Depreciation rates and saving rates are estimated from historical data and are kept constant after 1990. There is an exogenous assumption based on historical data for technical progress in the production function. For the lower growth scenario, the savings rate was reduced, resulting in about 10% lower GDP in 2060.

The economic growth rates predicted by the BLS in the reference scenario follow historical trends. For the period 1980 to 2060, the BLS produces a growth of 1.3%, 1.7%, and 2.4% annually for world total, developed, and developing countries, respectively, as compared to average population growth rates of 1.1%, 0.3%, and 1.3%.

C. Yield Trends

The annual yield trends used in the BLS for the period 1980–2000 are 1.2%, 1.0%, and 1.7% for global, developed country, and developing countries, respectively. In general, the rate of exogenous technical progress starts from historical values and for cereal crops approaches 0.5% per annum by 2060. Yields have been growing at an average of around 2% annually during the period 1961–1990, both for developed and developing (excluding China) countries (FAO, 1991). Here and elsewhere in the study, China is excluded from the aggregation for developing countries due to lack of data. In the 1980s, yields grew globally at an average yield increase of 1.3%, implying a falling trend in yield growth rates.

The falling growth rates utilized in the reference scenario of the BLS may be justified by several reasons. Historical trends suggest decreasing rates of increase, and yield improvements from biotechnology have yet to be realized. Much of the large yield increases in developed countries in the 1950s and 1960s, and in developing countries thereafter, have been due to intensification of chemical inputs and mechanization. Apart from economic reasons and environmental concerns, which suggest that maximum input levels have been reached in many developed countries, there would likely be diminishing rates of return for further input increases. In some developing countries, especially in Africa, increase of input levels and intensification of production are likely to continue for some time, but may also ultimately level off as well. Furthermore, since Africa has the lowest average cereal yields of all the regional groups and a high population growth rate, it will contribute an increasing share of cereal production, thereby reducing average global yield increases.

D. Arable Land

Availability of arable land for expansion of crop production is based on FAO data. In the BLS standard national models, a piece-wise linear time trend function is used to impose upper bounds (inequality constraints) on land use. In addition, this time trend function is modified with an elasticity term (usually 0.05 or less) that reacts to changes in shadow prices of land in comparison to 1980 levels. The upper limits imposed by the time trend function utilize the FAO data on potential arable land. The arable land limits are not adjusted due to climate change, even though they may be affected positively in some locations by extension of season length or drying of wet soils, or negatively by sea-level inundation or desertification.

E. Risk of Hunger Indicator

The indicator of number of people at risk from hunger used in the BLS is defined as the population with an income insufficient to either produce or procure their food requirements in developing countries (excluding China). The measure is derived from FAO estimates and methodology for developing market economies (FAO, 1984, 1987). The parameters of these distributions were estimated by FAO for each country based on country-specific data and cross-country comparisons. The estimate of the energy requirement of an individual is based on the basal metabolic rate (time in a fasting state and lying at complete rest in a warm environment). Body weight, age, and sex have an impact on this requirement. FAO presents two estimates of undernourished people, based on minimum maintenance requirements of 1.2 and 1.4 (the latter judged as more appropriate) times the basal metabolic rate. The BLS estimate for 1980, based on the 1.4 basal metabolic rate requirements, is 501 million undernourished people in the developing world, excluding China.

F. Limitations of World Food Trade Model

The economic adjustments simulated by the BLS are assumed not to alter the basic structure of the production functions. In fact, these relationships may change in a changed climatic regime and under conditions of elevated CO_2. For example, yield responses to nitrogen fertilization may be altered.

Recent changes in global geopolitics and related changes in agricultural production are not well represented in the BLS. Prices in previously planned economies were made more responsive compared to earlier versions, "plan targets" for allocation decisions were replaced, and some constraints were relaxed in the agricultural sector model. But better analysis depends on development of new models for these emerging capitalist economies.

In the analysis of BLS results presented in this chapter, consideration is limited to the major cereal food crops, even though shifts in the balance of arable and livestock agriculture are likely under changed climatic regimes. Livestock production is a significant component of the global food system and is also potentially sensitive to climatic change. Furthermore, the non-agriculture sector is poorly modeled in the BLS, leading to simplifications in the simulation of responses to climatic change.

G. The Reference Scenario

The reference scenario projects the agricultural system to the year 2060 with no climate change and with no major changes in the political or economic context of world food trade. It assumes:

- UN medium population estimates (about 10.3 billion by 2060) (UN, 1989; International Bank for Rural Development/World Bank, 1990).
- 50% trade liberalization in agriculture (e.g., removal of import restrictions) introduced gradually by 2020.
- Moderate economic growth (for the period 1980 to 2060, the BLS produces an average growth of 1.3%, ranging from 3.0%/year in 1980-2000 to 1.1%/year in 2040-2060).
- Technology projected to increase yields over time (1990-2060). Cereal yields of world total, developing countries, and developed countries assumed to increase annually by 0.7%, 0.9%, and 0.6%, respectively. (Recent increase (1965–1985) in annual productivity for less developed countries is about 1.5%/year.)

H. Climate Change Scenarios

These are projections of the world system including effects on agricultural yields under different climate scenarios (the "2xCO_2 scenarios" for the GISS, GFDL, and UKMO GCMs). The food trade simulations for these three scenarios were started in 1990 and assumed a linear change in yields until the doubled CO_2 changes are reached in 2060. Simulations were made both with and without the physiological effects of 555 ppm CO_2 on crop growth and yield for the equilibrium yield estimates. In these scenarios, internal adjustments in the model occur, such as increased agricultural investment, reallocation of agricultural resources according to economic returns (including crop switching), and reclamation of additional arable land as a response to higher cereal prices. These are based on shifts in supply and demand factors that alter the comparative advantage among countries and regions in the world food trade system. These economic adjustments are assumed not to feedback to the yield levels predicted by the crop modeling study.

I. Scenarios Including the Effects of Farm-level Adaptations

The food trade model was first run with yield changes assuming no external farm-level adaptation to climate change and was then re-run with different climate-induced changes in yield assuming the two levels of adaptation described above. Adaptation Level 1 includes those adaptations at the farm level that would not involve major changes in agricultural practices. It thus included changes in planting date, in amounts of irrigation, and in choice of crop varieties that are currently available. Adaptation Level 2 includes, in addition to the former, major changes in agricultural practices, e.g., shifts of planting date of greater than one month, the availability of new cultivars, expansion of irrigation systems, and increased fertilizer application. This level of adaptation would be likely to involve policy changes at the regional, national, and international level and would also be likely to involve significant costs. However, policy, cost, and water resource availability were not studied explicitly and were assumed not to be barriers to adaptation. Switching from one enterprise to another based on production and demand factors is included in the BLS.

V. WORLD FOOD TRADE RESULTS

A. The Reference Scenario: The Future without Climate Change

Assuming no effects of climate change on crop yields, but that population growth and economic growth are as specified above, world cereal production[7] is estimated at 3286 million metric tons (mmt) in 2060 (cf. 1795 mmt in 1990). *Per capita* cereal production in developed countries increases from 690 kg in 1980 to 984 kg in 2060. In developing countries (excluding China), cereal production increases from 179 to 282 kg/cap. Aggregated world *per capita* cereal production goes from 327 kg in 1980 to 319 kg in 2060. The declining aggregate trend for the future is caused by the relatively large difference in *per capita* cereal production in the developed and developing countries and the demographic changes assumed by the model.

Cereal prices are estimated at an index of 121 (1970 = 100) for the year 2060, reversing the trend of falling real cereal prices over the last 100 years. This

Table 5-6. Change in cereal production, prices, and number of people at risk of hunger in 2060 under GCM $2XCO_2$ climate change scenarios.

Region	Reference Scenario[a]	GISS	GFDL	UKMO
			% change from reference	
Cereal production (million metric tons)				
Global	3286.0	-1.2	-2.8	-7.6
Developed	1449.0	11.3	5.2	-3.5
Developing	1836.0	-11.0	-9.2	-10.8
Cereal prices (1970 = 100)				
	121.0	24.0	33.0	145.0
People at risk from hunger (millions)				
Developing[b]	641.0	10.0	17.0	58.0

[a] Reference scenario is for 2060 assuming no climate change.
[b] Estimates do not include China.

occurs because the BLS standard reference scenario has two phases of price development. During 1980 to 2020, while trade barriers and protection are still in place but are being reduced, there are increases in relative prices; price decreases follow when trade barriers are removed. The number of hungry people is estimated at about 640 million or about 6% of total population in 2060 (cf. 530 million in 1990, about 10% of total current population).

B. Effects of Climate Change with Economic Adjustment, but without Farm-level Adaptation

Cereal production, cereal prices, and people at risk of hunger estimated for the GCM $2XCO_2$ climate change scenarios (with the direct CO_2 effects taken into account) are given in Table 5-6. These estimations are based upon dynamic simulations by the BLS that allow the world food system to respond to climate-induced supply shortfalls of cereals and consequently higher commodity prices through increases in production factors (cultivated land, labor, and capital) and inputs such as fertilizer. The testing of climate change impacts without farm-level adaptation is unrealistic, but is done for the purpose of establishing a baseline with which to compare the effects of farmer response. We can safely assume that at least some farm-level adaptations will be adopted, especially techniques similar to those tested in Adaptation Level 1 that do not imply major changes to current agricultural systems.

Under the GISS scenario (which provides lower temperature increases) cereal production is estimated to decrease by just over 1%, while under the UKMO scenario (with the highest temperature increases) global production is estimated to decrease by more than 7%. The largest negative changes occur in developing regions, which average -9% to -11%, though the extent of decreased production varies greatly by country depending on the projected climate. By contrast, in developed countries production is estimated to increase under all but the UKMO scenario (+11% to -3%). Thus, disparities in crop production between developed and developing countries are estimated to increase.

Price increases resulting from climate-induced decreases in yield are estimated to range between about 25 and 150%. The 5.3% percent reduction in yields of the unadjusted GISS scenario causes a disequilibrium that is resolved via market mechanisms in the adjusted case. This results in a -1.2% consumer response and about a +4% (relative) producer response and leads to 24% higher relative prices for cereals. While this price response seems to be high, cereal prices only account for a modest fraction, perhaps one-third or less, of retail food prices. Hence, a 24% increase in world cereal prices does not imply a 24% increase in food prices.

These increases in price are likely to affect the number of people with insufficient resources to purchase adequate amounts of food. The estimated number of hungry people increases approximately 1% for each 2–2.5% increase in prices (depending on climate change scenario). People at risk of hunger increase by 10% to almost 60% in the climate change scenarios tested, resulting in an estimated increase of between 60 and 350 million people in this condition above the reference scenario projection of 640 million by 2060.

C. Effects of Climate Change with and without Direct CO_2 Effects on Yield and Under Different Levels of Farm-level Adaptation

Without the direct physiological effects of CO_2 on crop yields, simulated world cereal production in 2060 declines between 10 and 20%. These decreases are partially compensated by direct CO_2 effects which reduce the declines by about half. Globally, both minor and major levels of adaptation help restore world production levels with CO_2 effects, compared to the climate change scenarios with no adaptation (Figure 5-6). Averaged global cereal production decreases by up to about 160 mmt (0% to -5%) from the reference scenario projection of 3286 mmt with Level 1 adaptations. These involve shifts in farm activities that are not very disruptive to regional agricultural systems. With adaptations implying major changes, global cereal production responses range from a slight increase of 30 mmt to a slight decrease of about 80 mmt (+1% to -2.5%).

Level 1 adaptation largely offsets the negative climate change yield effects in developed countries, improving their comparative advantage in world markets (Figure 5-7). In these regions, cereal production increases by 4% to 14% over the reference scenario. However, developing countries are

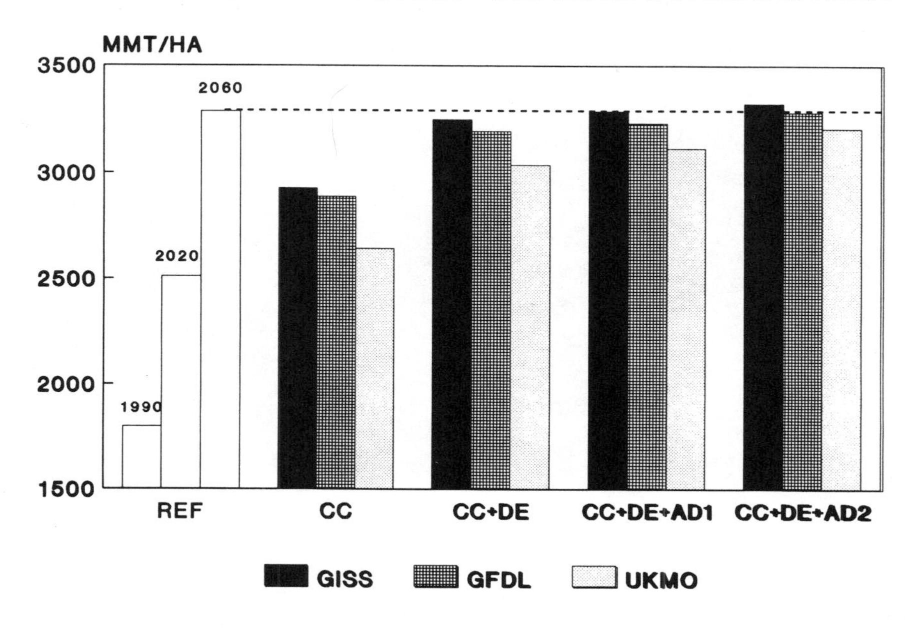

Figure 5-6. World cereal production projected by the BLS for the reference, GISS, GFDL, and UKMO doubled CO_2 climate change scenarios, with and without direct CO_2 effects on crop yields and with Adaptation Levels 1 and 2. Adaptation Level 1 implies minor changes to existing agricultural systems; Adaptation Level 2 implies major changes.

estimated to benefit little from this level of adaptation (-9% to -12% change in cereal production). More extensive adaptation virtually eliminates global negative cereal yield impacts derived under the GISS and GFDL climate scenarios and reduces impacts under the UKMO scenario to one-third.

As a consequence of climate change and Adaptation Level 1, the number of people at risk from hunger increases by about 40 to 300 million (6% to 50%) from the reference scenario of 641 million (Figure 5-8). With more significant farmer adaptation, the number of people at risk from hunger is altered by between -12 million for the GISS scenario and 120 million for the UKMO scenario (-2% and +20%). These results indicate that, except for the GISS scenario under Adaptation Level 2, the simulated farm-level adaptations did not mitigate entirely the negative effects of climate change on potential risk of hunger, even when economic adjustment, i.e., the production and price responses of the world food system, are taken into account.

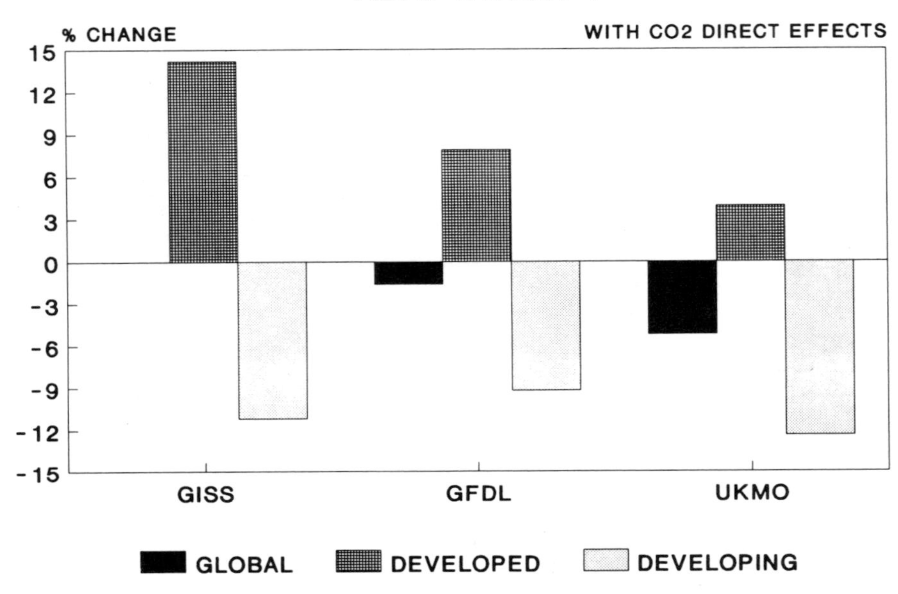

Figure 5-7. Change in global, developed country, and developing country cereal production under climate change scenarios in 2060 assuming Adaptation Level 1, implying minor changes to existing agricultural systems. Reference scenario for 2060 assumes no climate change and projects global production to be 3286 mmt, developed country production to be 1449 mmt, and developing country production to be 1836 mmt. Direct CO_2 effects on crop yields are taken into account.

VI. CONCLUSIONS

When three projections of climate change are imposed on the world food system up to the year 2060, given a continuation of current trends in economic development and population growth rates, it is estimated that, assuming a modest level of farm-level adaptation (e.g., minor shifts in planting dates, changes in crop variety), the net effect of climate change is to reduce global cereal production by up to 5%. This global reduction could be largely overcome by more major forms of adaptation such as installation of irrigation.

Climate change would increase the disparities in cereal production between developed and developing countries. Production in the developed world may well benefit from climate change, whereas production in developing nations may decline. Adaptation at the farm-level does little to reduce

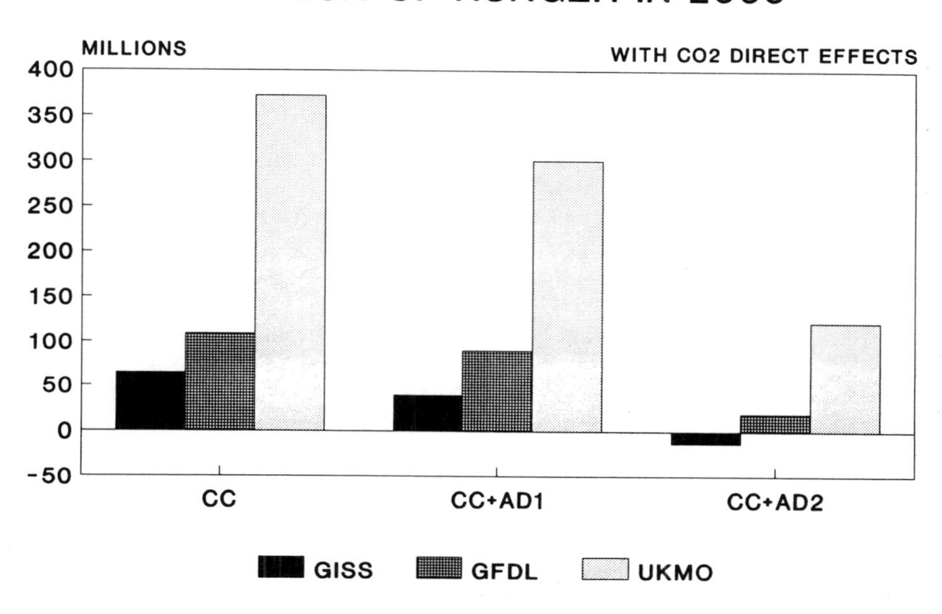

Figure 5-8. Change in number of people at risk of hunger in developing countries in 2060 calculated by the Basic Linked System for the GISS, GFDL, and UKMO doubled CO_2 climate change scenarios and two levels of farm-level adaptation. Reference scenario for 2060 assumes no climate change and projects 529 million people at risk of hunger in 1990 and 641 million in 2060.

the disparities, with the developing world suffering the losses. Cereal prices and thus the population at risk of hunger could increase despite adaptation. Even a high level of farm-level adaptation in the agricultural sector would not entirely prevent such adverse effects.

Overall, the study suggests that the worst situation arises from a scenario of severe climate change and little farm-level adaptation. In order to minimize possible adverse consequences—production losses, food price increases, and people at risk of hunger—the way forward is to encourage the agricultural sector to continue to develop crop breeding and management programs for heat and drought conditions (these will be immediately useful in improving productivity in marginal environments today). Measures taken to slow the growth of the human population of the world would lessen the demand for food and at the same time be consistent with efforts to slow emissions of greenhouse gases, the source of the problem, and thus the rate and eventual magnitude of global climate change.

NOTES

[1] Taken from a more detailed presentation of the study, Rosenzweig et al. (1993).

[2] The results of individual country studies on the potential effects of climate change on risk of hunger in Kenya, Senegal, Zimbabwe, and Chile are presented in Downing (1992).

[3] Individual country crop yield analyses are presented in Rosenzweig and Iglesias (1993).

[4] This approach, which mixes equilibrium climate and transient CO_2 projections, is the best that can be done given lack of availability of GCM transient climate change simulations for impact studies and the need for a dynamic time course for the economic model. However, it is not known what rates of future emissions of trace gases will be and when the full magnitude of their effects will be realized.

[5] Because they absorb longwave infrared radiation, greenhouse gases are predicted to cause changes in the balance between the energy absorbed by the earth and that emitted by it in the form of longwave infrared radiation, resulting in changes in global climate (IPCC, 1990).

[6] The BLS was designed at the International Institute for Applied Systems Analysis (IIASA) for food policy studies, but it also can be used to evaluate the effect of climate-induced changes in yield on world food supply and agricultural prices. For a description of the model see Fischer et al., 1988.

[7] The estimate for cereals includes wheat, rice, maize, millet, sorghum, and minor grains contained in the FAO AGROSTAT database; rice is included as rice milled equivalent (a factor of 0.67 is used to convert from rice paddy to milled rice).

LITERATURE CITED

Acock, B. and L.H. Allen, Jr. 1985. Crop responses to elevated carbon dioxide concentrations. In *Direct Effects of Increasing Carbon Dioxide on Vegetation*. eds. B.R. Strain and J.D. Cure, 33-97. U.S. Department of Energy, DOE/ER-0238: Washington, D.C.

Adams, R.M., C. Rosenzweig, R.M. Peart, J.T. Ritchie, B.A. McCarl, J.D. Glyer, R.B. Curry, J.W. Jones, K.J. Boote, and L.H. Allen, Jr. 1990. Global climate change and US agriculture. *Nature*. 345(6272):219-24.

Allen, L.H., Jr., K.J. Boote, J.W. Jones, P.H. Jones, R.R. Valle, B. Acock, H.H. Rogers, and R.C. Dahlman. 1987. Response of vegetation to rising carbon dioxide: Photosynthesis, biomass and seed yield of soybean. *Global Biogeochemical Cycles*. 1:1-14.

Cure, J.D. 1985. Carbon dioxide doubling responses: A crop survey. In *Direct Effects of Increasing Carbon Dioxide on Vegetation*. eds. B.R. Strain and J.D. Cure, 33-97. U.S. Department of Energy, DOE/ER-0238: Washington, D.C.

Cure, J.D. and B. Acock. 1986. Crop responses to carbon dioxide doubling: A literature survey. Ag. *and For. Meteor.* 38:127-45.

Downing, T.W. 1992. *Climate Change and Vulnerable Places: Global Food Security*. Environmental Change Unit, University of Oxford.

FAO. 1984. *Fourth World Food Survey*. United Nations Food and Agriculture Organization: Rome.

FAO. 1987. *Fifth World Food Survey*. United Nations Food and Agriculture Organization: Rome.

FAO. 1988. *1987 Production Yearbook*. Statistics Series No. 82. United Nations Food and Agriculture Organization: Rome. 351 pp.

FAO. 1991. *AGROSTAT/PC*. United Nations Food and Agriculture Organization: Rome.

Fischer, G., K. Frohberg, M.A. Keyzer, and K.S. Parikh. 1988. *Linked National Models: A Tool for International Food Policy Analysis*. Kluwer: Dordrecht. 227 pp.

Hansen, J., G. Russell, D. Rind, P. Stone, A. Lacis, S. Lebedeff, R. Ruedy, and L. Travis. 1983. Efficient three-dimensional global models for climate studies: Models I and II. *Monthly Weather Review*. 111(4):609-662.

Hansen, J., I. Fung, A. Lacis, D. Rind, G. Russell, S. Lebedeff, R. Ruedy, and P. Stone. 1988. Global climate changes as forecast by the GISS 3-D model. *Journal of Geophysical Research*. 93(D8):9341-9364.

IPCC. 1990a. *Climate Change: The IPCC Scientific Assessment*. eds. J.T. Houghton, G.J. Jenkins, and J.J. Ephraums. Cambridge University Press: Cambridge. 365 pp.

IPCC, 1990b. *Climate Change: The IPCC Impacts Assessment*. eds. W.J. McG. Tegart, G.W. Sheldon, and D.C. Griffiths. Australian Government Publishing Service: Canberra.

IPCC. 1992. *Climate Change 1992. The Supplementary Report to the IPCC Scientific Assessment*. eds. J.T. Houghton, B.A. Callandar, and S.K. Varney. Cambridge University Press: Cambridge. 200pp.

International Bank for Reconstruction and Development/World Bank. 1990. *World Population Projections*. John Hopkins University Press: Baltimore.

International Benchmark Sites Network for Agrotechnology Transfer (IBSNAT) Project. 1989. *Decision Support System for Agrotechnology Transfer Version 2.1 (DSSAT V2.1)*. Dept. of Agronomy and Soil Science. College of Tropical Agriculture and Human Resources. University of Hawaii: Honolulu.

Kane, S., J. Reilly, and J. Tobey. 1991. *Climate Change: Economic Implications for World Agriculture*. AER-No. 647. U.S. Department of Agriculture. Economic Research Service: Washington, D.C. 21 pp.

Kellogg, W.W. and Z.-C. Zhao. 1988. Sensitivity of soil moisture to doubling of carbon dioxide in climate model experiments. Part 1: North America. *J. of Climate*. 1:348-66.

Kimball, B.A. 1983. Carbon dioxide and agricultural yield. An assemblage and analysis of 430 prior observations. *Agronomy Journal*. 75:779-88.

Liverman, D., M. Dilley, K. O'Brien, and L. Menchaca. 1993. The Impacts of Climate Change on Maize in Mexico. In *Implications of Climate Change for International Agriculture: Crop Modeling Study*. eds. C. Rosenzweig and A. Iglesias. US Environmental Protection Agency: Washington, D.C. In press.

Magalhaes, A.R. 1992. *Impacts of Climatic Variations and Sustainable Development in Semi-arid Regions*. Proceedings of International Conference. ICID. Fortaleza, Brazil. 52 pp.

Manabe, S. and R.T. Wetherald. 1987. Large-scale changes in soil wetness induced by an increase in CO_2. *J. Atmos. Sci*. 44:1211-35.

Menzhulin, G., L. Koval, and A. Badenko. 1993. Potential effects of global warming and atmospheric carbon dioxide on wheat production in the Commonwealth of Independent States. In *Implications of Climate Change for International Agriculture: Crop Modeling Study*. eds. C. Rosenzweig and A. Iglesias. US Environmental Protection Agency: Washington, D.C. In press.

Parry, M.L., T.R. Carter, and N.T. Konijn. eds. 1988. *The Impact of Climatic Variations on Agriculture. Vol 1. Assessments in Cool Temperate and Cold Regions.* Kluwer: Dordecht, Netherlands. 876 pp.

Parry, M.L., T.R. Carter, and N.T. Konijn. eds. 1988. *The Impact of Climatic Variations on Agriculture, Volume 2. Assessments in Semi-Arid Areas.* Dordrecht: Kluwer. 764 pp.

Parry, M.L. 1990. *Climate Change and World Agriculture.* Earthscan: London.

Parry, M.L., M.B. de Rozari, A.L. Chong, and S. Panich. eds. 1992. *The Potential Socio-Economic Effects of Climate Change in South-East Asia.* Nairobi: U.N. Environment Programme.

Peart, R.M., J.W. Jones, R.B. Curry, K. Boote, and L.H. Allen, Jr. 1989. Impact of climate change on crop yield in the southeastern U.S.A. In *The Potential Effects of Global Climate Change on the United States.* eds. J.B. Smith and D.A. Tirpak. U.S. Environmental Protection Agency: Washington, D.C.

Rogers, H.H., G.E. Bingham, J.D. Cure, J.M. Smith, and K.A. Surano. 1983. Responses of selected plant species to elevated carbon dioxide in the field. J. *Environ. Qual.* 12:569-74.

Rosenberg, N.J. and P.R. Crosson. 1991. *Processes for Identifying Regional Influences of and Responses to Increasing Atmospheric CO_2 and Climate Change: the MINK Project, An Overview.* Resources for the Future. Dept. of Energy. DOE/RL/01830T-H5. Washington, D.C. 35 pp.

Rosenzweig, C. and A. Iglesias. eds. 1993. I*mplications of Climate Change for International Agriculture: Crop Modeling Study.* US Environmental Protection Agency: Washington, D.C. In press.

Rosenzweig, C., M. Parry, G. Fischer, and K. Frohberg. 1993. *Climate Change and World Food Supply.* Environmental Change Unit: Oxford University.

Smit, B. 1989. Climatic warming and Canada's comparative position in agricultural production and trade. In *Climate Change Digest.* Environment Canada. CCD 89-01.CCD 89-01. pp. 1-9.

Smith, J.B. and D.A. Tirpak. eds. 1989. *The Potential Effects of Global Climate Change on the United States.* Report to Congress. EPA-230-05-89-050. U.S. Environmental Protection Agency: Washington, D.C. 423 pp.

Strzepek, K.M., S.C. Onyeji, and M. Saleh. 1993. A socio-economic analysis of integrated climate change impacts on Egypt. In *As Climate Changes: International Impacts and Implications.* eds. K.M. Strzepek and J.B. Smith. Cambridge University Press. Cambridge. In press.

Tobey, J., S. Kane, and J. Reilly. 1993. An empirical study of the economic effects of climate change on world agriculture. *Climatic Change.* In press.

UN. 1989. *World Population Prospects 1988.* United Nations: New York.

Wilson, C.A. and J.F.B. Mitchell. 1987. A doubled CO_2 climate sensitivity experiment with a global climate model including a simple ocean. *Journal of Geophysical Research.* 92(13):315-43.

6

IMPACTS of CLIMATE CHANGE on the AGRICULTURE and ECONOMY of the MISSOURI, IOWA, NEBRASKA, and KANSAS (MINK) REGION

Pierre Crosson
Senior Fellow and Consultant, Resources for the Future, Washington, D.C.

I. INTRODUCTION

In the late 1980s, Resources for the Future (RFF) put together an interdisciplinary team consisting of several natural resource economists, two agroclimatologists, and an ecologist to study the effects of climate change on the economy of four midwestern states: Missouri, Iowa, Nebraska, and Kansas (hereafter referred to as the MINK region). The study was undertaken as part of RFF's Climate Resources Program. This paper presents some of the results from the study.[1]

The MINK region was selected for the study for several reasons. First, compared with other regions in the U.S. of similar size, the topography of MINK and land use in the region are relatively simple and homogeneous. Except for the hilly Ozark region of southeast Missouri where some forestry is practiced, the land is flat-to-rolling and is used primarily for agriculture. Second, relative to the rest of the country, the MINK economy is more dependent on natural resource-based sectors most likely to be impacted by climate change. This is not to say that the economy of the region is heavily dependent on agriculture. Data prepared by the U.S. Department of Commerce (1990) indicate that farm income in MINK accounts for only a little more than three percent of total regional income. When allowance is made for income in non-farm activities which serve farming, the direct and indirect contribution of agriculture to regional income is not more than 10–12 percent. However, in MINK, the share of farm income in total regional income is about three times the share of farm income in national income. The MINK economy, therefore, is likely to be more sensitive to climate change impacts on agriculture than the national economy.

Third, most general circulation models (GCMs) predict that the global "greenhouse effect" will result in significant warming in MINK, and some of the models also predict significantly less precipitation. These predictions suggest severe potential threats to agriculture and water resources in the region. Thus, study of climate change impacts in MINK should provide insights relevant to other regions under similar threats.

Most studies of the regional economic impacts of climate change have defined regional climates using GCM output (e.g., Parry et al., 1988; Smith and Tirpak, 1989). We chose instead to use the MINK climate of the relatively hot and dry 1930s as an analog of what the climate of the region might be like under global warming induced by a CO_2 equivalent doubling. There are both advantages and disadvantages to this approach.

The principal advantage is that the actual climate record does a far better job than GCMs in reflecting the spatial and temporal variability of the climate. The principal disadvantage is that in the 1930s mean summer temperatures in MINK averaged only 1-2°C above the climate from 1950 to 1980 (which we took to represent the current "normal" climate of the region), while the GCMs show that under a CO_2 equivalent doubling, temperatures in the region could be from 3°C to 8°C higher. Anyone who believes that the GCMs more accurately predict the future climate of the region will likely conclude that our results underestimate future climate impacts. However, there is much disagreement among the GCMs about regional climates, and much uncertainty about the results of all of them. (See the discussion by Santer, Chapter 3.) The hotter and drier 1930s—henceforth called the analog climate—were directionally the same as the GCM results for the region. Given this, we decided that the advantages of the analog approach outweighed the disadvantages.

This chapter focusses on impacts of climate change effects on regional agriculture and on the consequent feedback effects on the regional economy as a whole. The agricultural impacts were first estimated at the farm level and then extrapolated to the region as a whole. (See Easterling et al., (1991) for further detail.) In this paper, I take the farm level results as the point of departure for my discussion.

By the time the climate changes enough to have significant impacts, population growth and technological change will have transformed agriculture in most regions to something far different from what it is presently. In the MINK study, we estimated the impact of the analog climate on the region's agriculture in the 1980s to establish a baseline. Most of the emphasis in the study, however, is on the impacts on agriculture in the region as it might look in 2030. Construction of scenarios of the region's agriculture in 2030, with and without the analog climate, was thus a major part of the MINK research (see Crosson et al., 1991).

II. IMPACT OF THE ANALOG CLIMATE ON 1980s CROP PRODUCTION

We used the Erosion Productivity Impact Calculator (EPIC), a crop growth simulation model developed by U.S. Department of Agriculture scientists, to estimate the effects of the analog climate on yields of corn, wheat, soybeans, sorghum, and alfalfa hay on a number of representative farms in each of a number of Major Land Resource Areas (MLRAs) included in the study. The EPIC model was adjusted to show the yield effects of the analog climate. Since a higher concentration of CO_2 in the atmosphere not only is the primary driver of climate change but also stimulates faster crop growth, our simulations with EPIC show the crop yield effects of the analog climate both with and without CO_2 fertilization.

Four different crop yield scenarios were developed corresponding to a situation where the analog climate replaced the present climate (Easterling, et al., 1991). In the first scenario, EPIC was used to calculate the yield changes assuming no CO_2 fertilization effect and no management adjustments by farmers to compensate for the negative yield effects. In the second scenario, the effects of CO_2 fertilization were incorporated, but no management adjustments were assumed. In the third and fourth scenarios, we drew on the judgment of experts on MINK agriculture to simulate farm-level adjustments to climate change with (scenario 4) and without (scenario 3) CO_2 fertilization. The decline in yields under the analog climate would increase the costs of producing the various crops. The resources available to us did not permit estimation of the cost functions of the crops so we are unable to estimate the amounts of the cost increases and the subsequent declines in production. Instead, we assumed that in the no on-farm adjustment scenarios, costs of each crop would rise such that the decline in production would be proportional to the decline in yields. In the scenarios including on-farm adjustments, the declines in production reflect both the decline in yields and some shifting of land out of irrigated corn, because of increased scarcity of water, and into dryland production of wheat and sorghum.

The production effects for the scenarios with and without CO_2 fertilization but no management adjustments are given in Table 6-1. The table shows the changes in both the volume and the value of production. The changes in value were found by multiplying the changes in production volume by average prices of the various crops in 1982 dollars.

When the CO_2 fertilizer effect is not taken into account, and no management adjustments are permitted, the value of MINK production of the five crops under the analog climate declines 17.1 percent from the average 1984/87 value of production (1984/87 production was valued in 1982 prices for this calculation). With CO_2 fertilization, the decline in production is 8.4 percent

Table 6-1. Climate change impacts on crop production from average 1984–87 levels without farm-level adjustments.

	Volume (million metric tons)					Value (millions 1982 $)	
	Mo.	Iowa	Neb.	Kan.	Total	Total	% change in total
Without CO_2 fertilization							
Corn	-1.27	-10.24	-2.84	-0.68	-15.03	-$1644	-21.3
Wheat	+0.07	a	-0.02	-0.15	-0.10	-14	-0.8
Sorghum	-0.48	a	-0.43	-1.24	-2.15	-215	-17.1
Soybeans	-0.92	-1.92	-0.55	-0.38	-3.77	-789	-23.0
Hay	-0.40	a	-0.10	-0.30	-0.80	-47	-3.0
						-$2709	-17.1
With CO_2 fertilization							
Corn	-.84	-6.92	-1.34	-0.36	-9.46	-$1035	-13.4
Wheat	+0.19	+0.01	+0.12	+0.82	+1.14	+150	+8.2
Sorghum	-0.32	a	-0.20	-0.66	-1.18	-118	-9.4
Soybeans	-0.55	-0.96	-0.34	-0.24	-2.09	-438	-12.8
Hay	+0.40	+0.30	+0.10	+1.10	+1.90	+112	+7.0
						-$1329	-8.4

[a] Less than 10,000 tons

Note: Volume is percentage change in crop yields times average production in 1984–87; value is the change in production times prices in 1982. Details of the yield loss calculations and prices are in Easterling et al. (1991), sections 4 and 5 respectively.

Table 6-2. Climate change impacts on crop production from 1984/87 average, with farm-level adjustments and without and with CO_2 fertilizations.

Crop	Without CO_2 fertilization		With CO_2 fertilization	
	Millions	% change	Millions	% change
Corn	-1729	-22.4	-1236	-16.0
Wheat	+361	+19.7	+139	+7.6
Sorghum	-35	-2.8	+178	+14.1
Soybeans	-542	-15.8	-48	-1.4
Hay	+23	-1.5	+435	+27.3
Total	**-1922**	**-12.1**	**-532**	**-3.3**

Note: Volume is a combination of percentage changes in crop yields and shifts of land out of irrigated corn production into dryland production of wheat and sorghum. Value is the change in volume of production times crop prices in 1982.

instead of 17.1 percent (still with no management adjustments).

It is certain that farmers in MINK would seek ways to offset the negative effects of the analog climate. Consultation with agricultural economists and agronomists in the region indicated a number of known practices which farmers might adopt, such as earlier planting (except for wheat, which is planted in the fall), use of longer season varieties, and furrow diking for dryland row crops. Because water would become more scarce under the analog climate, some shifts of land in Nebraska and Kansas from irrigated corn production to dryland production of wheat and sorghum would be likely.

These adjustments were introduced into EPIC and yields were recalculated. The effects on the volume and value of crop production are shown in Table 6-2. Comparison of the without CO_2 fertilization cases in Tables 6-1 and 6-2 indicates that the simulated on-farm adjustments would reduce the loss of crop production from -\$2.7 billion to -\$1.9 billion, or 29 percent. The combination of CO_2 fertilization and on-farm adjustments reduces the loss further to -\$532 million (Table 6-2) or by 80 percent compared with the no adjustment, no CO_2 fertilization case. In these simulations, CO_2 fertilization is more important than on-farm adjustments in moderating the effects of the analog climate on crop yields. (For further discussion of the controversy surrounding CO_2 fertilization, see Chapter 8.)

III. IMPACTS ON THE 1980s REGIONAL ECONOMY BY WAY OF CROP PRODUCTION

We used IMPLAN, a regional input-output model developed for the U.S. Forest Service (Alward, 1986, 1989), to estimate the impacts of the analog climate on the MINK economy by way of agriculture. IMPLAN is used by the Forest Service to estimate the local impacts of their land management policies. Because of the widespread Forest Service land holdings, IMPLAN was made to be highly flexible in its geographic coverage, and can be used to describe and analyze the economies of any region in the United States. The basic unit of analysis is either the county or state, and regions can be defined by any set of counties or states. Aggregation of state data for Iowa, Kansas, Nebraska, and Missouri provided us with a model of the MINK region.[2]

A. Agricultural Output Multipliers

IMPLAN was used to calculate production, value added, employment, and personal income multipliers for the MINK economy. All four types of multipliers give essentially the same result in terms of the percentage impact on the MINK economy of a given percentage change in deliveries to final demand. We show here only the production multipliers because they capture reasonably well the range of estimates for the set of four multipliers.

Table 6-3 shows the production multipliers for the six largest (by value of production) farm level production activities in MINK. The $27.7 billion produced by these six activities was 87 percent of total on-farm production in 1982, according to IMPLAN data. Table 6-3 shows that of the $7.3 billion of feedgrains produced in MINK, $3.4 billion was delivered to final demand (purchases for consumption, investment, government, and export). Exports (to the rest of the U.S. and abroad) accounted for 88 percent of this total final demand for feedgrain production. The $3.9 billion of feedgrain production not delivered to final demand was delivered to cattle feedlots, producers of hogs and pigs, and other sectors in MINK using feedgrains as an input to their operations.

The production multiplier for feedgrains is 1.68 (Table 6-3). That is, every dollar of feedgrains delivered to final demand stimulates another 68 cents worth of production throughout the regional economy in activities directly and indirectly linked to feedgrain production. Because IMPLAN (like all input-output models) assumes that all inter-industry relationships in MINK are linear, the multipliers can be used to estimate changes in total regional production induced by changes in final demand for any combination of these given industries.

Table 6-3. Production, final demand, and production multipliers for selected activities.

Activity	Production	Final demand	Production multiplier
	billion $	billion $	
On-farm production			
Feedgrains	7.3	3.4	1.68
Cattle feedlots	7.0	0.5	1.93
Hogs and pigs	4.7	0.3	2.17
Oil-bearing crops	4.5	2.2	1.46
Hay and pasture	2.4	1.1	1.62
Food grains	1.8	1.3	1.67
Totals	**27.7**	**8.8**	**1.79**
Off-farm production			
Meat packing plants	14.2	12.3	2.57
Sausages & prepared meats	3.9	3.6	2.45
Prepared feeds n.e.c.	2.5	1.0	2.05
Creamery butter	1.5	1.1	2.45
Fluid milk	1.4	0.9	1.98
Fertilizer	0.8	0.2	1.63
Agric. chemicals n.e.c.	0.7	0.4	1.64
Farm machinery & equip.	3.2	2.5	1.32
Logging camps	0.1	0.1	1.45
Sawmills	0.1	0.1	1.29
Water transport	0.8	0.3	1.79
Totals	**29.2**	**22.4**	**2.26**

Source: U.S. Department of Agricuilture (1989).

Table 6-3 also shows production multipliers for a set of off-farm, agriculturally-related manufacturing activities. Note the importance of meat processing activities. Meatpacking plants and those producing sausages and prepared meat accounted for 62 percent of the $29.2 billion of production of the eleven activities listed, and for 71 percent of deliveries to final demand. Meatpacking alone accounted for 49 percent of production and 55 percent of deliveries to final demand. Meatpacking, in fact, is the largest single manufacturing activity in MINK, according to IMPLAN. Its multiplier of 2.57 is the highest of any activity in the region. The implication is that meatpacking plants directly and indirectly draw proportionally more of their inputs from within the MINK region than do any other agriculturally-related industries.

Clearly, in assessing the effects of climate change on the MINK economy special attention must be given to likely impacts on output of meatpacking plants. Reasons to expect direct effects on final demand do not readily come to mind. However, indirect effects by way of climate change impacts on feedgrain, soybean, and animal production would appear possible. If, for example, climate change is unfavorable for on-farm production of these commodities in MINK, the costs of meatpacking production might rise relative to costs elsewhere, leading to a decline in final demand for output of MINK's meatpacking plants.

B. Impacts of Declines in Crop Production

The negative impacts of the analog climate on crop yields in MINK and the consequent rise in production costs would affect the amount of regional crop production in two ways: 1) production would be cut back, because the previous level would no longer be profitable under the higher costs; 2) some farmers would shift entirely out of a no longer profitable crop and into a profitable one. These consequences are reflected in Tables 6-1 and 6-2. Table 6-4 shows the region-wide impacts of the production declines, using production multipliers from IMPLAN.

The declines in crop production would induce declines in total regional production, the amount depending on the production multipliers for the various crops. In the scenario without on-farm adjustments or CO_2 fertilization, the $1.6 billion decline in corn production, for example, would reduce total regional production by $2.8 billion because the production multiplier for feedgrains is 1.68 (Table 6-3). Table 6-4 shows that without on-farm adjustments or CO_2 enrichment the climate-induced changes in output of the five crops could lead to a $4.4 billion (1.4%) decline in overall regional output. Sixty-three percent of the decline in regional output is due to the drop in corn production; 26 percent is due to the decline in soybean production. Under the scenario with on-farm adjustments and the beneficial effects of elevated CO_2, the overall regional impact declines to $910 million, which is only a 0.3 percent decline in the overall value of regional output.

We noted above that the decline in feedgrain and soybean production could reduce animal production in the region, with significant implications for the important meat-packing industry. In fact, IMPLAN indicates a strong indirect linkage between the meatpacking industry and feedgrains, through

Table 6-4. Changes in the value of crop production and in total regional production under the analog climate, 1980s baseline (millions of 1982 $).

	Change in crop output[a,c]	Multiplier effect	Change in regional product[b]
			%
Without adjustments			
Corn	-1644	-2762	-0.9
Wheat	-14	-23	d
Sorghum	-215	-361	-0.1
Soybeans	-789	-1152	-0.4
Hay	-47	-76	d
Totals	**-2709**	**-4374**	**-1.4**
With adjustments			
Without CO_2 fertilization			
Corn	-1729	-2905	-0.9
Wheat	+361	+603	+0.2
Sorghum	-35	-59	d
Soybeans	-542	-791	-0.3
Hay	+23	+37	-1.0
Totals	**-1922**	**-3115**	**-1.0**
With CO_2 fertilization			
Corn	-1236	-2076	-0.7
Wheat	+139	+232	+0.1
Sorghum	+178	+299	-0.1
Soybeans	-48	-70	d
Hay	+35	+705	+0.2
Totals	**-532**	**-910**	**-0.3**

[a] From Table 6-1 (without CO_2 fertilization case)
[b] Percentages of total regional production ($309.2 billion in 1982, IMPLAN)
[c] With adjustment figures are from Table 6-2
[d] Less than 0.1 percent

Table 6-5. Effects of climate-change induced reductions in feedgrain production on overall regional production in MINK.

	Regional output change[a]	% change regional output
Production decline falls on exports		
Without adjustments and without CO_2 fertilization	-3.1	-1.0
With adjustments and CO_2 fertilization	-1.8	-0.6
Production decline falls on animal producers		
Without adjustments and without CO_2 fertilization	-29.9	-9.7
With adjustments and CO_2 fertilization	-17.6	-5.7

[a] Billions of 1982 dollars

Note: Based on the simulated declines in production of corn and sorghum shown in Table 6-4 and regional production multipliers derived from the IMPLAN model and shown in Table 6-3.

animal production. The meatpacking industry, as noted, is the largest single manufacturing activity in MINK, and also has the highest multiplier relationship to the rest of the economy.

In Table 6-5, two sets of estimates are presented of the multiplier effects of the decline in feedgrain (corn and sorghum) production imposed by the analog climate. In one set, we assume that the increase in costs of feedgrain production caused by the climate-induced losses of yield results in a reduction in export demand for the crops equal to the simulated decline in crop production. In this case, the supply of feedgrains available to animal producers in MINK is unchanged, so animal production and meatpacking in the region are not affected by the decline in grain output. The other set of multiplier effects are based on the assumption that the full amount of decline in grain production is borne by animal producers in MINK, exports of the crops being unaffected, with the shortfall in animals borne by the local meatpacking industry.

Neither of the two underlying assumptions is very realistic. Both export demand and internal MINK demand for grains would be reduced by the increased cost of producing these crops under the analog climate. However, given the resource limitations of this study, we had no way of estimating this more realistic outcome. The estimates obtained with the two underlying assumptions should be regarded as polar cases setting the limits within which the actual multiplier effects would fall.

Table 6-5 provides two estimates under each of the two underlying assumptions. One set of estimates is for the scenario with no CO_2 fertilization and no adjustments by farmers; the other is for the scenario with on-farm adjustments and with CO_2 fertilization. Under worst case conditions (no CO_2 fertilization, no on-farm adjustments, and the total burden of the decline in feedgrain production falling on the meatpacking industry in MINK), the climate-induced decline in feedgrain production would reduce total production of all sorts in the region by $29.9 billion, or 9.7 percent. Under the best case conditions (on-farm adaptations, CO_2 fertilization, and the total burden of the feedgrain production decline falling on exports), total regional production would decline by $1.8 billion, or 0.6 percent.

When the feedgrain production decline is absorbed internally it reduces animal production which in turn reduces output of the meatpacking industry. The size of this industry and its high multiplier, reflecting its relatively strong reliance on locally produced animal and other inputs, explains the difference between the two polar cases.

We believe that the impact of the analog climate on the MINK economy would be closer to the "less severe" than to the "more severe" polar case for three reasons. First, farmers would certainly make adjustments in response to changed climate. Second, the increase in atmospheric CO_2 concentrations likely will offset the negative effects of the analog climate on yields through the CO_2 fertilizer effect. Third, the assumption that the burden of reduced grain production would fall on exports is more realistic than that it would fall wholly on MINK animal producers. On a global scale, we assume that climate change would have offsetting effects on agricultural productivity (Crosson et al., 1991). That is, our analysis suggests that, after world-wide on-farm adjustments, climate change would not significantly affect world agricultural production totals, and hence would not affect prices of agricultural output. In this case, the higher grain production costs in MINK under the analog climate would weaken the region's competitive position in grain markets outside the region, leading to a decline in its exports. MINK animal producers at the same time could import cheaper feedgrain from outside the region under our assumption of no change in global feedgrain production and prices. Thus, the impact on animal production in the region, and hence on the meatpacking industry, would be less than in the worst case outcome. However, as noted above, over the long term some animal production might shift to other grain-producing regions less affected by climate change, which would induce a decline in meatpacking in MINK.

Estimates of the negative economy-wide impacts of the analog climate imply no adjustments to the impacts by people in non-farm sectors of the regional economy. Of course, just as farmers would adjust, so would non-farmers. People and other resources would search for employment elsewhere in MINK, and some surely would be successful. To that extent the estimates of economy-wide impacts in Tables 6-4 and 6-5 would be overstated. Displaced resources not finding employment elsewhere in MINK likely would move to other regions. To this extent the MINK economy would be permanently smaller, but other regions would be larger. The impact on the national economy would be smaller than the impact on MINK.

IV. IMPACTS IN 2030: AGRICULTURE

The objectives of this section are: (1) to develop a baseline scenario of MINK production and resource use with respect to corn, sorghum, wheat, and soybeans in 2030 on the assumption of no change in climate; (2) to consider how this scenario might change under the analog climate on the assumption that the change from the present climate to the analog is gradual over the period from 1984/87 to 2030.

A. Production Under the No Climate Change Scenario

MINK agriculture is a part of national and world markets for the various crops of interest. Consequently, the scenarios for MINK must be part of larger scenarios incorporating aspects of these world and national markets. These are complex issues which were explored in depth in Crosson et al. (1991). Only a brief summary is given here. In the more developed countries (MDCs), demand for feedgrains, wheat, and soybeans was assumed to grow only with population (projected by the United Nations [1989]) from the 1984–87 baseline to 2030, the income elasticity of demand for these crops at the farm level already being close to zero. Both population and *per capita* income growth drive demand for these crops in the less developed countries (LDCs). Over the projection period, global demand for the crops increases 2.4 times, with demand in the LDCs, higher by 3.3 times, accounting for over 90 percent of the increase.

Demand for MINK production of the crops in 2030 was derived from global demand in four steps. First, world trade in the crops was assumed to grow in proportion to global demand. Second, the U.S. share of world trade in the crops was assumed to remain at the 1980s levels. The main threat to this assumption is that some countries, e.g., Brazil, Argentina, India, China, Russia, Ukraine, could emerge as major exporters of these crops, claiming increasing market shares at the expense of the U.S. (and other MDC exporters). Third, steps 1 and 2 gave export demand for U.S. production of the crops. The U.S.

Table 6-6. Production in MINK in 2030 under no climate change.

	Billons of 1982 $
Corn	10.2
Wheat	3.7
Sorghum	2.1
Soybeans	7.1
Total	**23.1**

Source: Crosson et al. (1991).

population projection gave domestic demand. Fourth, and finally, MINK's shares of U.S. production were projected after taking account of factors that might change MINK production costs relative to those in the rest of the country. Interregional differences in land and water supply, new technology, and environmental policies were considered. Despite obviously high uncertainty about each of these factors, it was concluded that only increasing water scarcity (Frederick, 1991) would be likely to affect MINK's competitive position.

Corn production in Nebraska, and to a lesser extent in Kansas, is highly dependent on irrigation from groundwater. Frederick (1991) concluded that, even in the absence of climate change, continuation of current "mining" of groundwater would gradually increase the cost of the resource such that by 2030 a significant portion of presently irrigated land would shift to some non-irrigated use. Increased demand for in-stream ecological services of surface water also would diminish the supply for irrigation. On these arguments we assumed that MINK's share of national corn production would decline about 5 percent in the absence of climate change. (Over 60 percent of MINK corn production is on rain-fed land in Iowa and Missouri.) Production of each crop in 2030 is shown in Table 6-6.

B. Crop Yields Under the No Climate Change Scenario

As in the 1980s baseline situation, EPIC was used to simulate yields in 2030 for the case of no climate change. Specifically, five technical advances in grain and soybean production were incorporated in EPIC: improvements in photosynthetic efficiency, increases in the harvest index (ratio of plant economic product to plant biomass), advances in pest management, earlier leaf area development to increase the period of interception of sunlight, and increases in harvest efficiency, i.e., in techniques for reducing in-the-field losses at the time of harvest.

The effect of the five technical innovations was to raise the annual average rates of EPIC-simulated yields from 1984–87 to 2030 at 0.8% for corn, 1.2% for sorghum, 1.1% for wheat, and 1.1% for soybeans.

The innovations are described in detail in Easterling et al. (1991). Here I note only that a review of the literature on prospective advances in agricultural technology indicated that these innovations could be economically available to farmers generally, and therefore to farmers in MINK also, by 2030 if public and private agricultural research institutions, nationally and internationally, receive the support they will need to be able to respond adequately to globally rising demands for food and fiber.

C. Climate Change Scenarios

In considering the effects of the analog climate on crop yields and production in MINK, it was assumed that climate changes gradually from the 1980s baseline to take the analog form by 2030. It is plausible that in this case the agricultural research establishment in MINK would in time take note of the change in climate toward hotter and drier conditions and increase the resources devoted to research on technologies adapted to those conditions. To reflect this, we incorporated in EPIC not only the five technological improvements discussed above but also two others. One is increased irrigation efficiency, achieved in the with-climate-change scenario by additional research on techniques to reduce water losses to soil infiltration and evaporation. The other additional technological improvement is increased crop resistance to drought, modeled in EPIC as increased stomatal resistance in plant leaves.

An implication of adding the two technologies developed specifically in response to the analog climate to the five assumed to be developed in the absence of the analog is that the resources devoted to agricultural research in MINK would be greater with the analog climate than without it. We believe this is quite plausible. Hayami and Ruttan (1985) made a strong argument that agricultural research institutions respond to signals of increasing scarcity of agricultural resources by developing technologies which substitute more abundant for increasingly scarce resources. Climate is an agricultural resource; that is, it enters the agricultural production function as an input in the form of temperature and precipitation in the EPIC model. Deterioration of the quality of a resource increases its scarcity the same as a reduction in its quantity. The emergence of the analog climate in MINK between the mid-1980s and 2030 would represent a gradual deterioration in the quality of the climate as a resource for agricultural production.

In the discussion of the impact of the analog climate on the 1980s baseline situation, we developed four scenarios of consequent changes in crop yields

Table 6-7. EPIC-simulated percentage changes in 2030 baseline crop yields under the analog climate.

	Without adjustments	With adjustments	
Crop	Without CO_2 fertilization	Without CO_2 fertilization	With CO_2 fertilization
Corn			
Dryland	-28	-16	-6
Irrigated	-12	+2	+8
Sorghum			
Dryland	-21	+11	+25
Irrigated	-10	+7	+18
Wheat			
Dryland	-12	+1	+17
Irrigated	-1	+6	+18
Soybeans			
Dryland	-25	-11	+2
Irrigated	–	+1	+12

Source: Calculated from EPIC, Easterling et al. (1991).

and production. In the first, we used EPIC to simulate the impact of the analog climate on yields assuming no CO_2 fertilization and no on-farm adjustments to the changed climate. The second scenario assumed CO_2 fertilization but still no on-farm adjustments. The third simulated the impacts with farm adjustments and no CO_2 fertilization, while the fourth assumed adjustments with CO_2 fertilization.

We modified this procedure to analyze the impacts of the analog climate on the 2030 baseline situation. In one scenario we show the effects on 2030 baseline yields without either CO_2 fertilization or on-farm adjustments. In two other scenarios we assume that farmers adopt the same adjustments as in the 1980s case plus the two new technologies developed in response to climate change. We show these adjusted yields both with and without CO_2 fertilization.

Table 6-7 shows the yield effects of the analog climate in 2030 under the three scenarios. Because of the problematic future of irrigation in the region, we show the yield effects under irrigated and dryland conditions separately. Table 6-7 indicates a marked difference between the yield effects of the analog climate in scenarios both with and without adjustments. The differences are especially marked when the results for the no adjustments, no CO_2 fertilization

Table 6-8. Changes in the value of 2030 baseline crop production under the analog climate.

	Without adjustments		With adjustments			
	Without CO_2 fertilization		Without CO_2 fertilization		With CO_2 fertilization	
Crop	Billion 1982 $	Percent	Billion 1982 $	Percent	Billion 1982 $	Percent
Corn	-2.5	-23	-1.6	-16	-0.6	-6
Sorghum	-0.4	-20	+0.3	+15	+0.5	+26
Wheat	-0.4	-11	+0.1	+3	+0.6	+17
Soybeans	-1.8	-26	-0.8	-11	+0.1	+2
Total	**-5.1**	**-22**	**-2.0**	**-8**	**+0.6**	**+3**

Note: Percent yield changes from EPIC times the value of production shown in Table 6-6.

scenario are compared with those reflecting both CO_2 fertilization and farm-level adjustments. In fact, in the latter scenario the effect of the analog climate on yields is positive for sorghum, wheat, and soybeans, and much less negative for corn than in the no CO_2 fertilization, no adjustment scenario.

To translate the Table 6-7 yield changes into production changes, we used the same procedure as we did in estimating the production effects of the analog climate in the 1980 baseline situation, i.e., we assumed in the no-adjustments scenario that production would decline in proportion to yields. In the adjustment scenarios, the production change reflects both the change in yields and the shift of land in Nebraska and Kansas out of irrigated corn and into dryland wheat and sorghum, and some increase in irrigated corn production in Iowa and Missouri.

Table 6-8 shows the changes in the value of production of the various crops under the analog climate. The table, of course, reflects the marked differences in yield effects between the adjustment and no adjustment scenarios. We do not take the specific numbers in the table too seriously. However, we believe the swing from a large, negative impact in the absence of CO_2 fertilization or adjustments to a much smaller negative, if not positive, impact when account is taken of CO_2 fertilization and adjustments, is in the right direction.

Comparison of the no-adjustment, no-CO_2-fertilizer results in Table 6-8 with the comparable scenario in Table 6-1 indicates that the analog climate in 2030 would have a substantially more negative effect on production of the four crops (-$5.1 billion compared to -$2.7 billion)[3] than if it were imposed on the 1980s situation. Most of this difference is because production of the crops in the 2030 no-climate-change scenario is much greater than it was in 1984/87. That is to say, the percentage changes in yield under the analog climate in the 2030 scenario are not much different from the changes in the 1984/87 scenario.

Comparison of the no-CO_2 fertilizer effect with adjustments scenario in Table 6-8 with the similar scenario in Table 6-2 shows that the on-farm adjustments in 2030 would ease the production impact of the analog climate proportionately much more than they would in 1984/87. In the 2030 scenario, the adjustments reduce the production loss from -$5.1 billion to -$2.0 billion (Table 6-8), or 62 percent. In the 1984/87 scenario (excluding hay), the adjustments reduce the loss from -$2.7 billion (Table 6-1) to -$2.0 billion (Table 6-2), or 27 percent. The greater adjustment effect in the 2030 scenario reflects the incorporation in EPIC of the two technological improvements induced by the analog climate: increased irrigation efficiency and stomatal resistance.

D. Economy-Wide Impacts

In estimating the economy-wide impacts of the analog climate in the 1980s baseline by way of agriculture, we used production multipliers derived from the IMPLAN input-output model for MINK. These multipliers reflect technical and economic conditions determining MINK's economic structure. To develop projections of the multipliers for the year 2030 would require a detailed analysis of these underlying conditions, which was beyond the scope of this study. Something useful can be said nonetheless about the regional economic impacts of the climate change scenarios for 2030. The analysis of the scenarios for the 1980s situation showed that if all of the climate change-induced decline in crop production were reflected in reduced MINK exports, the effect would be to reduce total regional production by roughly 0.3 to 1.4 percent (Table 6-4). Noting this, and allowing for the growth of the MINK economy to 2030 (75 percent according to the Department of Commerce projections), suggests that even if the crop production multipliers were to double from the current to the 2030 baseline, the decline in crop production in the worst case 2030 scenario would have a small percentage impact on the regional economy, if all the decline occurred in crop exports.

However, if the decline in crop production resulted in an equal reduction of feedgrain supplies to animal producers in MINK, and if the crop production multipliers were to increase, then the analysis of the 1980s scenarios (Table 6-5) suggests that the worst case 2030 scenarios would imply a decline in total regional production of more than 10 percent, which is no longer a trivial amount. This worst case result for the economy as a whole probably is less likely than the best case result, for the same reason as in the 1980s scenario: the decline in crop production is more likely to show up as reduced exports than as reduced feedgrain supplies to local animal producers.

The worst case result in the 2030 scenarios also assumes that animal production in MINK remains closely related to feedgrain production, and that the animal-meatpacking relationship remains important. The high multiplier effect of the decline in crop production in the worst case result for the regional economy assumes continuation of the strong locational relationship in MINK between feedgrain production, animals, and meatpacking.

Of course, if crop production in MINK should increase, as in the scenario with CO_2 fertilization and on-farm adjustments, then the impact of the analog climate on the regional economy would be positive, at least so far as agriculture is concerned.

V. CONCLUSION

The MINK study suggests that, if over the next 40 years greenhouse gas-induced global warming should return the climate of the region to that of the 1930s, the impact on the region's agriculture and economy would be minor. There are two reasons: (1) farm-level adjustments to the changed climate, including adjustments made possible by innovative agricultural research, would offset much of the negative agricultural impact; and (2) agriculture presently is a small part of the total regional economy. Forty years from now it likely will be smaller still.

In a few words, the evolution of MINK's agriculture and economy over the next several decades is more likely to be driven by global and national trends in population, technology, economics, and policy unrelated to climate than by climate change.

NOTES

[1] The MINK study was published as seven separate reports. The general title of the project was "Processes for Identifying Regional Influences of and Responses to Increasing Atmospheric CO_2 and Climate Change—the MINK project. The specific titles are:

I. An Overview

IIA. Agricultural Production and Resource Use in the MINK Region Without and With Climate Change

IIB. A Farm-Level Simulation of the Effects of Climate Change on Crop Production in the MINK Region

III. Forest Resources

IV. Water Resources

VI. Consequences of Climate Change for the MINK Economy: Impacts and Responses

[2] For more detail on the IMPLAN model and its role in the MINK study see Bowes and Crosson (1991).

[3] For this comparison we exclude the effects on hay production in Table 6-1.

LITERATURE CITED

Adams, R. 1989. Global climate change and agriculture: An economic perspective. *American Journal of Agricultural Economics*. 71(5):1272–79.

Alward, G. 1986. *IMPLAN Version 2.0: Methods Used to Construct the Regional Economics Data Base.* USDA Forest Service General Technical Report RM-000 (draft). Rocky Mountain Forest and Range Experiment Station: Fort Collins, CO.

Alward, G. 1989. *Micro IMPLAN: Methods for Constructing Regional Economic Accounts.* A paper presented at the conference on Input-Output Modeling and Economic Development Application, Feb. 28–March 2, Kansas City, MO.

Bowes, M. and P. Crosson. 1991. *The MINK Project, Report VI, Consequences of Climate Change for the MINK Economy: Impacts and Responses.* TR052H, DOE/RL/01830T-H12. Department of Energy Carbon Dioxide Research Program: Washington, D.C.

Crosson, P., L. Katz, and J. Wingard. 1991. *The MINK Project, Report IIA, Agricultural Production and Resource Use in the MINK Region without and with Climate Change.* TRO52C, DOE/RL/01830T-H7. Department of Energy Carbon Dioxide Research Program: Washington, D.C.

Easterling, W., M. McKenney, N. Rosenberg, and K. Lemon. 1991. *The MINK Project, Report IIB, Farm-Level Simulation of the Effects of Climate Change on Crop Production in the MINK Region.* TRO52D, DOE/RL/01830T-H8. Department of Energy Carbon Dioxide Research Program: Washington, D.C.

Frederick, K. 1991. *The MINK Project, Report IV, Water Resources.* TR052F, DOE/RL/01830T-H10. Department of Energy Carbon Dioxide Research Program: Washington, D.C.

Hayami, Y. and V. Ruttan. 1985. *Agricultural Development.* Johns Hopkins University Press: Baltimore.

Parry, M., T. Carter, and N. Konijn. 1988. *The Impact of Climate Variations on Agriculture.* Kluwer Academic Publishers: Dordrecht.

Smith, J. and D. Tirpak. eds. 1989. *The Potential Effects of Global Climate Change on the United States.* U.S. Environmental Protection Agency, Office of Policy, Planning, and Evaluation: Washington, D.C.

United Nations. 1989. *World Population Prospects 1988.*

U.S. Department of Agriculture. 1989. *Micro IMPLAN Release 89-03 (A Complete Program).* USDA Forest Service, Land Management Planning: Fort Collins, CO.

U.S. Department of Commerce. 1990. *Regional Projections to 2040, Vol. 1: States.* Bureau of Economic Analysis: Washington, D.C.

7

ADAPTATION to GLOBAL CLIMATE CHANGE at the FARM LEVEL

Harry M. Kaiser
Department of Agricultural Economics, Cornell University

Susan J. Riha

Daniel S. Wilks
Department of Soil, Crop, and Atmospheric Science, Cornell University

Radha Sampath
Department of Agricultural Economics, Cornell University

I. INTRODUCTION

If even the most mild predictions of climate models hold true, the earth's climate will undergo unprecedented changes in the next 50 to 100 years. Current global general circulation models (GCMs), which are our best tools for predicting future climates, indicate that the earth's surface temperature could rise by an average of 1.5°C to 4.5°C under a doubling of carbon dioxide in the atmosphere (Intergovernmental Panel on Climate Change, 1990, 1992). These models also predict an increase in global precipitation, but there is more uncertainty among GCMs as to the potential distribution of precipitation changes.

Any change in climate will have implications for climate-sensitive systems such as forestry, other natural resources, and agriculture. With respect to agriculture, changes in temperature, precipitation, and solar radiation will produce agronomic effects such as changes in crop yields, scheduling of field operations, and grain moisture content at harvest. Climate change will also have a host of effects on the economics of agriculture, including changes in farm profitability, prices, supply, demand, trade, and regional comparative advantage. The agronomic and economic impacts of climate change will depend principally on two factors: (1) the magnitude of changes in climatic variables, and (2) how well agriculture can adapt to these changes.

The purpose of this paper is to examine agricultural adaptation issues in the context of climate change. Since previous papers in this section have examined climate change impacts on agriculture at global and regional levels, the focus here will be on the individual farm-level. The farm-level analysis will consist of examining the potential agronomic and economic impacts of several climate change and adaptation scenarios. The results are based on a protocol that links climatic, agronomic, and economic models to form an integrated model. Two climate change scenarios constructed from the global results of Hansen et al. (1988) are examined along with a baseline (no climate change) scenario. These climate scenarios are examined under two adaptation situations, one assuming farmers cannot adapt to the new climate by switching crops, cultivars, and/or timing of field operations and the other assuming farmers can use these adaptation strategies. The analysis is conducted for two representative farms of important grain regions of the midwestern United States: Southwestern Minnesota and Southeastern Nebraska. Two regions are investigated to examine whether there may be regional differences in agronomic and economic impacts from climate change.

II. THE INTEGRATED MODEL

The integrated model, which was developed by Kaiser et al., consists of a stochastic weather generator, a dynamic crop yield simulation model, and a farm-level linear programming model.[1] The stochastic weather generator (Wilks, 1992) produces daily values for minimum and maximum temperature, precipitation, and solar radiation based on a defined climate scenario. These daily climatic variables are used as inputs into the crop yield simulation model. The crop simulation model (Buttler and Riha, 1989), in turn, simulates crop yields, grain moisture contents, and available field time, which are parameters in the linear programming model. The linear programming model generates optimal crop mix, field scheduling, and net revenue, which is discounted by a risk term which measures variability in net revenue. The output of the unified model shows how crop yields, grain moisture contents, field time, crop mix, and farm net revenue are impacted by each climate scenario.

Unlike the comparative static approach used by most analyses of climate change impacts, climate change here is modeled as a gradual, dynamic process. The stochastic weather generator is used to simulate three climate scenarios over a 100 year simulation period, 1980–2079. The first climate scenario is a baseline case that reflects no change in climate. In climate scenario 2, which corresponds in temperature to the relatively mild "scenario B" of Hansen et al. (1988), average global temperature increases by 2.5°C by the year 2060 (time of equivalent doubling of CO_2), with half of the warming occurring between 2030 and 2060. Average precipitation, in this case, is assumed to increase linearly over time at a rate sufficient to increase average precipitation by 10% in 2060. The third climate scenario, which corresponds to the more severe "scenario A" of Hansen et al. (1988), includes an increase in average temperature of 4.2°C by 2060, again with half of the warming occurring between 2030 and 2060. Under this scenario, average precipitation is assumed to decrease linearly so that it is 20% drier by 2060.

One advantage of using a stochastic weather generator is that changes in variability of temperature and precipitation can also be imposed. For both climate scenarios 2 and 3, the standard deviations of average monthly temperature are decreased linearly with increases in average monthly temperature. This is based on the results from several GCM simulations of how temperature variability may change (Rind et al., 1989; Wilson and Mitchell, 1987). In addition, the diurnal temperature range (maximum minus minimum daily temperature) is decreased. The variance of average monthly precipitation is assumed to increase in the wetter scenario 2 and decrease in the drier scenario 3.

The daily climatic variables from the weather generator are used by dynamic crop simulation models to generate annual crop yields,[2] grain moisture contents, and daily field time availability over the 100 year period based on each climate scenario. In addition to the climatic variables, the crop simulation models require information on the type of soils in the region. For the Minnesota location, represented by a hypothetical 600 acre farm in Redwood County, the prominent soil series is Ves. This is a deep soil with excellent water-holding capacity. For the Nebraska location, represented by a hypothetical 800 acre farm in Lancaster County, the major soil series is Sharpsburgh, which is also a deep soil that has a good water-holding capacity.

The crop simulation models used in this study were developed independently by different groups of researchers. We used distinctly different crop models rather than a generic crop model for all crops in order to make maximum use of the knowledge available for the crops investigated. Teams of researchers particularly familiar with a specific crop are usually the developers of crop models. This means that crop models for different species are not necessarily similar in that they reflect the research agenda that has been pursued for a particular crop. In the case of maize (corn), the model used in this study was initially developed by Stockle and Campbell (1989) with a yield simulation added later by Riha and Rossiter (1991). The wheat model used in this study was also developed by Stockle and Campbell (1989) for spring wheat, and is modified in this study to simulate winter wheat by adding a winter survival component. The sorghum model used in this study is based on the SORKAM model of Rosenthal et al. (1989) while the soybean model is based on SOYGROW developed by Wilkerson et al. (1983).

While different crop models are used in this study, all crop models utilize the same procedures to simulate the soil and atmospheric physical environment (Buttler and Riha, 1992). This means that each simulated crop interacts with the same simulated environment. However, because each crop model was developed separately, the manner in which components of the physical environment (such as temperature and soil moisture status) affect crop growth is not necessarily the same across crop models.

Southwest Minnesota is dominated by maize and soybean production. Sorghum is not currently a major crop in Minnesota, but is included because it is fairly drought resistant and might displace maize if climate warming is severe enough. While not dominant in this region, winter wheat is included because it may become more prominent if climate change makes it relatively more profitable. While irrigation is used in Lancaster County Nebraska, all crops in this analysis are assumed to not be irrigated. Dryland production is dominated by sorghum in this region. This is followed in importance by wheat, soybeans, and maize. There is currently no double cropping done in this area, but double cropping of winter wheat followed by soybeans is included in the model because it may become possible in a longer growing season.

For each crop, yields and grain moisture contents are simulated for three cultivars, which are early-, mid-, and late-maturing. In general, later maturing varieties have a greater yield potential compared to early and mid varieties. However, later maturing varieties are also more vulnerable to yield reduction due to frost and drought, as well as more likely to have high grain moisture contents at harvest than early- and mid-maturing cultivars. Simulated crop yields and grain moisture contents are disaggregated by several planting and harvest dates, which are different for Minnesota and Nebraska. The field time variable is computed as daily hours available for performing field operations and is limited by excessive soil moisture.

The farm-level linear programming simulates the annual farm-level decision-making process, which includes crop mix and field operation scheduling decisions. It is assumed that the farmer makes these decision facing two sources of uncertainty affecting net revenue: yield levels and grain drying costs (due to uncertain grain moisture contents). To incorporate these two important sources of risk into the model, the agronomic model is used to generate yield and grain moisture content "states of nature." Each of the simulated states are assumed to have an equally likely chance of occurring. Expected net revenue per acre for each activity (crop, cultivar, and planting-harvest date combination) is then calculated as follows:

$$E(C_{ijk}) = PC\,[\sum_{t=1}^{N} YC_{ijkt} - VCC_t] \,/\, N, \text{ where:}$$

$E(C_{ijk})$ is expected gross revenue per acre for crop c, cultivar i, planting period j, and harvest period k; PC is the per bushel price of crop c (assumed to be known), YC_{ijkt} is the yield (bushels per acre) of crop c, cultivar i, planting period j, harvest period k, and state of nature t; and VCC_t are the variable costs for crop c, state of nature t. Note that for each crop, variable costs are assumed to be the same across cultivars and planting and harvest combinations; however, variable costs vary across states of nature since drying costs may differ.

Net revenue risk due to uncertain yields and grain drying costs is represented by a MOTAD framework (Hazell, 1971). It is assumed that the farmer maximizes expected net revenue, which is discounted by subtracting a risk term. The risk term is equal to a linear approximation of the standard deviation (total absolute deviations) of expected net revenue, multiplied by a risk aversion coefficient (a coefficient of 0 implies the decision maker does not care about risk, while any positive coefficients mean that the farmer dislikes risk). Discounted net revenue (hereafter referred to as net revenue) is maximized subject to a set of resource constraints including acreage and labor availability. The risk aversion coefficient in the objective function is set equal to 1, which is in the range which Brink and McCarl (1978) found representative for Cornbelt farmers.

To summarize the methodology, for each climate scenario the simulation begins with the generation of daily weather values for the 100 year period (1980–2079) using the stochastic weather generator. The daily values of the weather variables for each year are used by the crop simulation model to generate annual values for crop yields, grain moisture contents, and field time. Actual farm yields are generally lower than simulated yields since simulated crop yields represent potential yields under ideal management conditions. Consequently, the results of the crop models were adjusted for the economic model to represent actual farm-level yields. This was done for each crop by multiplying simulated yields by the ratio of the actual average yield for the 1980s decade to the average simulated crop yield for the 1980s. The agronomic results are tabulated on a decade-by-decade basis, resulting in ten sets of results for each scenario. These results are then used as parameters of the linear programming model to generate optimal crop mix and net revenue. All costs, prices, technical parameters, and resource endowments in the linear programming model are held at their 1980 values for the entire simulation period.[3] On the other hand, cultivar selection, crop yields, grain moisture contents, and grain drying costs vary by decade and by climate scenario based on the results of the crop simulation model.

III. ADAPTATION SCENARIOS

Agricultural adaptation to a changing climate will occur in several forms, including technical innovations, changes in agricultural land areas, and changes in use of irrigation. Technological innovations include the development of new plant cultivars (varieties) that may be bred to better match the changing climate. Changes in agricultural land areas in response to both changes in demand for agricultural products due to increasing population, as well as changes in supply due to climate change will be another form of adaptation. Finally, as the climate warms there will likely be shifts towards greater use of irrigation systems to grow crops. All of these adaptation strategies are discussed elsewhere (e.g., Adams et al., 1990; Council for Agricultural Science and Technology, 1992; Drabenstott, 1992) and will not be reviewed in detail here. Rather, farm-level adaptation strategies will be examined in this analysis.

Three types of farm-level adaptation to climate change are considered in this analysis. The first is the possibility of switching cultivars for a particular crop. In this study, the three cultivars included for each crop differ by length of time to maturity and potential yield. If climate change affects the length of the growing season, then farmers could use the adaptation strategy of switching to longer growing, higher yielding cultivars.

A second farm-level adaptation strategy is to alter crop mix in response to a change in climate. If climate change affects the relative yield and profitability of one crop in favor of another, then farmers should respond by making the appropriate change in crop mix.

Finally, farmers may also make adjustments in scheduling of field operations in response to climate change. If climate change increases or decreases the amount of time that farmers can be in the field, this will affect planting and harvest dates, and therefore yield levels indirectly. The change in climate will also affect grain moisture content at harvest, and therefore grain drying costs, which are an important component of total variable costs. Changes in scheduling of field operations, as well as in crop mix and cultivar selection, in response to climate change are reflected in the linear programming results, which provide optimal timing of field operations, crop mix, and cultivar selection.

IV. THE MINNESOTA RESULTS

Figure 7-1 shows average crop yields by decade[4] for climate scenario 2 as a percentage of their 1980 average for the simulation period, 1980–2070.[5] Recall that climate scenario 2 is the relatively mild climate change scenario (temperature increases by 2.5°C and precipitation increase by 10% in 2060). Soybean yields at the Minnesota site appear to be positively affected by the increase in temperature under scenario 2. By 2070, soybean yields are over 48% higher than they were in 1980. Winter wheat yields generally increase over the simulation period (except for 1990 and 2000). Towards the last

Climate Scenario 2

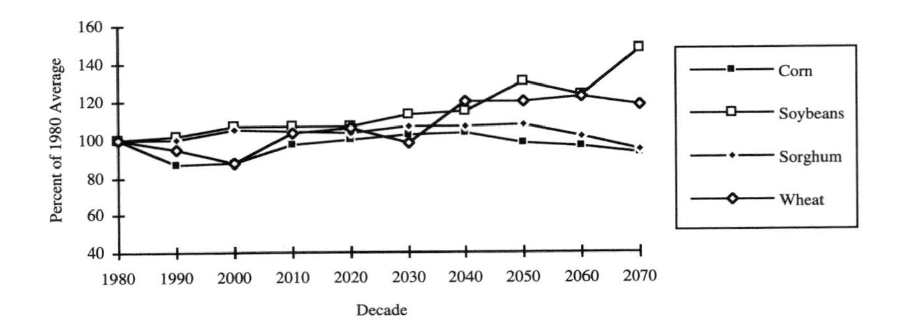

Climate Scenario 3

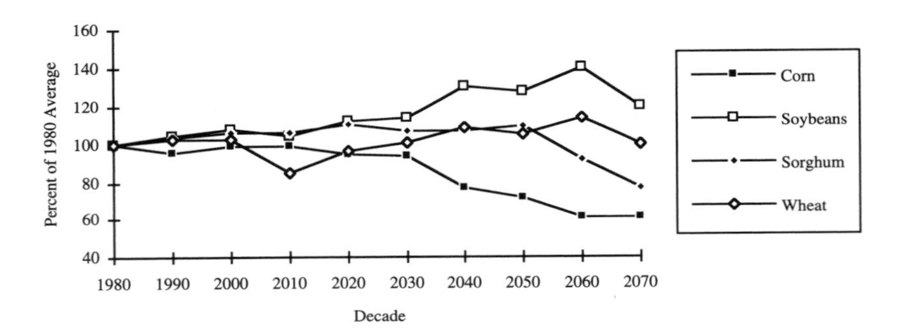

Figure 7-1. Minnesota Crop Yields as a Percentage of their 1980 Levels, 1980–2070.

several decades, winter wheat yields are about 20% higher than 1980 levels. Winter wheat yields benefit under this scenario because the increase in temperature reduces the likelihood of crop failure due to winter kill. Also, the increase in precipitation in this scenario lessens the chance of water stress. Sorghum yields are relatively stable throughout the 100 year period, with a minor decrease over the last 20 years. Maize yields have a somewhat odd initial result, actually decreasing by over 10% in 1990 and 2000 compared to 1980 levels. A detailed investigation of the weather generated revealed that there was less rainfall, on average, occurring in the particular samples generated for these two decades than what occurred in the 1980s. After reaching their 1980 level for the decades 2010 through 2040, maize yields decrease slightly for the remainder of the 100 year period due to increasing temperatures shortening the growing season.

In the more severe climate change scenario, yields for all crops are more negatively affected. For instance, maize yields decline consistently from 2010 through 2070, reaching 58% of their 1980 level by 2060. The steep decline in maize yields later on in the simulation is caused by water stress. Sorghum yields also decline, but not until after 2050 and not as much as maize yields. Winter wheat yields fluctuate somewhat over the 100 year period, but show no apparent trend either up or down relative to 1980 levels. Soybean yields, on the other hand, again appear to benefit by the hotter climate, but decline in the last decade of the simulation due to drought stress.

The key question is how the climate change scenarios affect net revenue. As is clear from the top graph in Figure 7-2, farm net revenue (discounted for risk) is actually improved under both climate change scenarios. In fact, net revenue is 87% higher in 2070 than it was in 1980 under scenario 2. Even in the more severe scenario, net revenue is as much as 33% higher (in 2060) than in 1980. The farm is responding to the positive impact on soybean yields under both climate change scenarios by increasing soybean acreage and decreasing maize acreage, as is shown in Figure 7-3. Optimal acreage of maize declines from just under 60% of the 600 acres for the decade of the 1980s to about 27% in 2070 for scenario 2, and 23% in 2070 for scenario 3. For both scenarios 2 and 3, sorghum and winter wheat are never optimal to grow.

What if farmers in this region could not adapt to climate change by switching cultivars, crop mix, and planting and harvesting dates? To examine this question, the model was re-solved by forcing the farm to grow the same crop mix using the same cultivars and planting and harvesting schedules found to be optimal in 1980 for the entire 100 year simulation. It is clear from comparing the no adaptation with the full adaptation results that farmers are able to lessen some of the negative yield effects of climate change by switching cultivars, as well as plant and harvest dates. For example, yields for soybeans and maize are as much as 20% and 25% less, respectively, in the no adaptation case compared to the full adaptation case. The percentage difference in net revenue for the no adaptation and full adaptation scenarios is presented in the bottom graph of Figure 7-2. If no adaptation is assumed, net revenue is as much as 80% less than the full adaptation case for the severe climate change scenario, and as much 25% less than the full adaptation case for the mild climate change scenario. These results illustrate how important the adaptation assumption is in predicting climate change impacts.

It appears that the relatively cool Minnesota region can adapt quite effectively to both of the climate change scenarios considered in this study. While both sorghum and maize yields were negatively impacted by the severe scenario, soybean yields actually increased, and maize yields increased or

Net Revenue

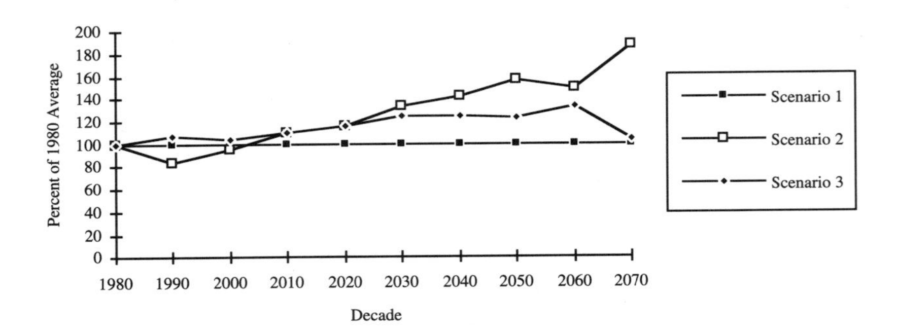

Percentage Difference Between Adaptation and No Adaptation

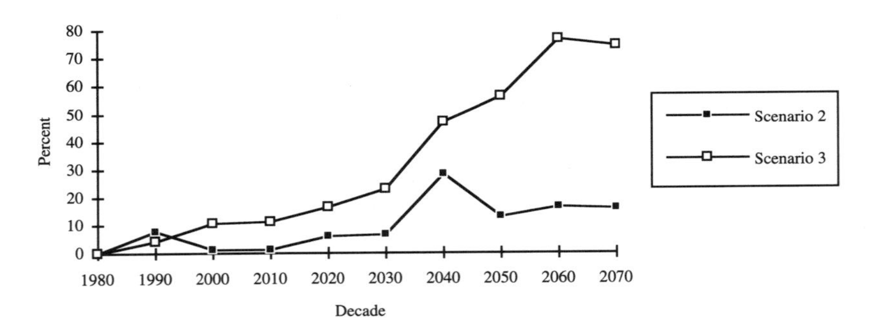

Figure 7-2. Net Revenue for Minnesota Farm as a Percentage of its 1980 Level by Climate Scenario, and Percentage Difference in Net Revenue Between the Adaptation and No Adaptation Cases, 1980–2070.

stayed the same. Net farm revenue was improved for both scenarios by growing more soybeans, less maize, and altering cultivars in response to the new climate.

V. THE NEBRASKA RESULTS

Figure 7-4 shows average crop yields by decade for the Nebraska location for climate scenario 2 as a percentage of their 1980 average for the simulation period, 1980–2070. Unlike the Minnesota results, soybean yields at the

Climate Scenario 2

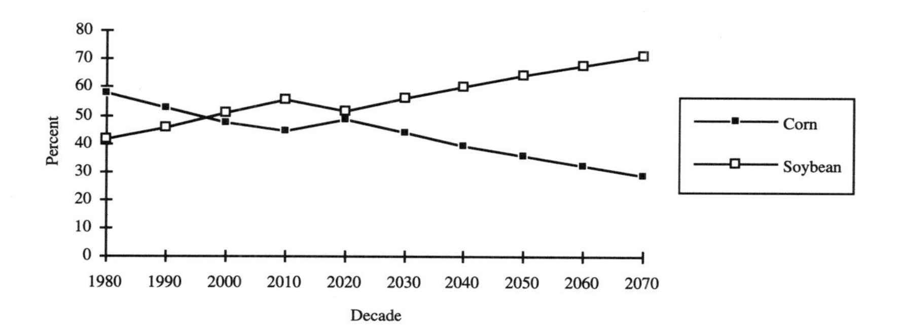

Climate Scenario 3

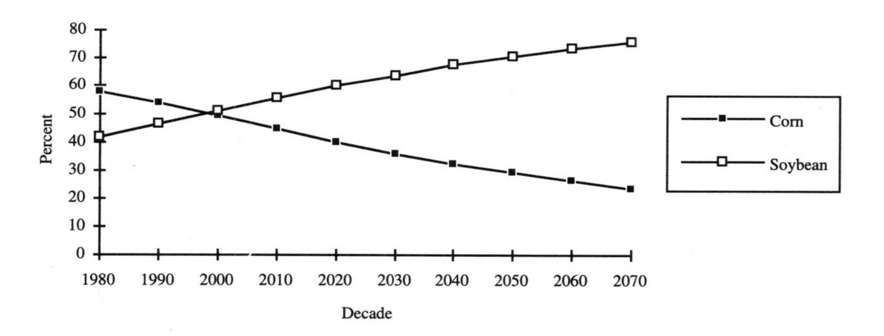

Figure 7-3. Optimal Mix of Crop Acreage for Minnesota Farm Under Climate Scenarios 2 and 3, 1980–2070.

Nebraska site appear to be virtually unaffected by scenario 2. By 2070, soybean yields are about the same as they were in 1980. Maize yields are also unaffected by this mild scenario because of increases in precipitation that occur throughout the 100 year period. At the same time, the increase in temperature over the simulation speeds the time of maturity and thereby decreases yield potential. These two trends balance each other out during this 100 year period resulting in little change in the yield of maize. Sorghum yields decline more than the other three crops in this scenario, falling to 80% of their 1980 levels by 2070. The decrease in sorghum yields is due to increases in

Climate Scenario 2

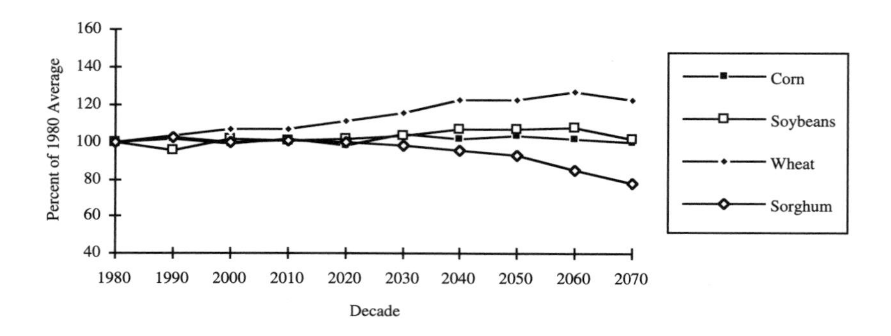

Climate Scenario 3

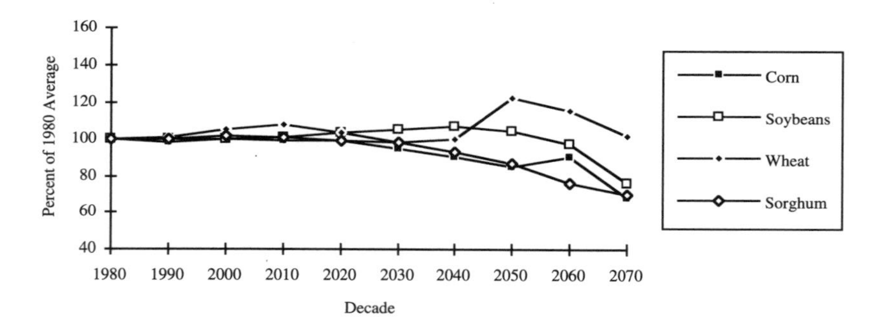

Figure 7-4. Nebraska Crop Yields as a Percentage of their 1980 Levels, 1980–2070.

temperature, which has the effect of decreasing the time to maturity and thereby decreasing the yield potential. Winter wheat yields benefit from climate scenario 2, increasing consistently throughout the 100 year simulation. The increase in temperature in this scenario has the effect of reducing the likelihood of crop failure due to winter kill. In addition, the increase in precipitation in this scenario reduces the likelihood of yield reduction due to water stress.

In climate scenario 3 (temperature increases by 4.2°C and precipitation decreases by 20% by 2060) soybeans, maize and sorghum yields are more negatively affected. Soybean yields increase slightly until 2040 and thereafter decrease. The predicted decrease in soybean yields after 2040 is due to increasing temperatures and is not due to drought stress. Maize yields decline throughout the simulation with most of the largest decrease occurring after 2060. The increase in temperature decreases the yield potential of maize because it speeds the time of maturity. In addition, the decrease in precipitation in this scenario decreases maize yields. Sorghum yields also decline under this climate scenario following a similar pattern to maize yields. The accelerated decrease in sorghum yields after 2040 is not due to drought stress, but rather to the minimum daily temperature during the yield formation stage averaging more than 18°C, which in the sorghum crop simulation model (SORKAM) decreases the yield potential. Development of later-maturing varieties or planting earlier might result in the maintenance of sorghum yield in the face of increasing temperatures.

The top graph in Figure 7-5 shows net revenue for the Nebraska location by climate scenario. Unlike the Minnesota location, net revenue decreases after 2010 for both of the climate change scenarios relative to the no climate change case. By 2070, net revenue for scenarios 2 and 3 are 85% and 60% of net revenue in the no climate change case. This location is responding to both climate change scenarios by growing less sorghum and more winter wheat, as is shown by Figure 7-6. Double cropping of winter wheat followed by soybeans is not optimal for either climate change scenario during the 100-year period.

If agriculture in this region could not adapt to climate change by switching cultivars, crop mix, and planting and harvesting dates, then the results would be even more negative. This is evident by the bottom graph in Figure 7-5, which presents the percentage difference in net revenue for the no adaptation and adaptation scenarios. If no adaptation is assumed, net revenue is as much as 16% less than the full adaptation case for the severe climate change scenario, and as much 10% less than the full adaptation case for the mild climate change scenario.

To summarize, it appears that while the Minnesota region could adapt quite effectively to both of the climate change scenarios considered in this study, the Nebraska location is worse off under both scenarios. The Nebraska region can lessen the negative effects of climate change by growing more winter wheat and less sorghum.

VI. SUMMARY

This chapter examined potential agronomic and economic impacts of several climate change and adaptation scenarios at the farm-level. The analysis was based on a protocol that links climatic, agronomic, and economic models to form an integrated model. The protocol was applied to two

Net Revenue

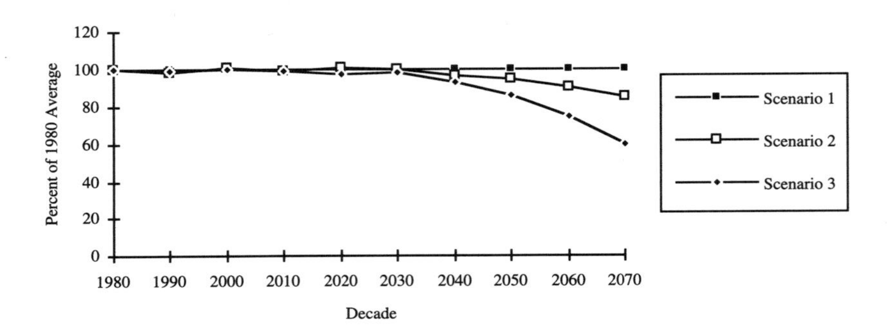

Percentage Difference Between Adaptation and No Adaptation

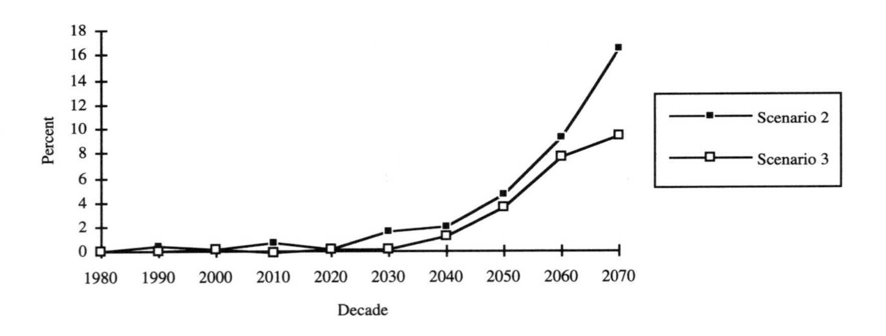

Figure 7-5. Net Revenue for Nebraska Farm as a Percentage of its 1980 Level by Climate Scenario, and Percentage Difference in Net Revenue Between the Adaptation and No Adaptation Cases, 1980–2070.

representative farms of important grain regions of the Midwest: Southwestern Minnesota and Southeastern Nebraska.

The results indicate that the two climate change scenarios had different impacts on the two regions. The Minnesota location was able to adapt quite effectively to both of the climate change scenarios considered in this study. While both sorghum and maize yields were seriously impacted by the more severe scenario, soybean yields actually increased. Farm net revenue was improved for both climate scenarios by growing more soybeans, less maize,

Climate Scenario 2

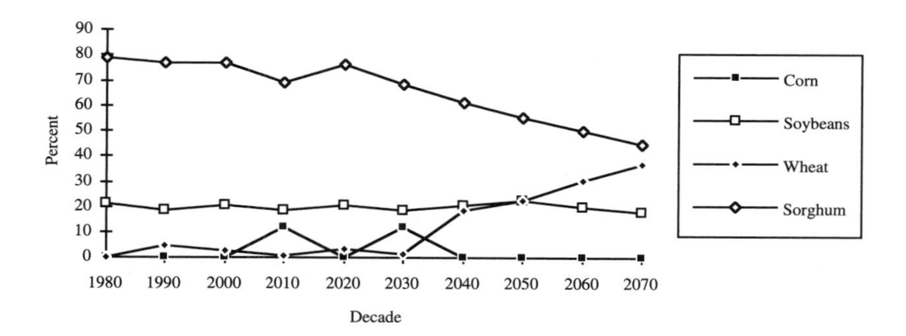

Climate Scenario 3

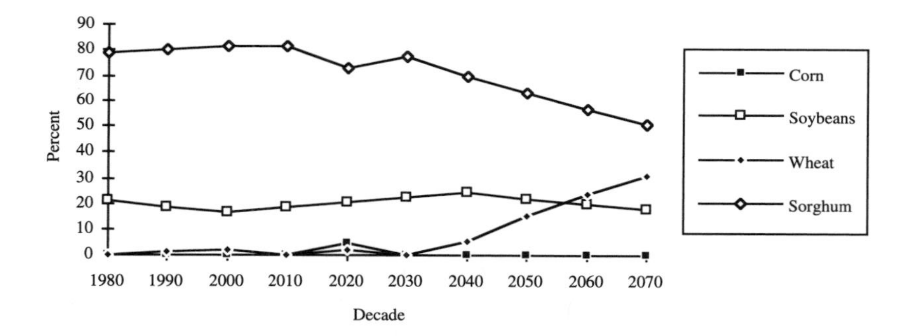

Figure 7-6. Optimal Mix of Crop Acreage for Nebraska Farm Under Climate Scenarios 2 and 3, 1980–2070.

and altering cultivars in response to the new climate. On the other hand, the Nebraska region was worse off under both climate change scenarios in later decades of the simulation. Even with such adaptation strategies as altering crop mix and crop cultivars, net farm revenue was lower under the two climate change scenarios than under no change in climate.

These results should not be used to draw general conclusions about climate change and agriculture. This paper has only reported the results of two climate change scenarios. More scenarios need to be investigated in order to get a better indication of the range of possible agronomic and economic outcomes of climate change. In addition, the analysis of other important grain regions would be useful to determine distributional effects of climate change. Warmer regions in the south, for example, may have more severe effects than the northern locations chosen for this study. Finally, since climate change will affect output prices, future research should make prices endogenous to climate change. The inclusion of more climate change scenarios, additional regions, and making prices endogenous should therefore be the subject of further study of this important topic.

NOTES

[1] For a more detailed description of the components of the integrated model, see Kaiser et al. (1993).

[2] While increased concentrations of CO_2 in the atmosphere should have some enhancing effect on crop yields, this so called "CO_2 fertilizer effect" is not considered in this study for several reasons. First, there is uncertainty among scientists over the magnitude of this enhancement. While the effects of elevated carbon dioxide in controlled greenhouse experiments are well documented, there are uncertainties over the magnitude of the CO_2 fertilizer effect under actual farm conditions where nutrients and other resources are limited. Second, there is evidence that there will be other effects accompanying the CO_2 fertilizer effect that will have a negative impact on yields, such as ozone depletion and companion gases created from fossil fuel combustion that have negative yield effects (such as sulfur dioxide, nitrogen oxides, and volatile organic hydrocarbons; see Erickson [1993] for detail). Hence, if the CO_2 fertilizer effect is to be considered, then so should be these other effects. Finally, the focus of this paper is on isolating the role of adaptation in response to climate change. Incorporating the CO_2 fertilizer effect into the analysis detracts from this focus.

[3] Data for the economic model that did not come from the crop simulation model includes variable costs, technical coefficients, and output prices. Variable cost data for all crops for Minnesota and Nebraska comes from Davenport (1988) and are equal to total cash expenses. Technical parameters that give the amount of labor required to produce an acre of each crop are based on the firm enterprise data systems (FEDS) coefficients for 1982 (U.S. Department of Agriculture). Output price data were obtained from selected issues of *Agricultural Prices* and *Nebraska Agricultural Statistics.*

[4] The yield averages for each decade are weighted averages of yields over the possible planting-harvest dates. The weights are derived by the linear programming results and are equal to the optimal proportion of how much of each crop is planted and harvested for each combination of planting-harvest dates. For example, if for a particular decade 75% of soybeans are planted in period 2 and harvested in period 5, and 25% of soybeans are planted in period 3 and harvested in period 6, then average soybean yield for that decade would equal .75 times the average soybean yield for planting period 2 and harvest period 5, plus .25 times the average soybean yield for planting period 3 and harvest period 6.

[5]Note that the results of the first climate scenario, which is the baseline no climate change case, are used as a reference for the two climate change scenarios in the figures that follow. That is, each decade's average yields are divided by the 1980 average yields. In other figures that do not use this indexing approach, the climate scenario 1 result is given by the vertical axis intercept for 1980.

LITERATURE CITED

Adams, R.M., C. Rosenzweig, R.M. Peart, J.T. Ritchie, B.A. McCarl, J.D. Glyer, R.B. Curry, J.W. Jones, K.J. Boote, and L.H. Allen, Jr. 1990. Global climate change and U.S. agriculture. *Nature.* 345:219–24.

Brink, L., and B. McCarl. 1978. The tradeoff between expected return and risk among cornbelt farmers. *Am. J. of Agri. Econ.* 60(1978):259–63.

Buttler, I.W. and S.J. Riha. 1992. Water fluxes in oxisols: a comparison of approaches. *Wat. Resources Res.* 28:221–29.

Buttler, I.W. and S.J.Riha. 1989. *GAPS: A General Purpose Simulation Model of the Soil-Plant-Atmosphere System (Version 1.1 User's Manual).* Department of Agronomy, Cornell University.

Council for Agricultural Science and Technology (CAST). 1992. *Preparing U.S. Agriculture for Global Climate Change.* Ames, Iowa.

Davenport, G. 1988. *State-Level Costs of Production, 1986.* U.S. Dept. of Agri. Washington, D.C.

Drabenstott, M. 1992. Agriculture's portfolio for an uncertain future: preparing for global warming. *Econ. Rev. Fed. Reserve Bank of Kansas City.* 77:5–20.

Erickson, Jon D. 1993. From ecology to economics: the case against CO_2 fertilization. *Ecol. Econ.* (in press).

Hansen, J., I. Fung, A. Lacis, D. Rind, S. Lebedeff, R. Ruedy, G. Russel, and P. Stone. 1988. Global climate changes as forecast by Goddard Institute for Space Studies three-dimensional model. J. *of Geophy. Res.* D93:9341–9364.

Hazell, P.B.R. 1971. A linear alternative to quadratic and semivariance programming for farm planning under uncertainty. *Am. J. of Agri. Econ.* 53:153–62.

Kaiser, H.M., S.J. Riha, D.S. Wilks, D.G. Rossiter, and R. Sampath. 1993. A farm-level analysis of the economic and agronomic impacts of gradual climate warming." *Am. J. of Agri. Econ.* (in press).

Karl, T.R., H. Diaz, and T. Barnett. 1990. *Climate Variations of the Past Century and the Greenhouse Effect.* Report based on the First Climate Trends Workshop, National Climate Program Office/NOAA, Rockville, MD.

Nebraska Agricultural Statistics Service. Various issues. *Nebraska Agricultural Statistics.* Lincoln, Nebraska.

Riha, S.J. and D.G. Rossiter. 1991. *GAPS: A General Purpose Simulation Model of the Soil-Plant-Atmosphere System User's Manual Version 2.0.* Department of Soil, Crop and Atmospheric Sciences, Cornell University.

Rind, D., R. Goldberg, and R. Ruedy. 1989. Change in climate variability in the 21st century. *Clim. Change.* 14:5–37.

Rosenthal, W.D., R.L. Vanderlip, B.S. Jackson, and G.F. Arkin. 1989. *SORKAM: A Grain Sorghum Crop Model.* Texas Agri. Exp. Station. College Station, Texas: Texas A&M University.

Stockle, C.O. and G.S. Campbell. 1989. Simulation of crop response to water and nitrogen: an example using spring wheat." *Transactions of the Am. Soc. of Agri. Engineers*. 32:66–74.

U.S. Department of Agriculture. 1979–82. *Firm Enterprise Data System (FEDS), Crop and Livestock Enterprise Research Budgets.* Unpublished Data Files of the Economic Research Service, Washington, D.C.

U.S. Department of Agriculture. Various issues. *Agricultural Prices*. Washington D.C.

Wilkerson, G.G., J.W. Jones, K.J. Boote, K.T. Ingram, and J.W. Mishoe. 1983. Modeling soybean growth for crop management. *Transactions of the Am. Soc. of Agri. Engineers*. 26:63–73.

Wilks, D.S. 1992. Adapting stochastic weather generator algorithms for climate change studies. *Clim. Change*. 22:67–84.

Wilson C.A. and J-F.B. Mitchell. 1987. Simulated climate and CO_2-induced climate change over Western Europe. *Clim. Change*. 10:11–42.

8

CARBON DIOXIDE EFFECTS on PLANTS: UNCERTAINTIES and IMPLICATIONS for MODELING CROP RESPONSE to CLIMATE CHANGE

David W. Wolfe
Department of Fruit and Vegetable Science, Cornell University

Jon D. Erickson
Department of Agricultural Economics, Cornell University

I. INTRODUCTION

Quantification of the direct physiological and yield response of plants to increased atmospheric CO_2 concentrations involves many assumptions that add to the uncertainty of economic forecasts regarding climate change. An assumption of a positive crop response to additional CO_2, a so-called "CO_2 fertilizer effect," can have a substantial impact on predictions of yield and economic welfare, and thus can influence policy decisions regarding resource management and international food security.

To account for the CO_2 effect, most crop models to date have incorporated optimistic multipliers or yield shifts, the magnitude of which are based on experiments conducted under controlled conditions with adequate water and nutrients, optimal temperatures, and nonexistent weed, disease, and insect pressures. Depending on the magnitude of the CO_2 fertilizer effect assumed in a crop model, climatic changes that inhibit growth may be compensated for, and forecasts can shift from damaging to favorable. Models are sometimes run with and without the CO_2 fertilizer effect for purposes of sensitivity analysis, but conclusions, publications, and recommendations are typically based on full realization of a strong positive crop response to elevated CO_2. This chapter explores the physiological complexities of this response in the context of likely environmental and economic constraints of the future, and offers specific suggestions for improving this aspect of crop models and subsequent economic welfare predictions.

II. PLANT PHYSIOLOGICAL RESPONSE TO CO_2

This section briefly discusses the basic mechanisms associated with the CO_2 fertilizer effect, including possible feedback mechanisms which make it difficult to scale up from individual leaf to whole plant and ecosystem level predictions.

A. Effects on Photosynthesis

Photosynthesis is the process in which CO_2 enters the plant through small openings in the leaves called stomates, is captured or "fixed" by photosynthetic enzymes, and is then converted into carbohydrates. When atmospheric CO_2 concentration goes up, more CO_2 will tend to enter the leaves of plants (i.e., photosynthetic rate increases) because of the increased CO_2 gradient between the leaf and air. Of even more importance is that an increase in external CO_2 concentration inhibits photorespiration, a process in which oxygen is absorbed and CO_2 released by the plant. At current atmospheric CO_2 levels, photorespiration can reduce the net carbon gain from photosynthesis by as much as 50% (Tolbert and Zelitch, 1983).

The biochemistry of photosynthesis differs among plant species and this greatly affects their relative response to CO_2. Most economically important crop and weed species can be classified as either a C3 or C4 type, the names referring to whether the early products of photosynthesis are compounds with three or four carbon atoms. In controlled experiments, the C3 species (including wheat, rice, soybean, most horticultural crops, and many weed species) tend to show significant gains in net photosynthesis from increased CO_2 because of photorespiration inhibition. In contrast, C4 plants (e.g., maize, sorghum, sugarcane, millet, and many pasture, forage, and weed species) show much less response to increased CO_2 due to a more efficient carbon assimilation process, Figure 8-1. The C4 plants obtain essentially no benefit from the CO_2 inhibition of photorespiration.

Table 8-1 shows the percent change in photosynthesis at 680 compared to 300–350 ppm CO_2 for several important crop species. These data are based primarily on experiments in which temperature, water, and nutrients were not limiting factors. The C3 crops listed had an average short term increase of 50% compared to 11.5% for the C4 species. The data in Table 8-1 also indicate that, regardless of photosynthetic pathway, the response is often much less when plants are acclimated to high CO_2 for at least one week prior to measurement (average long term increases of 25% and 5% for C3 and C4 species, respectively).

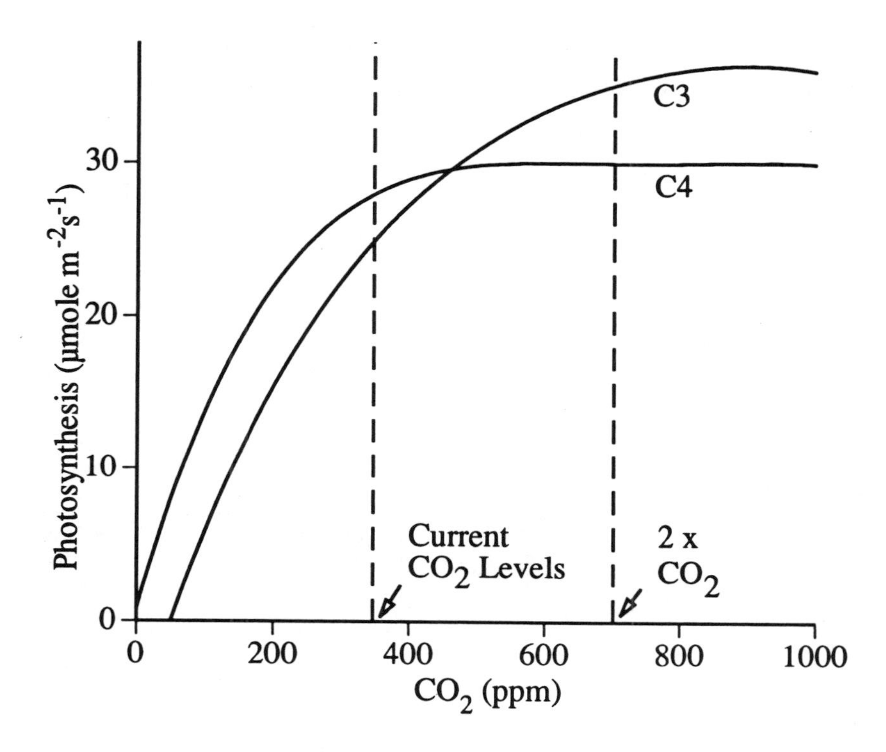

Figure 8-1. Photosynthesis per unit leaf area in relation to atmospheric CO_2 concentration. Typical curves for plants with the C3 and C4 photosynthetic pathway are compared. Although the specific photosynthetic values will vary among species, in general, C3 plants will show a greater relative benefit from a doubling of CO_2 compared to C4 plants.

B. Effects on Water Use By Plants

Another important direct effect of high CO_2 on plants is to cause a partial closure of the stomates. This restricts the escape of water vapor from the leaf (transpiration) more than it restricts photosynthesis. It is common for a doubling of CO_2 to decrease transpiration by 25–50 % (Cure and Acock, 1986). Water use efficiency (WUE), defined as the ratio of photosynthesis to transpiration, has been found to double with a doubling of CO_2 for some species (Pearcy and Björkman, 1983).

This stomatal response to CO_2 could moderate the increase in transpiration that is anticipated to occur with global warming, and could be of obvious benefit to plants growing under water-limited conditions. However, a more

Table 8-1. Photosynthectic response to CO_2-doubling for several crop species with either the C3 or C4 photosynthetic pathway.

		Percentage change in net photosynthesis after CO_2 doubling	
Photosynthetic pathway	Crop	Short term	Long term
C3	Barley	+50	+14
	Cotton	+60	+13
	Rice	+42	+46
	Soybeans	+78	+42
	Wheat	+41	+27
	Tomato	+30	+9
C4	Corn	+26	+4
	Sorghum	-3	+6

Note: Data are from several experiments and represent the percentage change at 680 ppm CO_2 compared with controls (300–350 ppm). Responses to short term and long term (>1 week) exposure to elevated CO_2 are presented.
Source: Yelle et al. (1989); Cure and Acock (1986).

efficient use of water per unit leaf area does not necessarily result in a reduction in total water requirements. The water savings benefit associated with partial stomatal closure can be counteracted entirely by larger plants with greater total leaf area (Rosenberg et al., 1990; Eamus, 1991). Allen et al. (1985), working with soybean, and Morison and Gifford (1984), comparing 16 crop species, found no difference in water use per plant between control and CO_2-enriched plants, although total plant leaf area was often greater in those grown at the higher CO_2 concentration.

A secondary effect of the stomatal closure response to increased CO_2 is to increase leaf temperatures, typically by 1–3° C (Idso et al., 1987; Jones et al., 1985). This occurs because there is less water evaporated at the leaf surface to cool it. A higher leaf temperature will increase the water vapor pressure within the leaf and *increase* transpiration. Also, if stomatal closure initially slows transpiration, humidity of the air surrounding the leaf becomes drier, which will tend to increase rather than decrease subsequent transpiration rates. These factors alone may counteract the water savings benefit from a CO_2 doubling (Jarvis and McNaughton, 1986).

C. Feedback Mechanisms and Acclimation to High CO_2

A major reason for uncertainty in predicting plant response to CO_2 is that physiological and morphological feedback mechanisms may limit the extent to which direct CO_2 effects are realized. The long term evolutionary response to higher CO_2 levels will also be very important, but little is known about this subject and it will not be addressed here.

i. Downregulation of Photosynthesis

The results in Table 8-1 indicate that the large increases in photosynthesis observed in short term experiments are seldom maintained when plants are exposed to high CO_2 for prolonged periods. Several explanations for this downward shift in photosynthesis, referred to as "downregulation," have been proposed. Many relate to the fact that, in the presence of high CO_2, the plant's ability to produce carbohydrates via photosynthesis may exceed the demand for carbohydrates in the remainder of the plant (Stitt, 1991). In some cases, excessive carbohydrates cause enlargement of starch grains within leaf cells, physically damaging organelles important in photosynthesis (DeLucia et al., 1985; Nafziger and Koller, 1976). Perhaps a more common mechanism is a feedback effect that reduces the activity of a key photosynthetic enzyme, Rubisco, by as much as 60% after prolonged exposure to high CO_2 (Sage et al., 1989).

A recent review (Stitt, 1991) concludes that the ability of a particular plant genotype to maintain photosynthetic rates at enhanced CO_2 levels will depend upon the supply-demand balance with regard to the products of photosynthesis (i.e., carbohydrates), and how this balance is regulated within the whole plant. Kramer (1981) suggested that indeterminate crop species, which have continued high demand for carbohydrates because they continue producing new fruit and leaves, respond more positively to increased CO_2 than determinate types.

ii. Leaf vs. Root and Fruit Growth

In general, how plants partition their biomass among leaves, roots, stems, seeds, and fruit has a tremendous impact on subsequent growth and yield and can easily overcome the influence of photosynthetic rate. A lack of correlation between photosynthesis and yield has long been recognized by plant breeders (Elmore, 1980). In terms of whole plant growth, the amount of leaf area available for light interception is just as important as photosynthesis per unit leaf area.

When considering the long term growth response in the field, shifts in biomass partitioning within the plant can have either a beneficial or negative effect, depending on environmental conditions. For example, an increase in leaf area due to high CO_2 would be of obvious benefit in terms of carbon- and energy-gaining ability, and could possibly compensate for the downregulation of photosynthesis phenomenon discussed above. However, large plants with greater leaf area will develop water and nutrient deficits more quickly if grown in a situation where these inputs are in limited supply.

The increase in root:shoot ratio sometimes observed at high CO_2 (Lawlor and Mitchell, 1991) may improve the ability to extract water and nutrients needed for growth, but if this occurs at the cost of reducing the amount of biomass allocated to the harvested portion of the plant, economic yield may be much less.

III. OPTIMUM YIELD RESPONSE TO CO_2 IN CONTROLLED ENVIRONMENTS

The preceding section described some of the reasons why an initial stimulation of photosynthesis due to enhanced CO_2 does not necessarily indicate an increase in yield potential. However, the experience in greenhouses and controlled environments has generally shown a benefit from additional CO_2, albeit less than the short term photosynthetic response, provided other growth conditions such as temperature, light, water, nutrients, air quality, and pest control are maintained near optimum. This information has formed the basis for quantification of the direct CO_2 effect as used in crop models.

A. Use of Supplemental CO_2 in Greenhouse Crop Production

It has been estimated that, worldwide, perhaps 50–75% of commercial greenhouses use some type of CO_2 enrichment to increase productivity (Wittwer, 1986). In The Netherlands, where the greenhouse industry is of considerable economic importance, this figure is closer to 80 or 90% for tomato, cucumber, sweet pepper, strawberry, and some flower and ornamental crops (Van Berkel, 1986). The widespread use of this technology is perhaps the best testimony of a positive CO_2 fertilizer effect in the greenhouse setting. Yield increases of 30–40% for some crops are not uncommon when CO_2 concentrations are maintained at three times ambient levels, i.e., near 1000 ppm (Wittwer, 1986). This is, of course, dependent on maintaining other environmental factors near the optimum for growth. It has been recognized for some time that an increase in water and fertilizer applications is necessary to obtain maximum CO_2 benefits in greenhouses (Wittwer and Robb, 1964).

There are a number of reasons for the strong positive response to CO_2 in greenhouses, in addition to the optimum growth conditions and lack of weed pressure. One is that many greenhouse crops have an indeterminate growth habit, which, as discussed in Section II-C(i), may make them more responsive to CO_2 enrichment than most field crops (Kramer, 1981). Another factor to note is that greenhouses without supplemental CO_2 often have CO_2 concentrations that have been depleted to very low levels (200–250 ppm; Goldsberry, 1986), and so the response to CO_2 doubling is more dramatic than we should expect from a doubling of atmospheric CO_2 from current levels of about 350 ppm, Figure 8-1.

B. Controlled Environment Experiments

The vast majority of our quantitative information regarding plant response to enhanced CO_2 is based on experiments in which CO_2 concentration alone was varied, while water, nutrients, temperature, and pest pressure were maintained near optimum for growth. Exhaustive reviews of this scientific literature by Kimball (1983) and Cure and Acock (1986) reveal that, under these circumstances, a doubling of CO_2 from about 350 to 700 ppm increases the productivity of C3 and C4 crop plants about 33% and 10%, respectively, on average. This information has had a tremendous impact on current opinion regarding the magnitude of the CO_2 fertilizer effect, and has been used extensively in crop models (see Section V).

Although the reviews by Kimball (1983) and Cure and Acock (1986) of controlled environment experiments are extremely valuable references that consolidate much of our current understanding regarding CO_2 effects, reliance on their calculated crop averages for predicting yield response to climate change on a global scale is far from ideal. Crop productivity in the field is never controlled by any one environmental variable exclusively; under field conditions temperature, light, water, and nutrient supply, etc. will have an interactive effect on the plant response to atmospheric CO_2 concentrations (see Section IV). Also, there are crucial differences in environmental conditions between experimental growth chambers and greenhouses and the field situation (Table 8-2).

Since the early 1980s a number of CO_2-enrichment experiments have been conducted in the field using open-top canopy chambers. Results have been highly variable, but in general they have confirmed results obtained in more controlled environments. Substantial yield increases of C3 crops occur (e.g., 20–80%, depending on crop and specific conditions) when temperatures are not too low or too high, and adequate water and fertilizer are supplied.

Table 8-2. Environmental differences between controlled environments and the field.

Environmental factor	Controlled environment	Field
Light	1) Often low intensity	1) Very high intensity in sunlight
	2) Constant	2) Highly variable
	3) Spectral differences from daylight	
Temperature	1) Usually optimum or high	1) In temperate regions often suboptimal early and late in growing season
	2) Often constant during day and night	2) Very variable during the day/night cycle, and during the growing season
Light x temperature	1) Poorly coupled	1) Strongly coupled
Water	1) Often high humidity	1) Very variable humidity
	2) Low wind speed	2) Wind speed variable, can be very high
	3) Regular application in small amounts	3) Application very erratic, if at all
Nutrition	1) Regular application in small amounts	1) Few applications in larger amounts, if at all
Rooting volume	1) Very small	1) Large
	2) Often soil-less growing media	2) Soil type and depth varies

Adapted from Lawlor and Mitchell (1991).

The field studies with open-top chambers are an improvement over laboratory and greenhouse experiments in that the plants are exposed to natural light conditions, and there are more realistic day-night and short term variations in temperature and humidity. Nevertheless, the environment is altered. Compared to the open field, temperatures are often a few degrees warmer, wind and air mixing around the plant foliage are less, and pest pressure may be less because of isolation from the rest of the crop (Lawlor and Mitchell, 1991). These factors will affect growth, and, in fact, increases in productivity due to the chamber effect alone have been reported. Open-top chambers *without* CO_2 enrichment increased the growth of soybean (Rogers et al., 1983) and maize (Rogers et al., 1986) compared to the open field.

Free-air CO_2 enrichment (FACE) is a relatively new approach for studies in the open field (Hendry, 1992). The technology for maintaining uniform CO_2 concentrations is still being tested, and few published results of crop response are available at this time. One preliminary study with cotton (Kimball, 1986b) found that, relative to a control at about 370 ppm CO_2, total yield increased 22% in a FACE treatment where CO_2 concentrations averaged about 1000 ppm in the morning hours to 500 ppm at midday. This yield response was less than the 50% increase reported for an open-top chamber treatment maintained near 650 ppm CO_2 in the same study.

IV. INTERACTIONS BETWEEN CO_2 AND OTHER ENVIRONMENTAL FACTORS

Plant growth is dependent on many factors in addition to CO_2, and if any one of these is in short supply, or at supraoptimal levels, growth will be slowed or plant survival itself may be jeopardized. The "principle of limiting factors" (Blackman, 1905) states that whichever factor is in shortest supply will be "limiting" and determine growth rate. We know that this model of plant response to the environment is simplistic, and there is reason to suspect that plant response to CO_2 may partially compensate for or exacerbate stresses associated with other factors. Also, it is unlikely that all plant responses to CO_2 would be limited by the levels of other environmental variables in exactly the same way. It is important that we learn to what extent yield response to CO_2, on both an absolute and relative basis, will be similar under stress compared to optimal environmental conditions.

The literature review by Cure and Acock (1986) found that data on interactions between CO_2 and other environmental factors were scarce and variable. They concluded that, until better quantitative information is available, prediction of response to CO_2 under specific environmental conditions will not be possible. Since that time, the USDA, EPA, DOE, and other agencies and research institutions have focused more attention on this issue. Our knowledge remains incomplete, but below we have attempted to describe some of the major findings regarding plant response to CO_2 in relation to changes in other biotic and abiotic factors controlling growth.

A. Temperature and CO_2

i. Photosynthetic Response

A strong interaction between temperature and photosynthetic response to CO_2 has been recognized for many years. At low temperatures there is little or no benefit from CO_2 enrichment, but as temperatures rise toward an optimum the proportionate stimulation of photosynthesis increases (Jolliffe and Tregunna, 1968; Berry and Björkman, 1980). The primary explanation for this is that the inhibitory effect of CO_2 on photorespiration (see Section II-A) increases with increasing temperature.

When plants have been acclimated to high CO_2 concentrations for some time, the CO_2 x temperature interaction becomes more complex and we may observe a *negative* effect from CO_2 at low temperatures. This is associated with the downregulation of photosynthesis phenomenon discussed in Section II-C(i), where the activity of the important photosynthetic enzyme, Rubisco, is reduced in plants exposed to high CO_2 for prolonged periods. When this occurs in CO_2-enriched plants whose photosynthetic biochemistry is also being slowed by low temperature, the result can be a *lower* photosynthetic rate than control plants.

The CO_2 x temperature interactive effect on photosynthesis is illustrated in Figure 8-2, which shows the photosynthetic response to temperature for control plants and for CO_2-enriched plants with and without an assumption of reduced Rubisco activity after acclimation to high CO_2 (Long, 1991). The benefit from increased CO_2 is much less at temperatures below 20° C than at warmer temperatures, even without an acclimation response. In simulations where it was assumed that acclimation to high CO_2 reduced Rubisco activity to 60 and 80%, the photosynthetic rate of CO_2-enriched plants became lower than the controls at temperatures of 22.5° and 12.5° C, respectively.

Another feature to note in the curves of Figure 8-2 is that the optimal temperature for photosynthesis is 4–5 degrees higher at the higher CO_2 concentration. Under some environmental conditions, this may compensate for the increase in leaf temperature that results from partial stomatal closure at elevated CO_2. An increase in leaf temperature will be of benefit for plants growing in an environment where air temperatures are below the photosynthesis optimum, but will have a negative effect when air temperatures are already high by shifting leaf temperature into the range where photosynthesis begins a sharp decline. Higher leaf temperatures resulting from CO_2 doubling could potentially place the crop productivity of more regions of the world in jeopardy from global warming than simply looking at air temperatures would indicate.

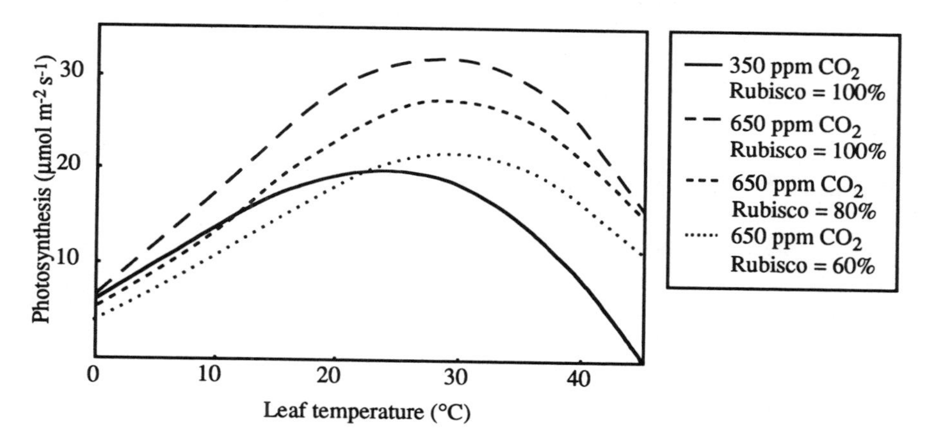

Figure 8-2. Photosynthesis per unit leaf area in relation to leaf temperature, CO_2 concentration, and activity level of the key photosynthetic enzyme, Rubisco. Plants at 350 ppm CO_2 are compared to plants at 650 ppm CO_2 with 60, 80 and 100% Rubisco activity levels. The reductions in Rubisco activity are observed in some species after prolonged exposure to high CO_2 concentration. Source: Long (1991).

ii. Plant Growth and Yield Response

Cure and Acock (1986) found some evidence to suggest that relative increases in plant growth due to CO_2 enrichment were generally greater at warmer temperatures. However, the studies reviewed were not specifically designed to investigate the temperature effect, and most involved only a narrow temperature range.

There is almost no information regarding the CO_2 x temperature interaction under field conditions. Little benefit from CO_2 enrichment was observed in natural tundra vegetation at a high latitude site with cool temperatures (Oechel and Reichers, 1987), whereas enrichment of warm wetland vegetation at a low latitude site produced significant increases in productivity (Drake and Leadley, 1991). Other factors, such as limited nutrient availability at the high latitude site, may also have been involved.

Idso et al. (1987) conducted a field study with open-top canopy chambers to examine the effect of temperature on weekly growth rates of five plant species grown with or without CO_2 doubling. The species were carrot, radish, water hyacinth, water fern, and cotton. All five species showed similar relationships between temperature and relative growth response to CO_2, despite their considerable genetic and morphological differences. The combined data were fitted to a single linear regression line as shown in Figure 8-3. The response to CO_2 enrichment ranged from a 60% *reduction* in growth at 12° C, to no CO_2 fertilizer effect at 18.5° C, to relative growth increases of 30% at 22° C, and 100% at 30° C.

The implications of the study by Idso et al. (1987) are that the simplistic assumption of an approximate 33% increase in C3 crop yields with a doubling of atmospheric CO_2 may substantially underestimate the CO_2 fertilizer effect at warm temperatures, while overestimating the effect when average temperatures are below 20° C. The strong negative effect at temperatures below 18° C is of particular concern since, despite global warming, the 24-hour average temperature will be well below 18° C for much of the critical crop production period in many temperate regions of the world. Some climate models suggest that the temperature increases with global warming will be greatest in the winter in northern latitudes, rather than spring through fall when most crops are produced (Bretherton, 1990).

The results of Figure 8-3, based on a limited data set and a single research approach, cannot be viewed as conclusive. Subsequent experiments with carrot and radish (Idso and Kimball, 1989) found a threshold temperature at which high CO_2 became inhibitory of 12° C rather than 18.5° C, and there was less benefit at high temperatures for radish than indicated in Figure 8-3. The specific slopes and threshold temperatures will undoubtedly be modified as we gather information for more species and temperature regimes. Nevertheless, our understanding of the biochemical mechanisms underlying the photosynthetic response to CO_2 and temperature (Figure 8-2) suggests that a general trend similar to that shown in Figure 8-3 will be found for many important crop species.

Considering our relative certainty that changes in temperature will accompany predicted increases in atmospheric CO_2 concentration, it is surprising that the interaction between these two environmental factors has not received more attention. Research on this subject, particularly under field conditions, should be a high priority.

B. Water and CO_2

Although a global annual increase in precipitation is expected with a CO_2-doubling, conditions may be drier during the summer months for large portions of North and Central America, western Europe, central Asia, eastern Brazil, and north and western Africa (Parry, 1990). One analysis indicates that droughts may begin to increase in frequency in the 1990s, and by the 2050s, severe droughts may occur about 50% of the time compared to 5% of the time under the current climate (Rosenzweig, 1989). Shifts in precipitation patterns will have a substantial impact on agriculture. Dryland farming (reliance on stored soil water without supplemental irrigation) is still common for production of wheat and some other food staples in the developed world, and

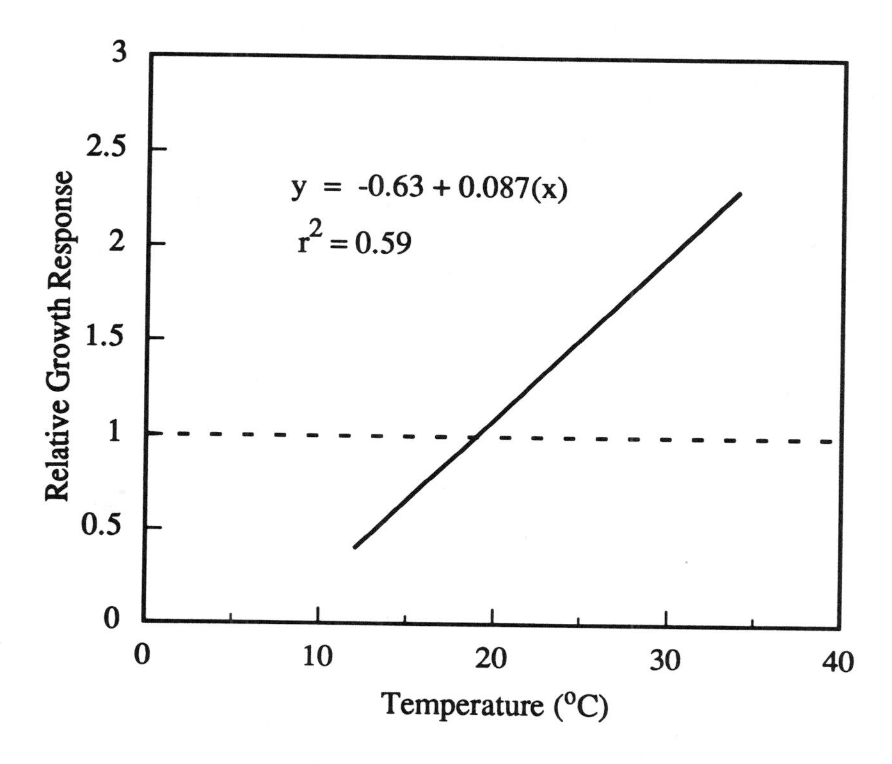

Figure 8-3. Relative growth response of plants to a 300 ppm increase in atmospheric CO_2 concentration in relation to average air temperature during the growth period. A value of 1.0 indicates no response to elevated CO_2, and values below or above 1.0 indicate negative or positive responses to CO_2, respectively. Results are based on data from five plant species. Source: Idso et al. (1987).

in most developing nations the majority of farmers do not have access to irrigation. These are compelling reasons for understanding the interaction between plant response to water stress and elevated CO_2.

The improved water use efficiency of CO_2-enriched plants may be of benefit under water-limited conditions (Section II-B). However, when CO_2 enrichment causes a larger leaf area and rooting volume, stored soil water can be depleted more rapidly despite a more efficient use of water, and this can lead to severe stress earlier in the plant life cycle than would occur without additional CO_2.

The results shown in Figure 8-4, published by Gifford (1979), are typical of many controlled environment experiments. A review by Kimball (1986a) concluded that, in general, the relative yield increase with CO_2 enrichment is

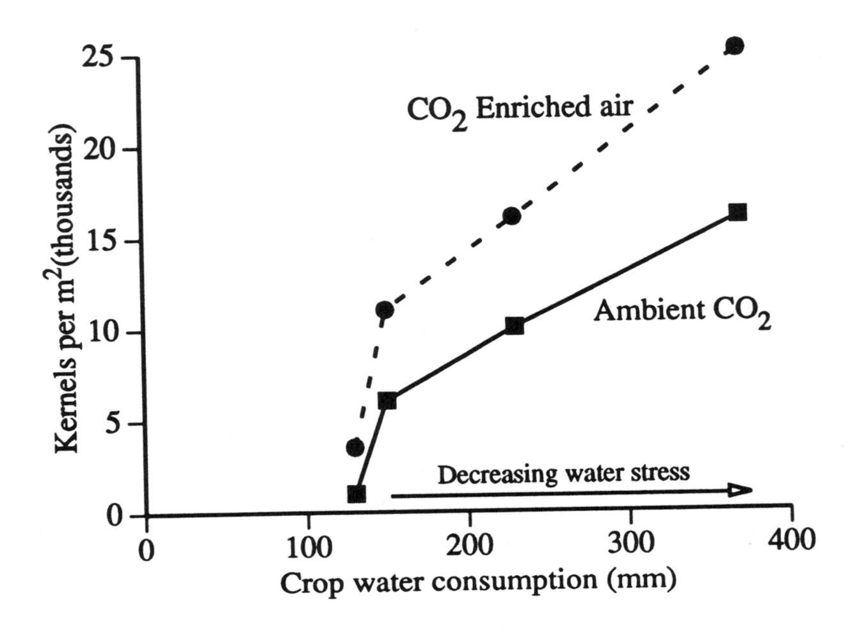

Figure 8-4. Yield response of wheat plants grown with or without CO_2 enrichment and exposed to various levels of drought stress. Increasing amounts of crop water consumption (i.e., transpiration) reflect increasing amounts of water applied and less plant water stress. Source: Gifford (1979).

as large or larger under water stress as under non-stressed conditions. Under some circumstances, the benefit from CO_2 enrichment may compensate for a *mild* water stress. However, under moderate to severe water stress, absolute yields are very low regardless of atmospheric CO_2 concentration, and the benefit from CO_2 enrichment is therefore minimal. This is illustrated in Figure 8-4, where the *relative* enhancement of yield by CO_2 enrichment increases with increasing water stress, but in *absolute* terms, yields are very low at both normal and enhanced CO_2 levels when water shortages become severe.

C. Nutrients and CO_2

The experience of commercial greenhouse producers, as well as CO_2 enrichment experiments, have demonstrated that the benefit from elevated CO_2 is much less when plant nutrients are in limited supply (Wittwer and Robb, 1964; Acock and Allen, 1985). Many investigations have focused on nitrogen and phosphorous, since these elements most frequently limit crop productivity. Wong (1979) grew cotton and maize in the greenhouse with varying nutrient solutions and found that the response to CO_2 was proportionally smaller in low-nitrogen treatments. In contrast, field experiments with cotton (Kimball, 1986b) and rice (Allen, 1991) found that, on a relative basis, the response to CO_2 enrichment improved at low compared to high nitrogen levels. In all cases, however, maximum yields with CO_2 enrichment occurred when nitrogen fertility was optimal for growth. Goudriaan and de Ruiter (1983) compared the effects of nitrogen and phosphorous deficiencies and found that some small positive CO_2 response was maintained even at low nitrogen levels, but there was no CO_2 fertilizer effect when phosphorous was in short supply. The experiment by Goudriaan and de Ruiter (1983) also confirmed earlier findings of Wong (1979) that maize, a C4 species, is less able than C3 species to take advantage of CO_2 enrichment when nitrogen is in short supply.

We have very little specific information about essential plant nutrients other than nitrogen and phosphorous, and we know even less about the effect of CO_2 on soil microbial processes that may affect nutrient availability. Allen (1991) found that CO_2 enrichment had a positive effect on nitrogen-fixing bacteria in the soil, and this improved nitrogen nutrition and yield of rice plants. In contrast, elevated CO_2 had the negative effect of doubling the quantity of nitrogen, phosphorous, and potassium leached below the root zone in an artificial tropical ecosystem (Körner and Arnone, 1992). This may have been associated with a stimulation of soil microbial activity or a more rapid turnover of fibrous roots at the higher CO_2 concentration. An accurate assessment of the effects of CO_2-doubling on crop productivity will require more information regarding these complex interactions between CO_2 and nutrient availability, and a better understanding of how specific nutrient deficiencies influence plant response to CO_2.

D. Weed, Disease, and Insect Pests and CO_2

i. Weed Competition with Crop Plants

An increase in atmospheric CO_2 will of course affect weed species as well as commercially important crop plants. Unfortunately, essentially all of our information regarding crop response to CO_2 is based on experiments in which competition from weeds was not a factor. Wittwer (1990) suggested that the rising CO_2 levels will generally favor crop production since the majority of our important food crops have the C3 photosynthetic pathway, while a high percentage of the major weed pests are C4 plants that will likely benefit less from CO_2 enrichment. This overlooks the fact that even an optimistic 30% increase in growth of C3 crops will not be sufficient to overcome the existing growth rate and competitive ability of some C4 weeds. Also, important C4 crops, such as maize and sugarcane, may experience yield reductions because of increased competition from C3 weeds. It should be noted that maize accounts for about 75% of all traded grain and is the major grain used in food relief programs in famine-prone regions (Parry, 1990).

Broad generalizations regarding CO_2 enrichment effects on crop-weed competition provide little insight into the specific weed control challenges that farmers will have to face in the coming century. The site-specific mix of weed and crop species, and the relative response of each of these species to environmental conditions in the future CO_2-rich world, will determine the economic outcome for both farmers and consumers.

ii. Disease and Insect Pressure

Evidence that the protein concentration of plants may be reduced as a result of increased CO_2 (Parry, 1990) implies not only lower nutritional value for human consumption, but also for leaf-feeding insects. This may induce greater insect mortality, but surviving insects could consume more foliage to compensate for poorer quality food (Fajer, 1989). Natural selection would tend to favor the evolution of insect genotypes that consume more plant material more rapidly. Bazzaz and Fajer (1992) suggested such augmented consumption by insects could negate any benefit gained from the hypothesized crop yield boon in a CO_2-rich environment (p. 72).

The climate changes that result from increased atmospheric CO_2 concentrations will undoubtedly influence the geographic range of insect and disease pests. Warmer temperatures in high latitude areas may allow more insects to overwinter in these areas. Also, crop damage from plant diseases is likely to increase in temperate regions because many fungal and bacterial diseases have a greater potential to reach severe levels when temperatures are warmer or when precipitation increases (Parry, 1990).

E. Air Pollutants, Ultraviolet (UV) Radiation and CO_2

i. Air Pollutants

The burning of fossil fuels, which is the primary cause of increasing atmospheric CO_2 concentrations, also introduces into the atmosphere several air pollutants or their precursors. The three air pollutants that cause the greatest damage to crops are ozone (O_3), sulfur dioxide (SO_2), and nitrogen dioxide (NO_2). Of these, O_3 probably has the greatest economic impact. Estimated yield losses at a 7-hr/day average O_3 concentration of 0.09 ppm are 12.5%, 30.7%, and 27.4% for corn, soybean, and wheat, respectively (Heck et al., 1984). The national composite average of O_3 (measured as second-highest daily maximum at 471 sites in the U.S.) crossed the 0.12 ppm mark in 1991

(U.S. EPA, 1991). Estimation of actual U.S. financial losses due to O_3-related yield reductions range from $1–5 billion per year (Fishman, 1990). The effects of SO_2 and NO_2, and resulting acid rain and dry deposition on crop yields, are less definitive and can have both positive and negative impacts. On the positive side, both nitrogen and sulfur are essential elements in plant nutrition, so that additions to the soil or leaves may benefit plant growth if these elements are in short supply. The negative effects include reductions in photosynthetic rate (Allen, 1990), reduced plant resistance to pathogens, and soil acidification (Canter, 1986). There is some evidence of a synergism between SO_2 and NO_2, such that the negative effect of each pollutant is greater when both pollutants are present (Carlson, 1983).

Since increased CO_2 concentrations cause partial stomatal closure, and most air pollutants enter the plant through the stomates, there is reason to believe that a CO_2 doubling will tend to lessen the yield reductions due to air pollutants. This does appear to be the case for some species, although CO_2 enrichment has seldom been found to completely overcome the adverse pollution effects. Elevated CO_2 concentration reduced O_3 leaf damage by 14% in bean and 66% in tobacco (Heck and Dunning, 1967). Leaf injury caused by SO_2 was reduced 60% at high CO_2 levels in another study (Hou et al., 1977). Allen (1990) used a simple model to show that stomatal closure with a CO_2 doubling could theoretically reduce the damaging effects of O_3 and SO_2 by 15%. Such partial alleviation of pollution stress by high CO_2 may not occur in all species. Carlson and Bazzaz (1982) found that several C4 species had the opposite reaction from the one described above for C3 crops. In their study, C4 plants exposed to SO_2 had reduced growth at high CO_2 concentrations, but not at 300 ppm CO_2.

The research conducted thus far indicates that the crop yield response to air pollutants in a CO_2-enriched world will depend on three key factors: (1) the concentration and potential toxicity of the air pollutant; (2) the degree to which partial stomatal closure limits contact of plant tissues with the pollutant; and (3) the relative magnitude of direct CO_2 effects on photosynthesis and plant growth. These factors will vary with the particular air pollutant, the CO_2 concentration, crop species and growth stage, and environmental factors.

ii. UV Radiation

Two trace gases contributing to global warming, the chlorofluorocarbons (CFCs) and oxides of nitrogen, also are associated with the degradation of the stratospheric ozone layer (at a rate of 3–5% per decade (Kerr, 1991)), and this is increasing the flux of UV radiation to the earth. An increase in UV radiation is not only a direct threat to human health, but also has a negative impact on photosynthesis and growth of many plant species. Approximately 300 plant species and varieties have been studied to date, and of these about half show physiological damage or growth reductions in response to UV radiation (Teramura, 1990). Many important crop species respond negatively, including maize, potato, soybean, tomato, cabbage, and squash. Results have been highly variable, dependent on environment and the particular variety used. Multi-year field studies measuring yield are scarce, but one such study with soybean found a 20–25% yield reduction in 4 of 6 years with UV-B radiation levels simulating a 25% ozone depletion (Teramura, 1990).

The CFCs released into the atmosphere remain active for many years, so that even as CFC production decreases because of new regulations, UV radiation levels are anticipated to continue rising well into the next century. For this reason, an understanding of the interaction between plant response to UV and CO_2 concentration will be important. Ultraviolet radiation affects a number of plant biochemical processes directly associated with photosynthesis and pigment formation, and also affects leaf morphology and biomass partitioning (Teramura, 1990). These effects are likely to alter plant response to CO_2, but at present we have essentially no information regarding this issue.

V. CROP MODELS AND THE CO_2 EFFECT

Some climate change researchers, such as Kaiser et al. (Chapter 7) do not include a CO_2 fertilizer effect in their crop growth models because of the many uncertainties involved. This is the exception; based primarily on data from controlled environment experiments conducted under optimum growing conditions, most modelers have incorporated global multipliers or yield shifts that assume very positive benefits from increased CO_2. The examples that follow illustrate the types of basic assumptions employed, and the impact these assumptions have on yield and economic welfare predictions.

A. Model Assumptions

i. Magnitude and Direction of the CO_2 Effect

Typically in climate change studies, temperature, precipitation, and other output from GCMs are used as input for the crop growth models. This can lead to an error in CO_2 concentrations assumed because the GCMs generate data based on an "*equivalent* CO_2 doubling," which includes the radiative forcing effects of various trace gases as well as CO_2. Trace gases may account for as much as 39% of the human-related forcing (Houghton, 1990). Under these circumstances, atmospheric CO_2 concentrations would be closer to 550 ppm at the time of equivalent CO_2 doubling rather than the 660 ppm (2 x 330) used in most CO_2 enrichment experiments and assumed in many crop models incorporating a CO_2 fertilizer effect.

A recent study by Rosenzweig et al. (1992) (Chapter 5) attempted to correct for this by estimating photosynthetic rates at 555 ppm CO_2. Their CO_2 fertilizer assumptions were nevertheless optimistic since they were based primarily on data from Cure and Acock (1986) and Kimball (1983), which summarize results from controlled environment experiments conducted under optimum growing conditions (see Table 8-2 and discussion in Section III-B). The 555 ppm CO_2/330 ppm CO_2 photosynthesis ratios used by Rosenzweig et al. were 1.21, 1.17, and 1.06 for soybean, wheat and rice, and maize, respectively. The authors indicated that this "... may overstate the positive effects of CO_2, because uncertainty exists concerning the extent to which the beneficial effects of increasing CO_2 will be seen in crops growing in variable, windy, and pest-infested (weed, insects, and diseases) fields under climate change conditions" (Rosenzweig et al., 1992, p.3).

In the MINK study (Chapter 6), Easterling et al. (1992) adopted the concept of radiation-use efficiency (RUE), which is the amount of biomass (i.e. total plant dry weight) produced per unit of incoming light energy. They then used the same Kimball (1983) data set utilized by Rosenzweig et al. (1992) and other groups to generate a curve predicting the increase in RUE with increases in CO_2 concentration. The RUE optimistically was assumed to increase 10%, 24%, and 33% with CO_2 doubling for maize, soybean, and wheat, respectively.

ii. Environmental and Economic Constraints

a. Environmental Factors

Most climate change research is focused on the effects of global warming, so modelers have focussed their attention on plant response to temperature. However, to our knowledge, the well-documented interaction between temperature and CO_2 illustrated in Figures 8-2 and 8-3 (Section IV-A) has not been incorporated into any of the models currently used in climate change studies. The CO_2 fertilizer effect is applied as a constant multiplier, independent of day-night and seasonal temperature fluctuations. The CO_2 fertilizer effect is always assumed to be positive, even when modeling within the low temperature range of 10–20° C, where all experimental evidence indicates the effect becomes relatively small or even negative. The robust and well-tested photosynthesis models that describe the underlying mechanisms of the CO_2 x temperature interaction (Long, 1991; Farquhar, 1980) have not been incorporated into crop models used in climate change research.

Several existing crop models have the capacity to simulate the supply-demand balance for water and nutrients. For the MINK study (Easterling et al., 1992; Chapter 6) nutrients were assumed non-limiting, but simulations with and without irrigation were conducted utilizing the water balance component of their crop model. They also added a linear function that predicts a 60% reduction in stomatal conductance (i.e., partial stomatal closure) with a doubling of CO_2. Although this function is based on controlled environment experiments and results in quite optimistic forecasts of water use efficiency at higher CO_2 concentrations, it is a step forward in terms of incorporation of well-established CO_2 x environment interactive effects.

Crop models used in Rosenzweig et al. (1992; Chapter 5) can predict crop growth as influenced by temperature, precipitation, and management practices such as irrigation and fertilizer application. The models were not modified to incorporate possible interactive effects, but crop yield predictions with and without irrigation and additional fertilizer were compared in their simulations.

To date, little effort has been made to incorporate the effects of weed, insect, or pest pressure, or effects from air pollutants and UV radiation, into crop models used in climate change research. These effects may completely negate any possible benefits from CO_2 enrichment. More research is needed to quantify these effects, and to quantify the interaction between plant response to these environmental factors and CO_2, so that the yield predictions for a CO_2-doubling become more realistic.

b. Economic Issues

The majority of the published literature on predicted crop response to climate change has assumed a maximum CO_2 beneficial effect, and most researchers agree that this would require an increase in water, fertilizer, and other inputs. Little effort has been made to determine whether increasing these expensive inputs in proportion to increases in CO_2 levels will be economically optimal. Also, the assumption has usually been that additional resources will be made available as needed, and as crop production zones shift. The reality is that the availability of water and other critical resources is extremely difficult to predict, and these resources may be in short supply in many underdeveloped as well as developed nations. Resource conservation and concerns regarding environmental impact are other factors that will constrain the use of increased amounts of water, fertilizer, and pesticides. The economic analysis by Rosenzweig et al. (1992; Chapter 5) and a recent review by Bazzaz and Fajer (1992) concluded that the poorer, underdeveloped nations will be at a greater disadvantage in a CO_2-rich world than they are today because of their limited ability to supply the inputs necessary for realization of the CO_2 fertilizer effect.

B. Sensitivity of Economic Forecasts to the CO_2 Effect

Rosenzweig et al. (1992; Chapter 5) found that, without the direct CO_2 fertilizer effect, forecasted crop yields were reduced due to climate change in all countries and GCM scenarios investigated. When the CO_2 fertilizer effect was incorporated, many of the crop yields increased, and all significantly improved. Economic analyses and analyses of risk from hunger were all based on yield forecasts that assumed a strong CO_2 fertilization effect.

Adams et al. (1990) reported that using the Geophysical Fluid Dynamics Laboratory (GFDL) model, for a doubling of CO_2, the impact of climate change on U.S. economic surpluses went from a negative $35.9 to a negative $10.5 billion, without and with the CO_2 fertilizer effect, respectively. Forecasts using the Goddard Institute of Space Studies (GISS) model showed that inclusion of the CO_2 fertilization effect converted a *negative* $6.5 billion change in economic surplus to a *positive* $9.9 billion change.

Easterling et al. (1992; Crosson, Chapter 6) modeled the CO_2 fertilizer effect alone as responsible for a 58% increase in value of production with the worst case scenario of no on-farm adjustments to climate change. In a scenario assuming use of adaptation technology, a positive CO_2 fertilizer effect shifted total forecasted *losses* of $2.03 billion to total *gains* of $645 million.

These results indicate why a critical analysis of the approach to modeling CO_2 effects is in order. As emphasized by Erickson (1993), regional, national, and even global agricultural and climate change policies may become dependent on the magnitude assigned to this one variable.

VI. SUMMARY

Most of our information regarding plant response to CO_2 is derived from controlled environment experiments where water and nutrients were in adequate supply, temperatures were near optimum, and weed, disease, and insect pests were not present. Under these circumstances, many C3 species (includes wheat, rice, soybean, and certain weeds) show a significant increase in photosynthetic rate and water use efficiency per unit leaf area at high CO_2 concentrations. Plant growth and yield may increase by as much as one-third with a doubling of CO_2. The C4 species, including the important crop plants maize, sugarcane, millet, and sorghum, usually show relatively little benefit from increased CO_2.

However, when plants grow in a field situation, the optimum conditions required for full realization of the benefits from CO_2 enrichment are seldom, if ever, maintained. This is particularly true for natural ecosystems and for agricultural ecosystems in underdeveloped countries where irrigation, fertilizer, herbicides, and pesticides are not available or are prohibitively expensive. Even in developed countries, the increase in water, nutrient, and chemical inputs necessary for maximum CO_2 benefit may not be cost-effective or may be limited by concerns regarding resource conservation or environmental quality. In some regions, the potential benefits from CO_2 may be negated by other factors associated with climate change, such as crop damage due to increases in air pollutants or UV radiation.

The specific temperature range for realization of a positive CO_2 effect will undoubtedly vary among crop genotypes, but we know that for most crops the lower temperature limit is somewhere between 10° and 20° C. This has important implications for many temperate regions of the world where, despite global warming, average temperatures during early and late portions of the growing season will be suboptimal and little benefit from a CO_2 doubling, or even a negative growth response, may occur.

In a recent review, Cline (1992) concluded, "It would seem risky to count on agriculture in general experiencing the same degree of benefits from carbon fertilization as has been observed in the laboratory experiments, especially in developing countries where the complementary water and fertilizers may be lacking" (p. 91). Nevertheless, most of our information is from controlled environment studies, and crop modelers have relied on these data for determining the magnitude of their CO_2 fertilizer variable. Some models are capable of separately simulating water and nutrient limitations, but yield and economic forecasts usually assume these growth factors are non-limiting. The many complex interactions between CO_2 and other environmental variables have, for the most part, not been incorporated into crop models used in climate research. Below are some specific suggestions for modeling and interpreting the CO_2 fertilizer effect:

1) Given the uncertainties regarding resource availability and crop response to CO_2 in a future CO_2-rich environment, a conservative policy approach would be to assume no CO_2 fertilizer effect. This may be particularly appropriate for some natural ecosystems and agricultural ecosystems in underdeveloped countries. For policy analysis, modelers should provide results without, as well as with, the CO_2 fertilizer effects. An effort should be made to incorporate economic and resource constraints into crop models as modifiers of the CO_2 variable.
2) The CO_2 x temperature interaction illustrated in Figs. 8-2 and 8-3 (Section IV-A) should be determined for major food crops and a wide temperature range, and incorporated into crop growth models. In the interim, it may be more accurate to assume no CO_2 fertilizer effect at suboptimal temperatures (e.g., less than 15° C).
3) The increase in leaf temperature at increasing CO_2 concentrations can be incorporated into crop models using existing leaf energy balance equations. Leaf temperature could then be linked to plant water use and photosynthesis components of growth models.

4) More field research investigating other environment x CO_2 interactions will be necessary to improve the accuracy of economic forecasts.
5) More information is needed regarding the effects of CO_2-doubling on competition from weeds, and damage from insect and disease pests.

LITERATURE CITED

Acock, B. and L.H. Allen, Jr. 1985. Crop responses to elevated carbon dioxide concentrations. In *Direct Effects of Increasing Carbon Dioxide on Vegetation*. eds. B.R. Strain and J.D. Cure, Chap. 4. U.S. Department of Energy, DOE/ER-0238: Washington, D.C.

Adams, R.M., C. Rosenzweig, R.M. Peart, J.T. Ritchie, B.A. McCarl, J.D. Glyer, R.B. Curry, J.W. Jones, K.J. Boote, and L.H. Allen, Jr. 1990. Global climate change and US agriculture. *Nature*. 345:219–24.

Allen, L.H., Jr. 1990. Plant responses to rising carbon dioxide and potential interactions with air pollutants. *J. Environ. Quality*. 19:15–34.

Allen, L.H., Jr. 1991. *Carbon dioxide effects on growth, photosynthesis, and evapotranspiration of rice at three nitrogen fertilizer levels*. Response of Vegetation to Carbon Dioxide, Report No. 62. Plant Stress and Protection Unit, USDA-ARS: Gainesville, Florida.

Allen, L.H., Jr., P. Jones, and J.W. Jones. 1985. Rising atmospheric CO_2 and evapotranspiration. In *Proceedings of the National Conference on Advances in Evapotranspiration. Amer. Soc. Agric. Engr.*. ASAE Pub. 14–85. St. Joseph: Michigan.

Bazzaz, F.A and E.D. Fajer. 1992. Plant life in a CO_2-rich world. *Scientific American*. 266:68–74.

Berry, J.A. and O. Björkman. 1980. Photosynthetic response and adaptation to temperature in higher plants. *Annual Review of Plant Physiol.* 31:491–543.

Blackman, F.F. 1905. Optima and limiting factors. *Annals Bot.* 19:281–86.

Bretherton, F.P., K. Bryan, and J.D. Woods. 1990. Time-dependent greenhouse gas-induced climate change. In *Climate Change: The IPCC Scientific Assessment.* eds. J.T. Houghton, G.J. Jenkins, and J.J. Ephraums. Cambridge University Press: Cambridge.

Canter, L.W. 1986. *Acid Rain and Dry Deposition.* Lewis Publishers: Chelsea, Michigan.

Carlson, R.W. 1983. Interaction between SO_2 and NO_2 and their effects on photosynthetic properties of soybean. *Environ. Pollut.* 32:11–38.

Carlson, R.W. and F.A. Bazzaz. 1982. Photosynthetic and growth response to fumigation with SO_2 at elevated CO_2 for C3 or C4 plants. *Oecologia*. 54:50–54.

Cline, W.R. 1992. *The Economics of Global Warming*. Institute for International Economics: Washington, D.C.

Cure, J.D. and B. Acock. 1986. Crop responses to carbon dioxide doubling: A literature survey. *Agric. and Forest Meteor.* 38:127–45.

DeLucia, E.H., T.W. Sasek, and B.R. Strain. 1985. Photosynthetic inhibition after long term exposure to elevated levels of atmospheric carbon dioxide. *Photosynth. Res.* 7:175–84.

Drake, B.G. and P.W. Leadley. 1991. Canopy photosynthesis of crops and native plant communities exposed to long term elevated CO_2 treatment. *Plant, Cell and Environ.* 14:853–60.

Eamus, D. 1991. The interaction of rising CO_2 and temperatures with water use efficiency. *Plant Cell and Environ.* 14:843–52.

Easterling, W.E., P.R. Crosson, N.J. Rosenberg, M.S. McKenney, and K.D. Frederick. 1992. Methodology for assessing regional economic impacts of responses to climate change: The MINK study. In *Economic Issues in Global Climate Change.* eds. J.M. Reilly and M. Anderson, 168–199. Westview Press: Boulder, Colorado.

Elmore, C.D. 1980. The paradox of no correlation between leaf photosynthetic rates and crop yields. In *Predicting Photosynthesis for Ecosystem Models.* eds. J.D. Hesketh and J.W. Jones, 155–67, Chap. 9. CRC Press, Inc.: Boca Raton, Florida.

Erickson, J.D. 1993. From ecology to economics: The case against CO_2 fertilization. *Ecol. Econ.* In press.

Fajer, E.D., M.D. Bowers, and F.A. Bazzaz. 1989. The effect of enriched carbon dioxide atmospheres on plant-insect herbivore interactions. *Science.* 243:1198–1200.

Farquhar, G.D, S. von Caemmerer, and J.A. Berry. 1980. A biochemical model of photosynthetic CO_2 assimilation in leaves of C3 species. *Planta.* 149:78–90.

Fishman, J. and R. Kalish. 1990. *Global Alert: The Ozone Pollution Crisis.* Plenum Press: New York.

Gifford, R.M. 1979. Growth and yield of CO_2-enriched wheat under water-limited conditions. *Aust. J. Plant Physiol.* 6:367–78.

Goldsberry, K.L. 1986. CO_2 fertilization of carnations and some other flower crops. In *Carbon Dioxide Enrichment of Greenhouse Crops, Vol. II.* eds. H.Z. Enoch and B.A. Kimball, Chap. 9. CRC Press, Inc.: Boca Raton, Florida.

Goudriaan, J. and H.E. de Ruiter. 1983. Plant growth in response to CO_2 enrichment at two levels of nitrogen and phosphorous supply. I. Dry matter, leaf area and development. *Neth. J. Agric. Sci.* 31:157–69.

Heck, W.W., J.D. Cure, J.O. Rawlings, L.J. Zaragoza, A.S. Heagle, H.E. Heggestad, R.J. Kohut, L.W. Kressk, and P.J. Temple. 1984. Assessing impacts of ozone on agricultural crops: II. Crop yield functions and alternative exposure statistics. *J. Air Pollut. Control Assoc.* 34:810–17.

Heck, W.W. and J.A. Dunning. 1967. The effects of ozone on tobacco and pinto bean as conditioned by several ecological factors. *J. Air Pollut. Control Assoc.* 17:112–14.

Hendrey, G.R., ed. 1992. *FACE: Free-Air CO_2 Enrichment for Plant Research in the Field.* CRC Press, Inc.: Boca Raton, Florida.

Hou, L., A.C. Hill, and A. Soleimani. 1977. Influence of CO_2 on the effects of SO_2 and NO_2 on alfalfa. *Environ. Pollut.* 12:7–16.

Houghton, J.T., G.J. Jenkins, and J.J. Ephraums, eds. 1990. *Climate Change: The IPCC Scientific Assessment.* Cambridge University Press: Cambridge.

Jarvis, P.G. and K.B. McNoughton. 1986. Stomatal control of transpiration: Scaling up from leaf to region. *Adv. Ecol. Res.* 15:1–10.

Idso, S.B. and B.A. Kimball. 1989. Growth response of carrot and radish to atmospheric CO_2 enrichment. *Environ. Exp. Bot.* 29:135–39.

Idso, S.B., B.A. Kimball, M.G. Anderson, and J.R. Mauney. 1987. Effects of atmospheric CO_2 enrichment on plant growth: The interactive role of air temperature. *Agric. Ecosys. Environ.* 20:1–10.

Idso, S.B., B.A. Kimball, and J.R. Mauney. 1987. Atmospheric carbon dioxide enrichment effects on cotton midday foliage temperature: Implications for plant water use and crop yield. *Agron. J.* 79:667-72.

Jolliffe, P.A. and E.B. Tregunna. 1968. Effect of temperature, CO_2 concentration, and light intensity on oxygen inhibition of photosynthesis in wheat leaves. *Plant Physiol.* 43:902–6.

Jones, P., L.H. Allen, Jr., and J.W. Jones. 1985. Responses of soybean canopy photosynthesis and transpiration to whole-day temperature changes in different CO_2 environments. *Agron. J.* 77:242–49.

Kaiser, H.M., S.J. Riha, D.G. Rossiter, and D.S. Wilks. 1992. Agronomic and economic impacts of global warming: A preliminary analysis of midwestern crop farming. In *Economic Issues in Global Climate Change.* eds. J.M. Reilly and M. Anderson, 117–31. Westview Press: Boulder, Colorado.

Kerr, R.A. 1991. Ozone destruction worsens. *Science.* 252:204.

Kimball, B.A. 1983. Carbon dioxide and agricultural yield: An assemblage and analysis of 430 prior observations. *Agron. J.* 75:779–88.

Kimball, B.A. 1986a. CO_2 stimulation of growth yield under environmental restraints. In *Carbon Dioxide Enrichment of Greenhouse Crops, Vol. II.* eds. H.Z. Enoch and B.A. Kimball, 55–67, Chap. 5. CRC Press, Inc.: Boca Raton, Florida.

Kimball, B.A. 1986b. *Effects of increasing atmospheric CO_2 on the growth, water relation, and physiology of plants under optimal and limiting levels of water and nitrogen.* Response of Vegetation to Carbon Dioxide, Report No. 39. U.S. Water Conservation Laboratory: Phoenix, Arizona.

Körner, C. and J.A. Arnone, III. 1992. Responses to elevated carbon dioxide in artificial tropical ecosystems. *Science.* 257:1672–75.

Kramer, J. 1981. Carbon dioxide concentration, photosynthesis and dry matter production. *Bioscience.* 31:29–33.

Lawlor, D.W. and A.C. Mitchell. 1991. The effects of increasing CO_2 on crop photosynthesis and productivity: A review of field studies. *Plant, Cell and Environ.* 14:807–18.

Long, S.P. 1991. Modification of the response of photosynthetic productivity to rising temperature by atmospheric CO_2 concentrations: Has its importance been underestimated? *Plant, Cell and Environ.* 14:729–39.

Morison, J.I.L. and R.M. Gifford. 1984. Plant growth and water use with limited water supply in high CO_2 concentrations. I. Leaf area, water use and transpiration. *Aust. J. Plant Physiol.* 11:361–74.

Nafziger, E.D. and R.M. Koller. 1976. Influence of leaf starch concentration on CO_2 assimilation in soybean. *Plant Physiol.* 57:560–63.

Oechel, W.C. and G.I. Reichers. 1987. *Responses of a Tundra Ecosystem to Elevated Atmospheric CO_2.* U.S. Department of Energy: Washington, D.C.

Parry, M. 1990. *Climate Change and World Agriculture.* Earthscan Publications, Ltd: London.

Pearcy, R.W. and O. Björkman. 1983. Physiological effects. In *CO_2 and Plants.* ed. E.R. Lemon, Chap. 4. Westview Press: Boulder, Colorado.

Rogers, H.H., G.E. Bingham, J.D. Cure, J.M. Smith, and K.A. Surano. 1983. Responses of selected plant species to elevated carbon dioxide in the field. *J. Environ. Quality.* 12:569–75.

Rogers, H.H., J.D. Cure, J.F. Thomas, and J.M. Smith. 1986. Soybean growth and yield response to elevated carbon dioxide. *Agric. Ecosys. Environ.* 16:113–28.

Rosenberg, N.J., B.A. Kimball, P. Martin, and C.F. Cooper. 1990. From climate and CO_2 enrichment to evapotranspiration. In *Climate Change and U.S. Water Resources.* ed. P.E. Waggoner, 151–75, Chap. 7. John Wiley & Sons: New York.

Rosenzweig, C. 1989. Global climate change: Predictions and observations. *Amer. J. Agr. Econ.* 71:1265–71.

Rosenzweig, L., M. Parry, G. Fischer, and K. Frohberg. 1992. *Climate Change and World Food Supply.* Preliminary U.S. EPA Report, Environmental Change Unit. University of Oxford: Oxford, 32 pp.

Sage, R.F., T.D. Sharkey, and J.R. Sieman. 1989. Acclimation of photosynthesis to elevated CO_2 in five C_3 species. *Plant Physiol.* 89:590–96.

Stitt, M. 1991. Rising CO_2 levels and their potential significance for carbon flow in photosynthetic cells. *Plant, Cell and Environ.* 14:741–62.

Teramura, A.H. 1990. Implications of stratospheric ozone depletion upon plant production. *Hortscience.* 25:1557–60.

Tolbert, N.E. and I. Zelitch. 1983. Carbon metabolism. In *CO_2 and Plants.* ed. E.R. Lemon, Chap. 3. Westview Press: Boulder, Colorado.

U.S. Environmental Protection Agency. 1991. *National Air Quality and Emissions Trends Report.* Office of Air Quality Planning and Standards: Research Triangle Park, North Carolina.

Van Berkel, N. 1986. CO_2 enrichment in the Netherlands. In *Carbon Dioxide Enrichment of Greenhouse Crops, Vol. I.* eds. H.Z. Enoch and B.A. Kimball, Chap. 2. CRC Press, Inc.: Boca Raton, Florida.

Wittwer, S.H. 1986. Worldwide status and history of CO_2 enrichment — An overview. In *Carbon Dioxide Enrichment of Greenhouse Crops, Vol. I.* eds. H.Z. Enoch and B.A. Kimball, Chap. 1. CRC Press, Inc.: Boca Raton, Florida.

Wittwer, S.H. 1990. Implications of the greenhouse effect on crop productivity. *HortScience.* 25:1560–67.

Wittwer, S.H. and W. Robb. 1964. Carbon dioxide enrichment of greenhouse atmospheres for food crop production. *Econ. Bot.* 18:34–56.

Wong, S.C. 1979. Elevated atmospheric partial pressure of CO_2 and plant growth. I. Interactions of nitrogen nutrition and photosynthetic capacity in C3 and C4 plants. *Oecologia.* 44:68–74.

Yelle, S., R.C. Beeson, Jr., M.J. Trudel, and A. Gosselin. 1989. Acclimation of two tomato species to high atmospheric CO_2. I. Sugar and starch concentrations. *Plant Physiol.* 90:1465–72.

9

DISTRIBUTIONAL EFFECTS of GLOBAL WARMING and EMISSIONS ABATEMENT[1]

Steven Kyle
Department of Agricultural Economics, Cornell University

I. INTRODUCTION

The problem of global warming resulting from the accumulation of various gases in the atmosphere has generated considerable concern as to the pattern of climate change that may occur and the distribution of costs and benefits of various policies which have been proposed to address the problem. This chapter discusses the extent to which the costs and benefits of climate change and of greenhouse gas abatement fall more heavily on some regions or sectors of society than others.

It is important to realize at the outset that most of what can be said about the impacts of climate change remains largely speculative. This is necessarily so given the level of resolution of most analytical tools used to simulate the effects of greenhouse gas (GHG) accumulation. (See Chapter 3 for further discussion of these issues.) That is, climate varies significantly across areas too small for current models to distinguish. A second problem is the long time period under consideration—on the order of 100 years or more—causing substantial uncertainty due to our inability to predict the myriad conditioning and contextual factors which are important in any evaluation of the economic and demographic growth which generate GHG emissions. A third factor is the still-evolving nature of our scientific understanding of the phenomena involved. Nevertheless, there are various observations that can be made both generally and on the basis of model results.

While there is uncertainty about how climate change will affect particular regions or groups, there is a great deal more that can be said with some degree of confidence about the effects of abatement policies. This is because abatement policies which are implemented in the present are subject to much less uncertainty than are projections over 100 years of perturbations of some arbitrary "base case." A second reason abatement policies can be analyzed with a greater degree of certainty is that we already have experience with many of the policies involved. Since much of the problem is ultimately derived from the burning of fossil fuels, previous work on the effects of fuel price changes or technological change provides a useful base from which to work.

The remainder of this chapter examines the distribution of effects of global climate change and of abatement policies at different levels of aggregation: at the household level, the national level, and the international level. The next section focuses on the pattern of climate change itself and is followed by a section discussing the distributional effects of GHG abatement policies. A final section compares the overall distributional picture and presents conclusions.

II. DISTRIBUTIONAL ASPECTS OF CLIMATE CHANGE

As noted above, there is much uncertainty about conditions 100 years in the future, but most agree that some level of warming is likely to take place (IPCC, 1992). Rosenzweig and Parry (Chapter 5) report on the potential effects of three different climate change scenarios, and provide the basis for much of the discussion that follows in terms of international effects. Kaiser et al. (Chapter 7) provide insights at the farm level. If there is significant change, then some level of adaptation will be necessary on the part of those directly dependent on weather for their incomes—mainly the agricultural sector.

While adaptation is virtually always modeled as being costly in economic models (see Kyle (1992) for a discussion), the ultimate effect of warming can be either good or bad for crop production, depending on whether heat or cold is the limiting factor for yields. It is entirely possible for some regions, countries, or households ultimately to benefit. Though weather itself is quite variable at the farm level, the existence of large areas with similar agro-ecological conditions and cropping patterns make generalizations at a broader level possible.

A. Farm Level Effects of Warming

It is clear that very little can be said generally with respect to the *direction* of change at the farm level in specific areas. Some farmers will ultimately be worse off if their weather changes adversely so as to tighten some binding constraint on production, or if land is rendered unsuited for agriculture. An extreme example of this is flooding, a substantial threat to agriculture given the fact that many of the most intensive areas for grain production are located in low-lying areas such as river deltas or basins. This is especially true of numerous important rice growing areas in Asia and elsewhere.

Rosenzweig and Parry (Chapter 5) present scenarios in which higher latitude producers can actually benefit from increased temperatures since this can result in longer growing seasons. An offsetting effect is the acceleration of plant growth in higher temperatures, allowing less time for grain filling. Increased moisture stress from higher temperatures (for a given level of rainfall) can also be a problem. Some authors such as Kane et al. (1989) project that a combination of all of these factors will cause a mixture of positive and negative effects of global warming on agriculture in higher latitudes.

In spite of this uncertainty, it is still the case that some producers are better able to *adapt* to change or uncertainty than are others. Adaptation can involve changes in the technology used to produce a given crop, a change to another crop, or a change to a non-agricultural activity. There are several factors which can contribute to the ability to adapt.

i. Income Level

This is perhaps the most important determinant of ability to adapt. Higher income farmers are better able to bear the risk that adaptation and change causes since they are in less danger of failing to achieve a subsistence level of production or income. The economic development literature has shown conclusively that poor peasants behave rationally in economic terms but may respond differently to technological or environmental change due to their much higher degree of risk aversion. In particular, they are likely to use fewer purchased inputs and to be less able to make large purchases of equipment. The premium put on assuring minimal levels of production of needed food crops may inhibit rapid shifts in farming systems. (See Schultz (1964) for a discussion of this.)

ii. Access To Credit

Adaptation, even to a new production mix generating higher income, almost invariably requires investment. Insofar as some farmers have easier and/or cheaper access to capital markets than do others, they will be able to make necessary changes to take advantage of opportunities or to avoid losses more quickly and fully than others. This clearly favors farmers who reside in countries with well-developed capital markets, and those in capital surplus countries with low real interest rates. Also favored would be relatively well-off farmers and/or those with secure title to their land since the ability to provide collateral has been shown to be an important determinant of access to credit.

iii. Diversification

Given an increase in the riskiness of particular crops and of farming in general, greater diversification in income sources will be a benefit. If a farm produces a variety of products it is less likely that all will fail at the same time. This consideration would tend to weigh heaviest on monoculture producers in vulnerable areas. Another consideration is diversification into off-farm income sources. This will be a more viable possibility in areas which have large shares of non-agricultural production, and where the off-farm economy is growing.

Many of the considerations above point to relatively greater difficulties for farmers in tropical developing countries than elsewhere. Not only do preliminary analyses of direct effects of global warming indicate greater adverse effects in poor countries, but these same countries are less able to achieve the changes required for adapting. Less developed countries also tend to have less developed capital markets, with access for farmers being a pervasive and chronic problem for many. High interest rates are common even where loans for the agricultural sector can be obtained (see Fry (1988) for a discussion of this). If the needed adjustments require investment, these factors will slow or eliminate farmers' ability to adapt.

B. Distributional Effects of Warming on the National Level

If, as discussed in the previous section, the process of global warming results in increased agricultural output in some areas and reduced output in others, this will affect both the rate and sectoral composition of economic growth in these countries. Insofar as adaptation places additional demands on capital markets, there will be less investment in other sectors and the overall composition of growth will change accordingly.

The effect on the level of growth depends on whether climate change has increased or reduced potential output. Given a presumption of adverse effects in tropical areas, developing countries can expect a slowdown of growth and a decrease in agricultural incomes *vis à vis* the rest of the economy. Given the large percentage of the labor force engaged in agriculture in low income countries, this effect is likely to increase overall income inequality within countries. Urban-rural disparities would also increase in such a scenario, with resulting increases in urban migration and overcrowding.

The extent to which growth is affected by the reallocation necessitated by climate change depends also on the relative capital and labor intensities of the various sectors, and the extent to which capital can be shifted from one sector to another. Ultimately the results depend on the direct effect of climate change, the extent to which adaptation requires additional investment, and the ease with which that investment can be financed. For countries with undeveloped domestic capital markets, this will depend critically on access to international capital markets. There is currently a wide range of debt burdens and perceived credit capacities across low and middle income countries. This fact could make an important difference in the ability of individual countries to take the steps needed to adapt.

C. Distributional Effects on the International Level

The pattern of redistribution between countries will depend both on the direct effect of climate change on agriculture and other activities, and on the changes in comparative advantage and trade flows that will result. In terms of direct effects, as discussed above, some countries will experience positive effects on the agricultural sector while others will suffer. Some low-lying countries may even disappear entirely if melting of polar ice causes significant rises in sea levels. Mauritius, a country comprised of an archipelago only a few meters above sea level, is in danger of disappearing entirely in some scenarios. In many areas of the world, particularly Asia, low lying river deltas are both densely populated and major grain producing zones.

For middle to high latitude countries which experience positive effects, there may be advantages not only from increased yields but also from increased export opportunities to tropical countries which suffer yield declines. However, it is important to note that even if yields do change according to this pattern, it cannot be taken for granted that additional revenue can be generated by exporting northern surpluses to grain deficit tropical countries; this will depend on the demand for imports in deficit countries as much as supply in surplus countries.

Increased export volume can fail to result in added revenue if demand is relatively inelastic. In fact, it is entirely possible for export revenue to *fall* if importing countries experience declines in income at the same time export volume increases in physical terms. Not only must potential importing countries have the needed income to purchase grain, they must also generate sufficient foreign exchange with exports of their own to effect the desired transfer, if they are to avoid increasing international indebtedness.

How can this be done? While there are numerous ways to generate income and/or foreign exchange, it stands to reason that in the aggregate, countries which lose comparative advantage in the agricultural sector must be able to realize their new comparative advantage in non-agricultural production. For most small open economies at mid to lower *per capita* incomes, this means an increase in manufacturing exports. To successfully exploit this possibility, it is necessary for developing countries to have access to developed country markets. That is, trade liberalization is a prerequisite for maximizing possible gains.

While an absence of trade liberalization would prevent high-latitude grain surplus countries from realizing all of their potential gains, it would likely prove to be harder on low-income importing countries. If, after suffering a permanent negative agricultural shock, they experience low or no growth in other sectors, stagnation could result. The higher proportion of the labor force dependent on agriculture in low income countries, together with their typically greater dependence on foreign trade, results in a greater relative effect of global climate change on growth and welfare.

The above might provide an example of a justifiable claim under Article 4.4 of the Framework Convention on Climate Change which states that developed countries "shall also assist the developing country Parties that are particularly vulnerable to the adverse effects of climate change in meeting costs of adaptation to those adverse effects." Assistance for developing countries could help promote their participation in international GHG control agreements.

III. DISTRIBUTIONAL ISSUES AND POLICIES FOR EMISSION ABATEMENT

While the distribution of costs and benefits of climate change and GHG abatement varies at different levels of aggregation, a natural unit of analysis for policy questions is the nation; global negotiations take place between national governments and policies designed to discourage emissions will be implemented on a national level. In order to achieve a viable agreement, it will be necessary for each country to see a positive outcome in the long run, which is an appropriate time horizon to consider when analyzing policies intended to have long run effects in addressing what is in fact a slowly developing problem. This section addresses the likely consequences of some emission abatement policies on economic growth in different countries and the resulting change in the ability to achieve important goals such as poverty alleviation or equitable income distribution. Though poor countries currently generate relatively small amounts of GHGs, largely because they are poor, their primary goal for their citizens is to raise their welfare above subsistence levels. Doing this without exacerbating environmental problems will be the true challenge in devising an acceptable long run strategy.

However, in addition to analyzing the long run, as has been the rule in previous studies of this issue, it is also important to analyze the short run impacts of policies on different sectors of the economy. This short run effect can reveal a great deal about the political feasibility for current leaders of signing on to an agreement. These leaders must out of necessity take a much shorter view on any matter than does the typical economic analysis since political requirements for staying in office do not hinge on policy results in the next century.

Even conservative estimates make it clear that carbon dioxide (CO_2) is the largest contributor to the greenhouse effect (Drennen and Chapman, 1992). The next most important contributor, chlorofluorocarbons (CFCs), are already subject to an international treaty and are in the process of being replaced by substitutes. Given the importance of CO_2, and the fact that the excessive accumulations are primarily a by-product of the use of carbon-based fuels, one of the most commonly cited policies which could be used to reduce emissions is a carbon tax.[2] This tax would be applied to fuels based on their carbon content and would result in prices more reflective of the true costs of burning them. Such a tax would, of course, have important economic consequences since fuel prices would have to rise considerably if substantial reductions in use are to be achieved. Several analysts have estimated that tax rates on the order of 500% for coal and 100% for gasoline would be required to achieve a 50% reduction in carbon emissions (Hoeller et al., 1990). Taxes this high are nevertheless still cheaper to implement than a "command and control" regulatory approach would be, since economic incentives are more efficient than bureaucratic regulation and allocation.

The costs to each country of reducing emissions varies widely and depends not only on the type of policy proposed, but also on the structure of each economy and the level of development achieved. Since any international agreement will be negotiated by national governments, success requires that each country see that the final accord is both in their interest—in the sense that the national calculation of costs and benefits is positive—and that it is equitable—in the sense that everyone's contribution is fair when compared to other countries.

The question posed here—the differential effects of carbon emission abatement in different economies—assumes that each country will in fact promote conservation by providing appropriate incentives to its internal economy. This takes the issue one step further than proposals such as tradable emission permits. In such schemes, government trade permits and international transfers result. However, the question is left open as to whether recipient governments would actually act on the incentives created and implement policies to reduce carbon emissions. It is far from clear that they can significantly affect such activities as deforestation, but it is clear that they can affect incentives for energy use. This section discusses the costs and benefits for implementing such policies.

Previous work on carbon tax proposals have focused mainly on the taxes required to achieve a specified level of emissions in a future year, where baseline emissions are determined by a calculation of the energy requirements (given current and/or foreseeable trends and technologies) needed to achieve desired growth rates in national income. (See, for example, Hoeller et al., 1990; Manne, 1990; Manne and Richels, 1990a, 1990b; Marks et al., 1990; and Sathaye and Ketoff, 1990.) However, it is likely that higher energy prices in and of themselves are likely to have a negative impact on economic growth rates in the short run. Hogan and Jorgenson (1990) address this issue in the case of the U.S., but do not go into any detail regarding the likely differences in adverse growth effects between countries of different structures or income levels. It is clear that policy makers in low income countries must consider both problems: the sectoral distribution of costs and benefits in developing countries (as in Sathaye and Ketoff [1991]) and the long run effects on productivity and growth. The majority of studies taking a global view do not disaggregate sufficiently for the differences in structure in developing countries to be distinguished.

A self imposed carbon tax, or equivalent abatement measures, amount to an internal transfer within each country. Acknowledging that this is an internal transfer highlights the true nature of the costs of adjustment; the problem is one of adjustment to a new set of relative prices where the costs which are internal to the country are those of adjustment (e.g., retooling or replacing capital stock to take best advantage of changed prices) and altered calculations of profit and comparative advantage which can affect the level and sectoral composition of growth. Whether taxes are used to promote adjustment by investing in new technology and conservation or are spent in other ways is an important determinant of the ultimate outcome. Overall, though conservation and abatement measures impose costs in the short run, there is reason to believe that long run effects may be beneficial.

A. Distribution and Energy Saving Measures in the Short Run

The distributional effects of energy saving measures would initially fall most heavily on nations which emit the most carbon, mostly from energy use. It is clear that, at the outset, this would mean that the industrialized countries in the OECD and in the former Eastern bloc would be most severely affected in absolute terms. The fuel mixes for different regions differ widely in their reliance on different types of fuels. Figures 9-1a and 9-1d show that consumption of carbon bearing commercial energy products are a far greater proportion of total energy in many poor countries than in rich ones. Many developing countries in general rely heavily on oil, especially for some activities where there are few substitutes, as in transport. China is an exception to this, possessing and consuming large amounts of coal. The large hydro potential in Latin America is a plus from the point of view of a "GHG

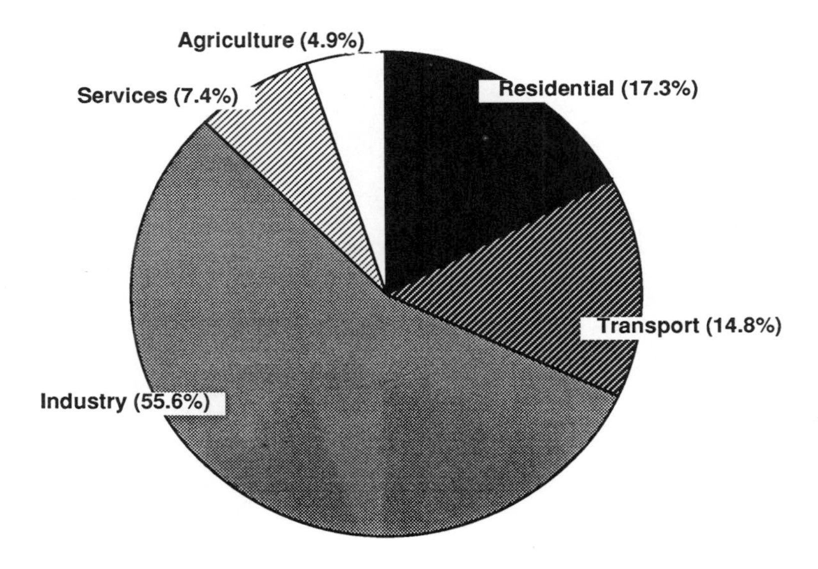

Figure 9-2. Sectoral shares of emissions for nine countries.

In summary, measures to discourage energy use by making it more expensive are more onerous for:

- countries which rely disproportionately on fossil fuels
- countries with large transport sectors
- more urbanized countries
- higher income groups
- countries with large heavy industry sectors
- countries where more expensive commercial fuels would promote deforestation caused by using firewood.

Those groups which are affected most within a country are most likely to be active in opposing imposition of carbon taxes at the national level. Accordingly, these are the groups that international transfers could target in order to promote acceptance of a carbon tax plan.

B. Energy Efficiency, Conservation Measures, and Distribution in the Long Run

In comparing long run effects on different countries, it should be noted that the energy intensity of GNP tends to increase with the level of *per capita* income, although there is considerable variation at higher income levels,

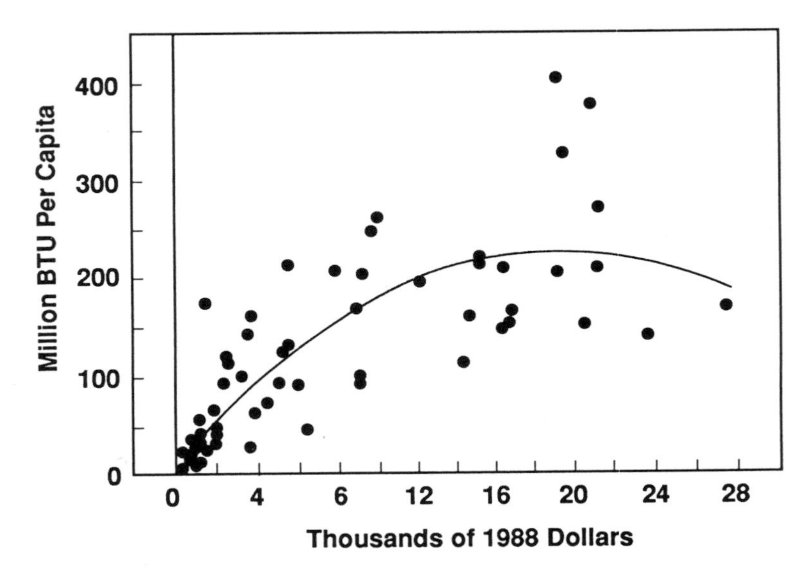

Figure 9-3. Relationship between energy and *per capita* income.

Figure 9-3. Those countries using a lot of energy *per capita* would feel the greatest shock from a tax, while those low and middle income countries engaged in building an energy intensive infrastructure, such as a manufacturing base and modern transportation system, will experience the greatest adverse effects in terms of growth, at least in the medium term before new technologies can be developed.

More expensive energy means that energy intensive activities will be undertaken at a lower level and are likely to be delayed somewhat in terms of their "typical" place in the growth path. All of this means that the cost in terms of growth foregone will vary according to countries' income levels. Given the common finding in econometric studies that energy and capital are complementary inputs, a tax on energy will reduce the rate of return on capital and so restrain capital accumulation and growth.[3]

Various studies have explored this relationship. Dahl (1991) surveys studies of energy/output elasticities, and a summary of these results is shown in Table 9-1. It can be seen that estimates of the short run elasticity of commercial energy with respect to output is 0.47, but rises to 1.19 in the long run. For industrialized countries the long run elasticity is a bit lower at 0.93. Both developing and industrialized countries show substantially higher output elasticities for residential than for industrial use, supporting the point made above that there are likely to be strong effects on the pattern of

Table 9-1. Summary of average energy elasticities.

	P_{SR}	P_{IR}	P_{LR}	Y_{SR}	Y_{IR}	Y_{LR}
Developing countries						
Total energy	-0.12	-0.38	-0.54	0.45	0.82	1.19
Industrial use	-0.19	-0.34	-0.50	0.48	1.11	1.15
Residential use	-0.14	-0.80	-0.27	1.33	1.49	2.48
Oil	-0.06	-0.26	-0.17	0.46	1.10	1.03
Industrialized countries						
Total energy	-0.22	-0.50	-0.45	–	1.13	0.93
Industrial use	0.24	-0.57	-0.35	–	0.88	0.84
Residential use	-0.23	0.59	-0.74	–	1.45	0.99
Oil	-0.35	-0.88	-1.01	–	1.59	1.35

P_{SR}, P_{IR}, P_{LR} = Short, intermediate and long-run price elasticities.
Y_{SR}, Y_{IR}, Y_{LR} = Short, intermediate and long-run output elasticities.
Source: Dahl, 1991.

production in the event of the imposition of a sizable tax. However, the price elasticities for industry are higher than for residential uses in developing countries in the long run.[4]

The most striking aspect of Table 9-1, and one which should give supporters of carbon taxes some pause, is that energy demand seems to be generally price inelastic but income elastic. This means that the adverse effects on growth are likely to be substantial compared to gains in reducing use of carbon fuels. This somewhat bleak outlook is tempered to some extent by the average elasticities for oil, where for industrialized countries usage is both price and income elastic in the long run, suggesting that there would be a strong tendency to substitute non carbon fuels such as nuclear or hydroelectric wherever possible. This last observation points to the possibility that the costs of energy conservation may not be as onerous in the long run as these elasticities might suggest, and may even be negative. That is, econometric estimates such as these, based on relatively short time series (20–25 years), fail to capture the cost saving potential of new technologies and fuel substitution.

C. Benefits from Abatement Policies

Clearly, the most important benefit of abatement would be avoidance of the adverse effects of global warming. However, there are some additional benefits that can be foreseen which do not depend on necessarily uncertain projections of climate change on a regional basis. First, there would be substantial increases in the efficiency of energy use, and consequently the operating costs of energy using machinery would be lower, at least in pretax terms. While forecasting technological changes is chancy at best, it is a safe bet that higher effective energy prices would quicken the pace of technical change.

Second, an increase in energy efficiency and decrease in rates of use would result in a stretch-out of the life of known fossil fuel reserves. This would push back the need to find replacement fuels for finite resources even as research into those replacements is being encouraged because of the abatement policies. In addition, there would be less pressure to look for fossil fuel reserves in ecologically fragile areas such as wildlife refuges or sensitive off-shore sites.

Finally, a decrease in fossil fuel needs would mean a decrease in dependence on imports for the majority of countries which have inadequate or no reserves of their own. This would reduce pressure on the balance of payments and permit increases in imports of, for example, capital goods or other needed inputs. The value of this benefit would be proportional to a country's current fuel imports.

One common characteristic of all of these benefits is that, though substantial, they are only evident in the long run. In the short run there are likely to be substantial costs needed in order to be in a position to benefit in the future. For example, in order to gain the benefits of higher fuel efficiency, lower import bills, and avoidance of adverse climactic effects, it is necessary to replace current fuel using capital with new more efficient capital. This capital cost is incurred in the short run, and can be quite substantial. As was noted above, even if the present value of these investments is extremely attractive, costs will likely be higher than benefits in the short run.

IV. PROSPECTS FOR AN INTERNATIONAL ABATEMENT TREATY

When all of the considerations discussed above are added up there are several major obstacles to achieving a global agreement to slow global climate change, even allowing for the possibility that the long run costs estimated by economists may be benefits instead. First and foremost is the fact that there are wide differences between countries in the costs and benefits of joining an agreement. This could be dealt with through inter-country transfers, but even then there are some additional problems.

First, even if all of the costs and benefits are added up, and the calculation looks good from a particular country's point of view, the timing of the costs and benefits are not likely to be attractive. In other words, even a positive present discounted value can be rather uninspiring when there are large costs up front and the benefits are delayed and are in any case speculative. Given the inability of most developing countries to access world capital markets to begin with, this could pose a major problem.

Second, the fact that there may be competitive advantages to be gained from not implementing a tax that falls heavily on energy intensive industries in trading partners gives this problem some of the qualities of the "prisoner's dilemma." We may all be better off in the long run if we can all agree to cooperate on the solution. In the short run, though, there may be advantages to being outside of a general agreement.

Third, many of the costs are out-of-pocket while the benefits are less tangible (e.g., non-monetary) or are merely costs to be avoided. Governments, who are the ones at the negotiating table, worry more about direct expenses and less about ones that don't appear in their budgets. Taxes, from this point of view, are nice for governments since they add to revenue, but are bad for business since they add to costs. Thus, there is likely to be a large difference of opinion within countries as to the benefits of joining a carbon tax agreement.

If, as is argued above, there are likely to be costs in the short run regardless of one's view of eventual net benefits, there will be a financial burden on countries trying to implement abatement policies. As noted above, these costs are likely to be out-of-pocket since a large fraction of the expense consists of replacing outmoded capital equipment. What is especially important from a policy point of view is that increased capital expenditures need to be financed, and this implies an increase in debt (foreign debt if capital equipment is imported). There will also be pressure to cushion the impact of negative growth rates by borrowing during the process of adjustment.

Most developing countries are already constrained in world capital markets in that they cannot borrow readily. Additional funds could be provided on a concerted basis by industrialized countries if a political decision were made to do so; voluntary private flows of capital are not likely to respond. Can the developed countries realistically hope to provide this much in credit in order to make it possible for low and middle income countries to make needed investments? The position of the U.S. during the Bush Administration was that it was willing to spend only $75 million for this purpose. Clearly, there is a large gap between this figure and what might realistically be required. This observation naturally invites speculation regarding the possibility of retiring international debt in return for implementation of a carbon tax. While this idea may have some merit, there are some fundamental problems with earmarking debt retirement for compensation for the costs of GHG abatement. First, if one takes the view that the debt is unpayable to begin with, then it will sooner or later be retired or eliminated without need to agree to a carbon treaty. Second, the distribution of debt is not the same as the distribution of effects resulting from a carbon tax. Third, insofar as retired debt is owed to private creditors, it is unlikely that they will voluntarily renounce contractual claims of any value without compensation. To transfer hard currency to creditor banks to retire unpayable debt rather than transferring it to developing countries for whom foreign exchange is a serious constraint seems perverse.

Perhaps more important is the likelihood that most countries would remain unable to access credit markets even if outstanding debt were eliminated. Many countries are currently making no net payments on their debt; they are at best paying interest and often incurring new debt even to do this. The upshot is that there is no net inflow of capital. Were international lenders to agree to write off existing debt it is extremely unlikely that they would then embark on new loans for the purpose of capital replacement pursuant to a carbon tax or for almost any other purpose. Though writing off debt would clearly be a benefit, it would not by itself provide the wherewithal to make needed investments so long as countries remain shut out of world capital markets.

A further problem with debt write-offs, and one that is shared with other redistributive schemes, is that it is neither assured nor even likely that recipient countries would regard GHG abatement as the best way to spend large inflows of resources. There may very well be other investments aimed at promoting economic development which political leaders would regard as far more pressing.

If it is indeed true, as argued here, that the short run costs for developing countries of implementing GHG abatement efforts are likely to be viewed as high by many governments, and that even if provided the necessary transfers they would not necessarily use them for this purpose, what remains of an international agreement? Pessimistic scenarios include, of course, the specter of general chaos, with coastal flooding and massive disruption of society. An optimistic scenario might include as comprehensive an agreement as could be reached among, for example, high income countries. The optimists' hope would be that this agreement would provide incentives for the development of new technologies which could prove more important for GHG emission patterns in 100 years than any government policy. The pace of technological change over the past 100 years provides some comfort for this view and highlights the uncertainty of the technological assumptions associated with long range projections.

Perhaps more important, such an agreement could help alter the expectations upon which economic agents base their decisions when investing in capital. Since capital costs constitute the major portion of transition costs, these expectations can play a major role in promoting a more "GHG conscious" capital stock in the future. For example, a small carbon tax which rises over time would achieve both a revision of expectations and would avoid the most onerous costs associated with rapid imposition of a large tax. Thus, a go-slow approach may be a good idea from an economic point of view as well as being politically feasible.

V. CONCLUSIONS

In summary, the distributional effects of global warming are very uncertain, even in terms of the direction of the effect. Nevertheless, low income countries have less ability to support the investments needed to take best advantage of changed conditions. Their greater dependence on agriculture makes them more sensitive and vulnerable to the effects of large climate changes. In scenarios where a general warming occurs, there is some presumption that developing countries will suffer more since crops are already grown at temperatures close to the maximum plants can sustain. In contrast to the uncertainty and long run nature of estimates of the effects of global warming, the distribution of costs of GHG abatement schemes are relatively clear, with energy using sectors being hit hardest. Reconciling the interests of those suffering short run costs with the larger interest of achieving the long run goal of avoiding climate change will provide the main challenge for reaching international accords to address the problem.

NOTES

[1] Based on issues raised during a panel discussion at the Conference on Agricultural Dimensions of Global Climate Change at Cornell on October 8, 1992. Panelists included: R. Barker (Cornell University), C. Jolly (Auburn University), N. Rosenberg (Battelle), C. Rosenzweig (Columbia University), and H. Shue (Cornell University). However, the author takes full responsibility for any statements made herein.

[2] Though many analysts have pointed to biological activities such as agriculture as sources of CO_2, these studies incorrectly focus only on those parts of the biological cycle which result in *emissions*. In order to emit, a process must also *withdraw* carbon from its surroundings. So, in cases where the activity truly is a cycle, and not the destruction of a cycle as would be the case if, e.g., a forest were converted to a sterile desert, such emissions must be balanced with uptake; the resulting net releases are likely to be many orders of magnitude smaller than those estimated solely on the basis of a partial view (see Drennen and Chapman, 1992).

[3] Among the studies emphasizing the relation between growth and energy use are: Ang (1987), Desai (1986), Erol and Yu (1988), Fry (1988), Salib (1984), and Yu and Choi (1985). Studies which take a more disaggregated view emphasizing the relation between energy use and economic structure include: Bohi (1981), Boyd and Uri (1991), Dahl (1991), Fiebig et al. (1987), Jannuzzi (1989), Jones (1991), Longva et al. (1988), and Seale et al. (1991).

[4] It should be noted that the estimates that these figures are based on depend almost exclusively on data series of relatively short duration. Developed country estimates rely on post-World War II data at best, while developing country estimates are almost universally based on much shorter time series. Many "long run" estimates are in fact

based on cross sectional studies and are subject to problems when interpreted as long run relations. Nevertheless, these estimates do provide evidence of the likely effects of an energy or carbon tax in the shorter time period of interest to policy makers.

LITERATURE CITED

Ang, B.W. 1987. A cross-sectional analysis of energy-output correlation. *Energy Economics.* 9(4):274–86.

Bohi, D. 1981. *Analyzing Demand Behavior: A Study of Energy Demand Elasticities.* Resources for the Future. Johns Hopkins University Press: Baltimore.

Boyd, R. and N. Uri. 1991. The impact of a broad based energy tax on the U.S. economy. *Energy Economics.* 13(4):258–73.

Dahl, C. 1991. *Survey of Energy Demand Elasticities in Developing Countries.* Energy Modeling Forum. WP 11.11. IIASA: International Institute for Applied Systems Analysis, Vienna.

Desai, D. 1986. Energy–GDP relationship and capital intensity in LDCs. *Energy Economics.* 8(2):113–17.

Drennen, T. and D. Chapman. 1992. Biological emissions and North-South politics. In *Economic Issues in Global Climate Change.* eds. J. Reilly and M. Anderson, Ch. 12. Westview Press: Boulder.

Erol, U. and E.S.H. Yu. 1988. On the causal relationship between energy and income for industrialized countries. *The Journal of Energy and Development.* 13(1):113–22.

Fiebig, D.G., J. Seale, and H. Theil. 1987. The demand for energy: Evidence from a cross-country demand system. *Energy Economics.* 9(2):149–53.

Framework Convention on Climate Change. 1992. UN, A/AC.237/18, May 15.

Fry, M. 1988. *Money, Interest, and Banking in Economic Development.* Johns Hopkins University Press: Baltimore.

Hoeller, P., A. Dean, and J. Nicolaisen. 1990. *A Survey of Studies of the Costs of Reducing Greenhouse Gas Emissions.* WP #89. OECD Department of Economics and Statistics: Paris.

Hogan, W. and D. Jorgenson. 1991. Productivity trends and the cost of reducing CO_2 emissions. *The Energy Journal.* 12(2):67–82.

IPCC. 1992. *Climate Change 1992. The Supplementary Report to the IPCC Scientific Assessment.* eds. J.T. Houghton, B.A. Callandar, and S.K. Varney. Cambridge University Press: Cambridge.

Jannuzzi, G. de M. 1989. Residential energy demand in Brazil by income classes: Issues for the energy sector. *Energy Economics.* 11(3):254–63.

Jones, D. 1991. How urbanization affects energy use in developing countries. *Energy Policy.* Sept. 1991:621–30.

Kaiser, H.M., S.J. Riha, D.G. Rossiter, and D.S. Wilks. 1990. Agronomic and economic impacts of gradual global warming: A preliminary analysis of midwestern crop farming. Paper presented at a conference *Global Change: Economic Issues in Agriculture, Forestry, and Natural Resources.* Nov. 19, 1990. Cornell University. Sponsored by USDA-ERS.

Kane, S, J. Reilly, and R. Bucklin. 1989. *Implications of the Greenhouse Effect for World Agricultural Commodity Markets.* USDA: Washington, D.C.

Kyle, S. 1992. *The Effects of Carbon Taxes on Economic Growth and Structure: Will Developing Countries Join a Global Agreement?* Cornell University Department of Agricultural Economics. Working Paper No. 92–6.

Longva, S., O. Olsen, and S. Strom. 1988. Total elasticities of energy demand analyzed within a general equilibrium model. *Energy Economics.* 10(4):298–308.

Manne, A. 1990. *Greenhouse Economics for the Developing Countries: Evaluating the Costs of Imposing Emission Limitations.* Mimeo, World Bank, July 1990.

Manne, A. and R. Richels. 1991. Global CO_2 emission reductions—the impacts of rising energy costs. *The Energy Journal.* 12(1):87–107.

Marks, R., P. Swan, P. McLennan, R. Schodde, P. Dixon, and D. Johnson. 1991. The cost of Australian carbon dioxide abatement. *The Energy Journal.* 12(2):135–52.

Rosenzweig, C., M. Parry, G. Fischer, and K. Frohberg. 1993. *Climate Change and World Food Supply.* Environmental Change Unit: Oxford University.

Salib, A.B. 1984. Energy, GDP, and the structure of demand: An international comparison using input-output techniques. *Journal of Energy and Development.* 9(1):55–61.

Sathaye, J. and A. Ketoff. 1991. CO_2 emissions from major developing countries: Better understanding the role of energy in the long term. *The Energy Journal.* 12(1):161–96.

Schultz, T.W. 1964. *Transforming Traditional Agriculture.* Yale University Press: New Haven.

Seale, J.L., W.E. Walker, and I. Kim. 1991. The demand for energy: Cross country evidence using the Florida model. *Energy Economics.* 13(1):33–40.

Yu, E.S.H. and J.Y. Choi. 1985. The causal relationship between energy and GNP: An international comparison. *Journal of Energy and Development.* 10(2):249–72.

10

AFTER RIO: The STATUS of CLIMATE CHANGE NEGOTIATIONS

Thomas E. Drennen
Department of Agricultural Economics, Cornell University

I. INTRODUCTION

The Framework Convention on Climate Change (1992), signed by 153 nations at the 1992 U.N. Conference on the Environment and Development (UNCED) in Rio, begins a process which should, at a minimum, lead to a stabilization of emissions from developed countries of industrial-related greenhouse gases at 1990 levels. While it may lead to much deeper and more meaningful reductions over time, several obstacles to successful implementation remain and, as currently drafted, its overall impact on limiting projected temperature increases will be minimal.

This chapter reviews the basic obligations of the Framework Convention for both developed and developing countries, including commitments to reduce emissions, requirements to prepare inventories, and a mechanism to finance the terms of the Convention. This is followed by a look at questions unresolved in negotiations, such as the comprehensive nature of the agreement, the weighting of the various gases, and how to resolve uncertainties regarding sources and sinks. Finally, an estimate of the initial effectiveness of the Framework Convention is presented.

II. THE AGREEMENT

The ultimate objective of the Framework Convention on Climate Change (1992) (hereafter referred to as the Convention) is the "stabilization of greenhouse gas concentrations in the atmosphere at a level that would prevent dangerous anthropogenic interference with the climate system" (Art. 2). While the text does not specify what level of atmospheric concentrations might cause such interference, it seems clear that the objective of the

Framework Convention is to prevent continued exponential buildup of greenhouse gases. To this end, the Framework Convention (Art. 4) requires all Parties to supply detailed inventories of sources and sinks for all greenhouse gases not covered by the Montreal Protocol. Developed country Parties[1] are further required to "adopt national policies" (Art. 4.2.a) which detail plans for limiting greenhouse gas emissions by sources and increasing removal by sinks with the specific goal (Art. 4.2.b) of returning, either "individually or jointly to their 1990 levels of these anthropogenic emissions of CO_2 and other greenhouse gases not controlled by the Montreal Protocol."

The Convention enters into force only after ratification by 50 countries. Developed countries are given six months thereafter to submit the inventories and national plans; developing countries have three years. Article 7 establishes a Conference of the Parties (CoP) to review implementation efforts, such as the accuracy of required submissions and the overall adequacy in terms of achieving the stated objective of stabilization of atmospheric concentrations. The Convention requires an initial review within 12 months of the date of entry into force of the Convention. Based on this review, the CoP has the authority (Art. 4.2.d) to "take appropriate action which may include the adoption of amendments to the commitments."

Developed countries must assist developing countries with the data reporting requirements of the Convention through both financial and technological assistance. Such funding resources must be additional to existing funding, meaning that foreign aid cannot simply be diverted to fulfilling the requirements of the Framework Convention. Failure to provide such funding theoretically exempts (Art. 12.5) developing countries from their data reporting requirements until such funding is made available.

In addition to meeting the full agreed incremental costs of developing countries associated with meeting their data requirements, Article 4 also requires that developed countries assist those "developing country Parties that are particularly vulnerable to the adverse effects of climate change in meeting costs of adaptation to those adverse effects." This is a broad commitment on the part of developed countries and is noteworthy in that this is exactly the type of long term commitment that several countries, notably the U.S., were wary of when first establishing a multilateral fund to deal with an environmental problem, in that case the protection of the ozone layer.[2] Examples of items potentially covered by this adaptation clause include: assistance with efforts to protect low-lying coastal areas; increased disaster relief to those suffering from the increased incidence of natural disasters such as drought; and even the costs to install air conditioning in areas that become too hot for comfort. The Global Environment Facility (GEF) is designated as the funding agency on an "interim basis," until such time as the CoP can agree on a funding agency acceptable to all Parties. Initially created in 1990 as a three year pilot program to provide investment and assistance for developing country projects dealing with global warming, biodiversity, international waters, or ozone depletion, the GEF is jointly operated by the World Bank, UNDP, and UNEP. However, its use as the funding agency for the Framework Convention was bitterly contested. Developing countries objected to the expanded use of the GEF, arguing that this mechanism did not provide them with adequate control over the distribution of funds since it is the World Bank that manages its investment projects. Developed countries maintained their insistence on using the GEF. Compromise was finally reached by naming the GEF as the interim agency (Art. 21.3) and requiring that it "be appropriately restructured and its membership made universal."

III. OBSTACLES TO IMPLEMENTATION

Negotiations on the Framework Convention took place over an 18 month period, culminating in a final session in New York City one month before UNCED. Heading into this final session, the text of the Convention (INC, 1992a) was far from agreed upon, and included hundreds of bracketed items (items not agreed to by all delegates). Exerting strong leadership, the Chairman of the Session, Jean Ripert, presented a completely rewritten text to the delegates on the first day of this final session. The chairman's text (INC, 1992b), while containing no brackets, was a watered-down version of the heavily debated earlier drafts. Ripert deflected criticism of this text by emphasizing his opinion that acceptance of this text provided the only real hope for finishing in time for the Rio Summit because of a "lack of agreement among the industrialized countries" regarding the scope of the commitments (Ripert, 1992).

It is hardly surprising that, faced with the possibility of failure to produce an agreement in time for Rio, a more general and sometimes vague text emerged from this final session. It also means that many issues will have to be resolved at further negotiating sessions. Nevertheless, the Framework Convention begins a process for reporting data and adopting national action plans, and establishes an administrative structure to facilitate future amendments to the Convention.

When compared to another recent Framework Convention, the Vienna Convention for the Protection of the Ozone Layer (1985), this Convention compares favorably. In terms of commitments, the Vienna Convention (Art. 2.1) only required Parties to "take appropriate measures...to protect human health and the environment against adverse effects resulting or likely to result from human activities which modify or are likely to modify the ozone layer." The "appropriate measures" were not further elaborated on. Yet despite the Vienna Convention's vague language, it began a process leading within two years to a partial phaseout of ozone destroying chemicals (Montreal Protocol, 1987) and within five years to a complete phaseout (London Amendments, 1990).

Today that process is heralded as the ultimate success in international treaty making and yet the Vienna Convention was signed initially by only 20 nations and the European Community and was criticized as doing too little, too late. While many differences between these two global environmental issues exist, the lessons of the Vienna Convention should not be ignored. Perhaps the most important lesson is that once the process begins, it can take on a life of its own.

Many items stand in the way of success on the scale of the Vienna Convention however, including: the existence of many loopholes and ambiguities of text; the actual scope of the Convention; uncertainty regarding the ability of the major greenhouse gases to affect climate change, and the magnitude of sources and sinks. These items are discussed in turn.

A. Loopholes and Ambiguities

While establishing goals for developed and developing countries in terms of data reporting requirements and setting national policies, the Convention contains ambiguous language which may provide countries with possible rationales for delaying implementation. The overall effect of these loopholes and ambiguities is unclear; they may be counterbalanced by the effect of public scrutiny of submitted plans and inventories and the likely resulting pressure from various governmental and non-governmental groups.

As an example of a loophole, consider language in Article 4 regarding the formulation of national policies. In adopting these policies, developed countries may take into account (Art. 4.2.a.): "the differences in [developed country Parties'] starting points and approaches, economic structures and resource bases, the need to maintain strong and sustainable economic growth, available technologies and other individual circumstances." This wording could lead various European countries to argue that because of other existing national policies, such as high gasoline taxes, widespread mass transit, or commitment to non-carbon emitting fuels (such as France's nuclear program), their effort need not be as great as required by the U.S. whose heavy reliance on fossil fuels makes it a prime greenhouse gas contributor. Likewise, the U.S. action plan could call for a very gradual adoption of policies, citing the need to "maintain strong economic growth." Or the U.S. could claim that existing policies put the U.S. ahead of other countries in combating climate change.

Claiming credit for existing programs is exactly what the U.S. did at the final negotiating session prior to the Rio Summit in an attempt to shed its image as a "spoiler" with respect to specific commitments. The U.S. delegation released a document (U.S. Delegation, 1992) outlining actions currently under way, such as the 1990 Clean Air Act and the adoption of a National Energy Strategy. Such action, the document states, will result in emission reductions of 7–11% of projected emissions for the year 2000. The plan (p. 8) also stresses the massive funding that the U.S. has already contributed to climate change research, claiming that it accounts for roughly half of the total spent by all countries on climate change research.

Other regions benefit from ambiguous wording also. Developing countries, while given three years to submit plans (Art. 12.5), may actually delay their submission if they are considered "least developed countries." Such countries are allowed to make their initial communication "at their discretion." And for those countries currently "undergoing the process of transition to a market economy," flexibility is allowed in implementation of their commitments (Art. 4.6). This arose due to concern by countries (such as Hungary) whose 1990 emissions were less than they were in preceding years due to a restructuring of their economies. These countries can therefore use an earlier year, when emissions were greater, as their reference year.

B. Comprehensive or Not?

The Framework Convention did not completely resolve an ideologic battle regarding actual commitments to reduce greenhouse gases. Several OECD countries had pushed for specific commitments in regards to CO_2; Germany is committed to a 25–30% reduction from 1990 levels by the year 2005.[3] The U.S., however, insisted on the concept of a comprehensive approach, covering all greenhouse gases, and without specific time schedules until further uncertainty is resolved. It could be claimed that the comprehensive approach won out, since the Framework Convention covers the sources and sinks for all greenhouse gases not covered by the Montreal Protocol. However, the wording of the Convention is ambiguous enough to suggest that this question has not been fully resolved.

A comprehensive approach, as envisioned by the U.S., would allow each country broad latitude in deciding which gases to reduce (see U.S. Department of State, 1990; and Stewart and Wiener, 1992). Under such an approach, it might be possible to actually increase emissions of one gas, such as CO_2, while limiting emissions of another gas, such as CH_4. Implicit in this approach is the concept of adopting a weighting scheme for the gases which takes into account the effectiveness of each individual gas as a contributor to climate change.

Stewart and Wiener (1992) argue the importance of the comprehensive approach in creating a level playing field among nations concerning emission levels. Under a truly comprehensive approach, countries with very different portfolios of greenhouse gas emissions would have the flexibility "to select that combination of greenhouse gas source and sink controls that is least economically and socially costly" (Stewart and Wiener, 1992, p. 94). As an example, the authors note that CO_2 emissions in the U.S. are approximately six times greater than CH_4 emissions, whereas in India, CH_4 emissions are twice as great as CO_2 emissions (Ibid., p. 93).

The adopted text (Art. 4.2.b) states that national policies should aim to return emissions to 1990 levels of "anthropogenic emissions of CO_2 and other greenhouse gases not controlled by the Montreal Protocol:" The next paragraph states that calculations undertaken in regard to the above should take into account the respective contributions of such gases. Does this mean that countries can use emission reductions in one gas to offset the contributions of other gases? Or does the text mean that each gas should be returned to 1990 levels? The U.K. formally noted that their interpretation of this phrase was the latter, that each gas is to be considered separately (Grubb, 1992).

Earlier language (proposed for Art. 4; Framework Convention, 1992), although heavily bracketed, was more specific as to the overall scope of the comprehensive approach: "These developed country policies and measures will have the effect of, as a first step, stabilizing individually or jointly emissions of the total of greenhouse gases..." (INC, 1992a). By this wording, it seems clear that it would be the total of all greenhouse gases that would have be stabilized, not necessarily individual gases.

Obviously, questions of equity are directly tied to resolution of this issue (see Chapman and Drennen, 1990; and Shue, 1992b). While Stewart and Wiener (1992) note the differing portfolios of greenhouse gas emissions between various countries, such as the U.S. and India, they make no mention of either *per capita* emission levels, or of whether the Indian emissions result from agricultural practices at a subsistence level. In contrast to these Indian emissions, the majority of U.S. emissions emanate from what some term luxury uses when compared to developing country levels (see, for example, Shue 1992a). The statement that Indian emissions of CH_4 are twice the emissions of CO_2 is irrelevant; the bottom line must be whether these emissions are in excess of what is minimally necessary at some basic level. As Shue (1992a) notes: "The central point about equity is that it is not equitable to ask some people to surrender necessities so that other people can retain luxuries."

C. Global Warming Potentials (GWPs)

The resolution of the above issues raised concerning the comprehensive nature of the agreement and equitable treatment of various emissions is closely linked to resolution of a problem relating to the individual effectiveness of the various greenhouse gases.

All greenhouse gases are not created equal. Each gas absorbs infrared

Table 10-1. Greenhouse gas weighing schemes (per unit mass basis).

	Radiative forcing	IPCC 1990	IPCC 1992	Lashof and Ahuja	Reilly
CO_2	1	1	1	1	1
CH_4	58	21	11	10	21
N_20	206	290	270	180	250
CFC-11	3970	3500	3400	1300	4000
CFC-12	5750	7300	7100	3700	6800

Sources: IPCC, (1990, 1992), Lashof and Ahuja, (1990) and Reilly and Richards (1992).

radiation differently, referred to as instantaneous radiative forcing. By this measure, CH_4 is 58 times more effective on a per unit mass basis than is CO_2; CFC-11 is 3970 times more effective; and N_2O is 206 times more effective. These weights are often used to argue the importance of a comprehensive approach controlling not just CO_2, but all gases.

But instantaneous forcing is only one determinant of the overall importance of a gas. The gases have different lifetimes as well; methane lasts just 10 years before breaking down, CO_2 lasts about 200 years. Scientists have attempted to take the entire life cycle of the gases into account in determining a weighting scheme, often referred to as the Global Warming Potential (GWP). Unfortunately, disagreement on several factors exists, including the actual lifetime of the gases, the reactions leading to the breakdown of the chemicals, and the importance of the resulting gases. Agreement on such a weighting scheme would be an important component of any comprehensive agreement.

Table 10-1 lists five potential weighting schemes. Both the IPCC (1990, 1992) and the Lashof and Ahuja (1990) weights attempt to calculate the importance of the gases taking both instantaneous and long term effects into account:

$$GWP = \frac{\int_0^n a_i c_i dt}{\int_0^n a_{co_2} c_{co_2} dt}$$

Table 10-2. Contributions of 1990 emissions of manmade emissions to greenhouse warming under five alternative weighing schemes (in percent).

	Radiative forcing	IPCC 1990	IPCC 1992	Lashof and Ahuja	Reilly
CO_2	58.2	76.4	84.1	86.4	76.5
CH_4	39.0	18.5	11.0	10.0	18.4
N_20	2.8	5.1	5.2	3.6	5.0

Sources: IPCC, (1990, 1992), Lashof and Ahuja, (1990) and Reilly and Richards (1992).

where a_i is the instantaneous forcing of the trace gas, i, and c_i is the concentration of the trace gas remaining at time, t. The values for CO_2 are in the denominator. Implicit in this calculation are any indirect effects of other gases formed after the breakdown of the original trace gas. For example, over a 100 year period (n=100), methane is counted as methane for 10 years (its estimated life), then as CO_2 and water vapor for the remaining 90 years.

The Reilly index (Reilly and Richards, 1992) attempts to include not only the physical components of the GWP, but also both climate and non-related economic effects of increasing concentrations of greenhouse gases. Examples of these effects used by Reilly include beneficial effects due to CO_2 fertilization and damaging effects to crops due to increased ultraviolet radiation resulting from a decreased level of stratospheric ozone. Reilly thus introduces two additional sources of uncertainty with this expanded GWP: whether these effects are beneficial or damaging in nature, and the likely costs of these additional factors. For the sake of discussion, Reilly uses an estimate of $28 billion for CO_2 fertilization and $456 billion for damages for an effective CO_2 equivalent doubling. Reilly presents several damage functions; the case of a quadratic damage function is assumed here.[4]

The IPCC (1992, p. 22) recently revised its estimates for several of the GWPs and notes that they can no longer recommend values which include the indirect effect of gases "because of our incomplete understanding of chemical processes." The indirect effects are those effects resulting from the breakdown of the original trace gases.

Applying these various weights to 1990 manmade emissions of three key greenhouse gases, CO_2, CH_4, and N_2O, illustrates the importance of resolving remaining uncertainty (Table 10-2). The CFCs are not included in this analysis since they are not affected by the Framework Convention.

Selection of a weighting scheme is very important for determining the

Table 10-3. Annual emissions of methane from various sources, in Teragrams (10^{12} grams or millions of metric tons).

Source	Quantity	Percent of Total
Natural wetlands (includes bogs, swamps, tundras)	115	21.5
Rice paddies	110	20.5
Enteric fermentation (ruminant animals)	80	15.0
Biomass burning (includes fuel wood, agricultural burning, forest fires)	55	10.3
Gas drilling, venting, transmission	45	8.4
Termites	40	7.5
Landfills	40	7.5
Coal mining	35	6.5
Oceans	10	1.9
Fresh waters	5	0.9
Total	**535**	**100.0**

Source: Cicerone and Oremland (1988).

relative importance of each gas. Unfortunately, selection of a GWP could become highly politicized. Countries highly dependent on fossil fuels would benefit from the use of the instantaneous radiative forcing index, as this would shift the blame away from CO_2. Developing countries, highly dependent on agriculture, would benefit from the Lashof and Ahuja index, since CO_2 accounts for 87% of the total effect under this weight.

D. Uncertainty Regarding Sources and Sinks

Implementation of the Framework Convention will require resolution of two additional issues, namely: how to deal with the uncertainty regarding magnitudes and locations of existing sources and sinks for the various greenhouse gases; and the fundamental difference between biological and industrial emissions of gases. Neither issue will be resolved in the near term and both touch on issues other than the strictly scientific, raising subtle questions of politics and global justice.

Consider first the problem of sources and sinks. Reaching agreement on meaningful reduction strategies for any greenhouse gas requires a thorough understanding of the sources and sinks for that gas. Sources that are easily quantified include CO_2 emissions from fossil fuel use. More difficult are emissions from biological sources, due to the great variety and size of sources. Still more difficult to quantify are sinks for the various gases.

Consider the sources of methane (Table 10-3). The largest source is natural wetlands and bogs where methane is continuously formed through the breakdown of organic matter by bacteria, referred to as anaerobic decomposition. Other biologic sources include: rice paddies, ruminant animals, termites, and biomass burning. None of the biological sources seem amenable to accurate data estimates of emissions, to effective regulation, or to monitoring of plans for emissions reductions. Non-biological sources include releases directly from the processes of mining coal and the drilling, venting, and transmission of natural gas.

The degree of uncertainty may be even greater for emissions of nitrous oxide. The primary source of nitrous oxides is now assumed to be the decomposition of nitrogen fertilizers. Until recently, it was assumed that the release of nitrogen from fossil fuel combustion was responsible for the annual increase, but it has been recently shown that this estimate was based on faulty sampling techniques (IPCC, 1990, p. 26). This is a significant revelation as it is now assumed that future emissions of this gas would not be significantly curtailed by controls of fossil fuel plants, but would have to come from changes in agricultural practices, a more difficult task.

As uncertain as our knowledge is with respect to sources, the problems are only compounded regarding quantifying the sinks for the gases. As an indication of the difficulty in resolving remaining uncertainty, consider that the magnitudes of the global carbon cycle have not yet been resolved. Resolving magnitudes on a regional level will only be more difficult. The IPCC (1990, p. 13) quantifies sources of 7.0 +/- 0.5 GtC/yr and sinks of 5.4 +/- 1.0 GtC/yr, indicating a missing sink of 1.6 +/- 1.4 GtC/yr. It is possible that the preparation of the required inventories (Art. 4) will lead to a resolution of the global carbon budget; it is equally plausible that countries will claim part of the missing sink lies within their borders.

Another consideration often overlooked is the difference between biological and industrial emissions in terms of net greenhouse effect. The overall importance of biological sources of methane has been overestimated since previous estimates have ignored the recycling of carbon. Methane released from fossil fuel sources is adding carbon to the atmosphere which was removed tens of thousands of years ago, whereas biological sources of methane, such as bovines, are simply recycling carbon. For example, to calculate the greenhouse impact of methane emitted from bovines, one must also consider the CO_2 that is removed from the atmosphere during the feed growing process. A similar principle applies to every biological source of methane: rice production, termites, and wild animals. If only the emission is considered, and the ecological cycle of atmospheric CO_2 removal is ignored, then the apparent contribution of biological sources to the greenhouse effect will be overstated.

A simple example illustrates the difference that the consideration of both atmospheric lifetimes and cycling makes in terms of determining relative greenhouse effect. The average U.S. automobile releases .28 kg of CO_2 for each kilometer driven (1 lb of CO_2 per mile).[5] For a CO_2 weighting factor of 1, this implies a greenhouse equivalence of .28. A typical cow releases .125 kg per day of CH_4 (Drennen and Chapman, 1992). Traditional calculations of the effect of methane emissions from bovine animals would weight the CH_4 emissions by the instantaneous radiative forcing weight (58) for a greenhouse equivalent rating of 7.25. Comparing the two sources of greenhouse gases, cars vs. bovines, shows that a cow on a daily basis has the same effect as driving 26 km.

Alternately, taking the atmospheric lifetimes into account by applying the GWP proposed by Lashof and Ahuja (where 1 kg of CH_4 is weighted at 10 times 1 kg of CO_2), the cow emits 1.25 greenhouse equivalent units, comparable to driving a car 4.5 km per day. And by factoring in the effect of cycling, the cow emits just .905 greenhouse equivalent units, comparable to driving a car just 3.2 km per day.[6] For this example, ignoring the differences in atmospheric lifetimes and the effects of cycling results in an 800% error. Ignoring just the effects of cycling results in a 40% error.

IV. EFFECTIVENESS OF THE AGREEMENT

These various uncertainties regarding whether the agreement is truly comprehensive, the weighting of the various gases, the benefits and costs of emissions (i.e., CO_2 fertilization), the magnitude of sources and sinks, the difference between biological and industrial emissions, and equity-related concerns will make it rather difficult to proceed with implementation efforts. The result will be that for the foreseeable future, amendments or Protocols to the Convention will likely focus on carbon dioxide, or on those emissions of greenhouse gases that are industrial-related.[7] The latter would still require the selection of an appropriate weighting scheme but at least these emissions are more amenable to measurement and verification than are emissions from biological sources. Finally, an industrial-based comprehensive approach would keep the burden of achieving reductions on the industrialized countries, and would not require emission reductions from the agricultural sectors in developing countries. Such a policy makes sense from a point of view of responsibility for causing the problem and the ability to pay (Drennen et al., 1992).

So how effective is the Convention likely to be in the near future? While stabilization of concentrations is at the heart of the Objective of the Convention, Article 4 (Commitments) deals indirectly with an emissions freeze, which would require far less drastic action than would stabilization.[8] Article 4 suggests that developed countries adopt national policies to limit emissions of greenhouse gases, that, as a first step, return emissions to "earlier levels of anthropogenic emissions" by "the end of the present decade." At least as a first step, these "earlier levels" apparently refer to 1990 levels.

The question that must be asked is how far such a step, which basically implies an emissions freeze (although it does not say for how long), would lead towards meeting the spirit of the objective? Two alternative scenarios, originating from the Model of Economic Development and Climate Change (Drennen, 1993), are considered. The first assumes the successful negotiation and immediate implementation of an emissions freeze of industrial-related CO_2 and CH_4 emissions by all countries. The second scenario considers a freeze of those same emissions by industrialized countries only. Both scenarios assume that consumption of CFCs will be phased out in accordance with the terms of the London Amendments to the Montreal Protocol. Further, neither scenario deals with N_2O emissions due to the high level of uncertainty regarding their sources.

The imposition of a freeze on industrial-related CO_2 and CH_4 by all countries of the world would have a large impact on future temperature commitments, Figure 10-1. Temperature change commitment by 2036 under the freeze is 1.5°C, as compared to 2.35°C for the reference case.[9] The difference is even more pronounced over the time frame of 100 years, 2.3°C for the case of the freeze as compared to 5.3°C for the reference case.

For the freeze by industrial region countries only, the results indicate a relatively small impact on temperature change commitments, Figure 10-2. Temperature change commitment under this more modest emissions freeze by 2036 is 2.17°C, just .17 degrees less than for the reference case. The importance of this result is that it concretely demonstrates that an emissions freeze by developed countries only would prove ineffectual at achieving the long term results required to slow climate change. However, pinning hopes on a global freeze are unrealistic and would indeed be unfair to those countries currently consuming very low levels of fossil fuels.

V. CONCLUSIONS

The Framework Convention on Climate Change, as opened for signature at the U.N. Conference for the Environment and Development, is generally vague in regard to actual commitments of the Parties but begins a process that may ultimately lead to the adoption of meaningful amendments or Protocols.

While the ultimate objective of the Convention (Article 2) is the "stabilization of greenhouse gas concentrations in the atmosphere," the actual commitments (Article 4), suggest an emissions freeze. Theoretically, such a freeze would include all greenhouse gases not covered by the Montreal Protocol. However, due to ambiguities in language, uncertainties regarding

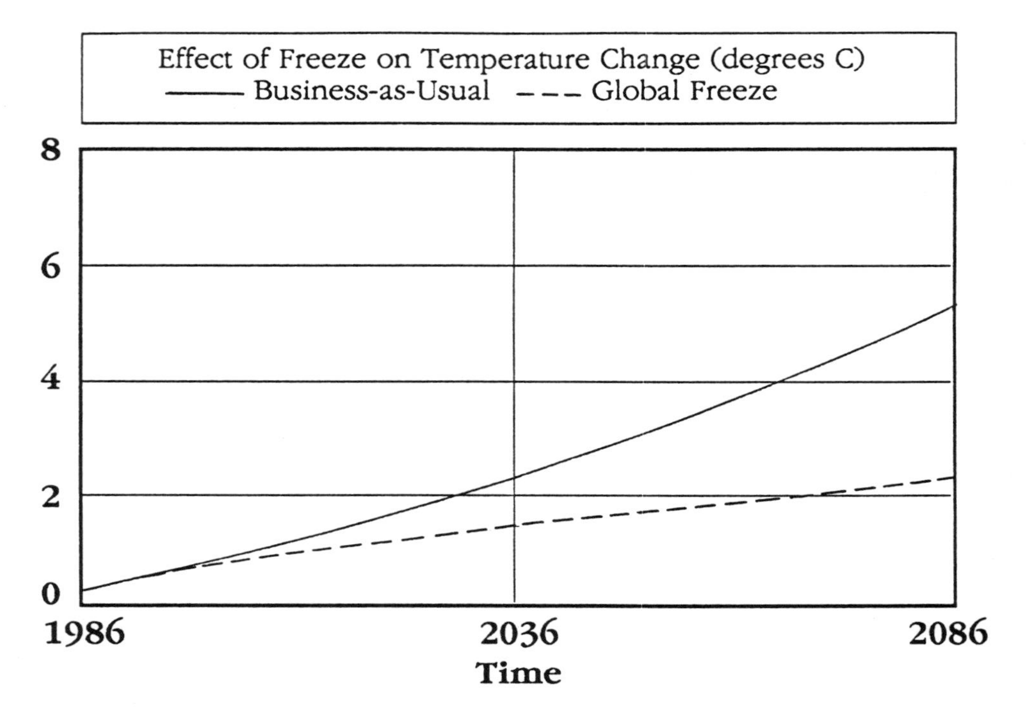

Figure 10-1. Projected temperature change for two scenarios: 1)business-as-usual; and 2) global emissions freeze of industrial-related greenhouse gases.

the magnitudes and locations of sources and sinks, the difference between biological and industrial emissions, the uncertainty regarding benefits and costs, the selection of an appropriate weighting scheme, and concerns about equity, actions in the foreseeable future will likely cover either CO_2 alone or just those gases resulting from industrial uses, namely the use of fossil fuels.

While the imposition of a freeze on industrial-related emissions of CO_2 and CH_4 by all countries of the world would have a large impact on future temperature change commitments, such a global action is unlikely due to issues relating to equity. However, a freeze by industrialized countries only would have a limited effect on future temperature change commitments. Adherence to the implied terms of the Framework Convention on Climate Change would be ineffectual at reducing future temperature change commitment. Meaningful action will require much more than an emissions freeze, and would include actions aimed at reducing dependence on fossil fuels, slowing population growth, and increasing forested areas.

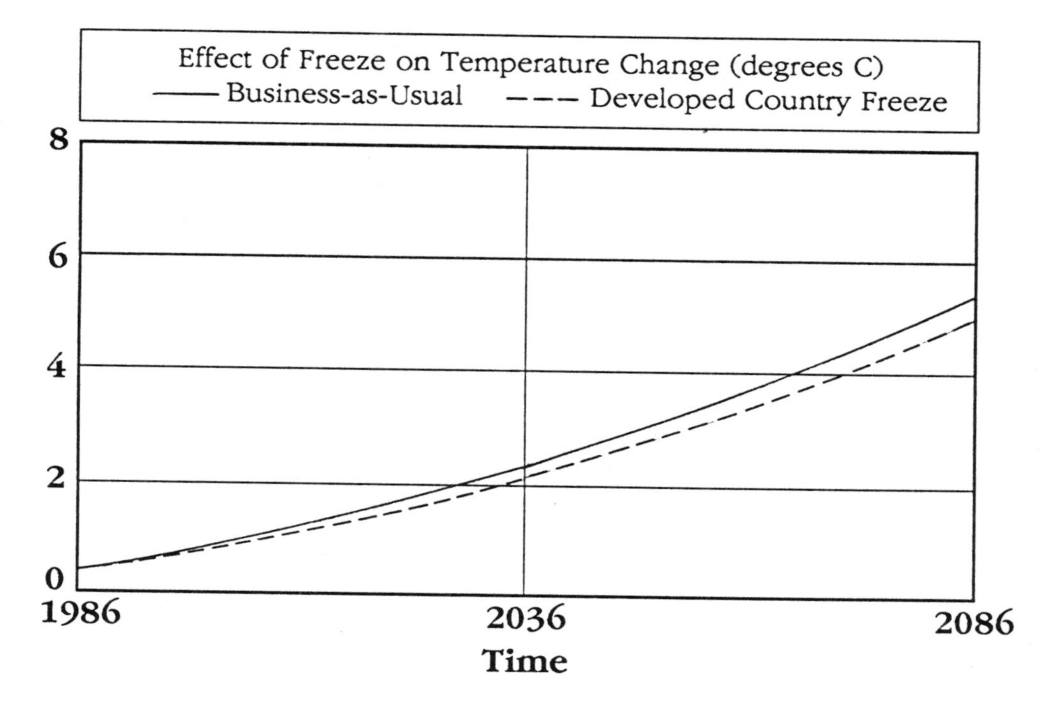

Figure 10-2. Projected temperature change for two scenarios: 1) business-as usual; and 2) industrialized country emissions freeze of industrial-related greenhouse gases.

NOTES

[1] These Parties include: Australia, Austria, Belarus, Belgium, Bulgaria, Canada, Czechoslovakia, Denmark, European Community, Estonia, Finland, France, Germany, Greece, Hungary, Iceland, Ireland, Italy, Japan, Latvia, Lithuania, Luxembourg, Netherlands, New Zealand, Norway, Poland, Portugal, Romania, Russian Federation, Spain, Sweden, Switzerland, Turkey, Ukraine, United Kingdom of Great Britain and Northern Ireland, and the United States of America.

[2] The U.S. was initially hesitant about establishing a fund for protection of the ozone layer, but facing pressure from several fronts, agreed to the concept. A phrase was added to the text (London Amendments to the Montreal Protocol, Article 10, Para. 10, 1991) stating that the "financial mechanism set out in this Article is without prejudice to any future arrangements that may be developed with respect to other environmental issues." (See Drennen, 1993).

[3] Statement by the German Delegation at the Intergovernmental Negotiating Committee, New York, May 2, 1992.

[4] The Reilly Index presented here is different from that presented elsewhere (Reilly, 1992; Drennen, 1993) and contains corrected information.

[5] Assumes 5.35 lbs of carbon per U.S. gallon and an average fuel efficiency rating of 20 mpg.

[6] This result is based on an input-output analysis of the daily cycle of a 500 kg bovine (See Drennen and Chapman, 1992).

[7] This does not mean industrial sector emissions only, but rather those gases resulting from man-induced, non-agricultural processes. Such an approach would primarily cover CO_2 and CH_4 emissions from fossil fuels. It might also include future substitutes for CFCs as they come into use.

[8] The IPCC (1990) estimates that to stabilize atmospheric concentrations of CO_2 would require emission reductions of 60% from current levels.

[9] The reference case assumes moderate *per capita* income growth of 1.0%, 0.5%, 1.35%, and 2.0%, respectively, for industrialized countries, developing countries, Commonwealth of Independent States and Eastern Europe, and China. Population growth rates are assumed to decline in all regions over time, consistent with projections of the United Nations (1991). Under these assumptions, population grows from 5 billion to 9.3 billion in 50 years. Finally, these scenarios assume an adequate supply of fossil fuels, no expansion in the nuclear sector, and a phaseout of CFC consumption (see Drennen, 1993).

* Portions of this chapter have appeared previously in *Law and Policy*. Blackwell Publishers.

LITERATURE CITED

Chapman, D. and T. Drennen. 1990. Equity and effectiveness of possible CO_2 Treaty Proposals. *Contemporary Policy Issues* 8(3):16–28.

Cicerone, R.J. and R.S. Oremland. 1988. Biogeological aspects of atmospheric methane. *Global Biogeological Cycles* 2(4):299–327.

Drennen, T. and D. Chapman. 1992. Negotiating a response to climate change: The role of biological emissions. *Contemporary Policy Issues* 10(3):49–58.

Drennen, T., G. Nicholson, and H. Shue. 1992. Equitable Cooperation on Environment and Development. A Background Paper for the 21st Century Leadership Development Program of the Rockefeller Foundation. Draft, May 5.

Drennen, T. 1993. *Economic Development and Climate Change: Analyzing the International Response*. Ph.D. Dissertation. Cornell University.

Framework Convention on Climate Change. 1992. UN, A/AC.237/18, May 15.

Grubb, M. 1992. The climate change convention: An assessment. *International Environmental Reporter*. August 12, 540–42.

Intergovernmental Negotiating Committee for a Framework Convention on Climate Change (INC). 1992a. *Report of the Intergovernmental Negotiating Committee for a Framework Convention on Climate Change on the Work of the First Part of the Fifth Session*. United Nations Document A/AC.237/18, March 10.

Intergovernmental Negotiating Committee for a Framework Convention on Climate Change (INC). 1992b. *Completion of a Framework Convention on Climate Change: Working Papers by the Chairman*. United Nations Document A/AC.237/CRP.1, April 30.

Intergovernmental Panel on Climate Change. 1990. *Climate Change: The IPCC Scientific Assessment.* Eds. J.T. Houghton et al. Cambridge: Cambridge University Press.

Intergovernmental Panel on Climate Change. 1992. *The Supplemental Report to the IPCC Scientific Assessment.* Cambridge: Cambridge University Press.

Lashof, D. and D. Ahuja. 1990. Relative contributions of greenhouse gas emissions to global warming. *Nature* 334:529–31.

London Amendments. Report of the Second Meeting of the Parties to the Montreal Protocol on Substances that Deplete the Ozone Layer. 1990. UNEP/OzL.Pro.2/3, June 29. Reprinted in *Ozone Diplomacy: New Directions in Safeguarding the Planet,* 1991, R. Benedick. Cambridge: Harvard University Press.

Montreal Protocol on Substances that Deplete the Ozone Layer. 1987. Reprinted in *Ozone Diplomacy: New Directions in Safeguarding the Planet,* 1991, R. Benedick. Cambridge: Harvard University Press.

Reilly, J. 1992. Climate-change damage and the trace-gas strategy. In *Global Climate Change: Agriculture, Forestry, and Natural Resources,* eds. J. Reilly and M. Anderson, 72–86. Washington: Westview Press.

Reilly, J. and K. Richards. 1992. *Climate-change Damage and the Trace-gas strategy.* Economic Research Service, Washington, DC. Draft, July 8.

Ripert, J. 1992. Comment at the Intergovernmental Negotiating Committee. New York.

Shue, H. 1992a. Subsistence Emissions and Luxury Emissions. Presented at the 1992 Law and Society Association Meetings, May 30. Panel entitled: Above the Boundaries: Ozone Depletion, Equity, and Climate Change.

Shue, H. 1992b. The unavoidability of justice. In *The International Politics of the Environment,* eds. A. Hurrell and B. Kingsbury. Oxford: Oxford University Press, Ch. 14.

Stewart, R. and J. Wiener. 1992. The comprehensive approach to global climate policy: Issues of design and practicality. *University of Arizona Journal of International and Comparative Law* 7:83–112.

United Nations. 1991. *World Population Prospects: 1990.* New York: United Nations.

U.S. Delegation. 1992. *U.S. Views on Global Climate Change.* Handout of the U.S. Delegation at the Intergovernmental Negotiating Committee for a Framework Convention on Climate Change. New York, April 30–May 9.

U. S. Department of State. 1990. Materials for the Informal Seminar on U.S. Experience with 'Comprehensive' and 'Emissions Trading' Approaches to Environmental Policy. February 3. Washington, D.C.

Vienna Convention for the Protection of the Ozone Layer, Final Act. 1985. UNEP. Nairobi. Reprinted in *Ozone Diplomacy: New Directions in Safeguarding the Planet,* 1991, R. Benedick. Cambridge: Harvard University Press.

11

FOUR QUESTIONS of JUSTICE

Henry Shue
Ethics & Public Life, Cornell University

I. FOUR KINDS OF QUESTIONS

It would be easier if global warming raised only one question about justice, but several questions are not only unavoidable individually but are entangled with each other. In addition, each question can be given not simply alternative answers but answers of different kinds. Leaving aside the many important questions about justice that do not have to be raised in order to decide how to tackle threats to the global environment, we will find four questions that are deeply involved in every choice of a multilateral plan for action. Our leaders can confront these four questions explicitly and thoughtfully, and thereby hope to deal with them more wisely, or they can leave them implicit and unexamined and simply blunder into positions on them while thinking only about the other standard economic and political considerations that always come up.

What they cannot do is to evade taking actions that will in fact be just or unjust. The subject of justice will not go away. Issues of justice are inherent in the kinds of choices that must ultimately be made. Fortunately, these four issues that are intertwined in practice can be separated for analysis (Table 11-1).

A. Allocating the Costs of Prevention

Whatever sums are spent in the attempt to prevent additional warming of the climate must somehow be divided up among those who are trying to deal with the problem. The one question of justice that people most readily see is this one: who should pay for whatever is done to keep global warming from becoming any worse than necessary?

Stabilizing emissions at 1990 levels will not stabilize temperature, as shown in the previous chapter by Drennen. In order to stabilize temperature, emissions of the various greenhouse gases cannot be added by human processes faster than natural processes can handle them. The current

Table 11-1. Four questions of justice.

1. What is a fair allocation of the costs of preventing the global warming that is still avoidable?
2. What is a fair allocation of the costs of coping with the social consequences of the global warming that will not, in fact, be avoided?
3. What background allocation of wealth would allow international bargaining (about issues like 1. and 2.) to be a fair process?
4. What is a fair allocation of greenhouse gas emissions over the long term and during transition to the long term allocation?

scientific consensus is that in order to stabilize the atmospheric concentration of CO_2, emissions would have to be reduced below 1990 levels by more than 60% (IPCC, 1990). Even if this international scientific consensus somehow were a wild exaggeration and the reduction needed to be, say, a reduction of only 20% from 1990 levels, we would still face a major challenge. And of course every day that we continue to add to the measurably growing concentration, we increase the size of the reduction from current emissions necessary to stabilize the concentration at any given total smaller than the total whenever serious reductions begin.

The need to reduce emissions, not merely to stabilize them at an already historically high level like the 1990 level, is only part of the bad news for the industrial countries. The other part is that the CO_2 emissions of most countries, containing large percentages of the human population, will be rising for some time. I believe that the emissions from these poor, economically less developed countries also ought to rise insofar as this rise is necessary to their providing a minimally decent standard of living for their now impoverished people. This is, of course, already a (very weak) judgment about what is fair: namely, that those living in desperate poverty ought not to be required to restrain their emissions, thereby remaining in poverty, in order that those living in luxury should not have to restrain their emissions. Any strategy of maintaining affluence for some people by keeping other people at or below subsistence is, I take it, patently unfair because it is so extraordinarily unequal—intolerably unequal.[1]

Be the fairness as it may, the poor countries of the globe are in fact not going to restrain themselves voluntarily from taking the measures necessary for creating a decent standard of living for themselves in order that the affluent can avoid discomfort. For instance, the Chinese government, presiding over 22 percent of humanity, is not about to adopt an economic policy of no-growth for the convenience of Europeans and North Americans already living much better than the vast majority of Chinese, whatever others think about fairness. Economic growth means growth in energy consumption, because economic activity uses energy. And growth in energy consumption, in the foreseeable future, means growth in CO_2 emissions. China specifically has far more domestic coal, the dirtiest fuel of all in CO_2 emissions, and no economically viable way in the short run of switching to completely clean technologies or importing, on its own, the cleaner fossil fuels, like natural gas, or even the cleaner technologies, which do exist in wealthier countries, for burning its own coal. In May 1992, Chen Wangxiang, general secretary of China's Electricity Council, said that coal-fired plants would account for 71–74.5 percent of the 240,000 megawatts of generating capacity China plans to have by 2000 (IER, 1992). So, until other arrangements are made and financed, China will most likely be burning vast and rapidly increasing quantities of coal with, for the most part, far from the best available coal-burning technology, not to mention the best energy technology overall. The only alternative China actually has with its current resources is to choose to restrain its economic growth, which it will surely not do.

Fundamentally, then, the challenge of preventing additional avoidable global warming takes this shape: how does one reduce emissions for the world as a whole while accommodating increased emissions by some parts of the world? The only possible answer is: by reducing the emissions in other parts of the world by a greater amount than the increase in those parts of the world that are increasing their emissions. The battle to reduce total emissions should be fought on two fronts. First, the increase in emissions by the poor nations should be held to the minimum necessary for the economic development to which they are entitled. From the point of view of the rich nations, this would serve to minimize the increase that their own reductions must exceed. Nevertheless, the rich nations must also reduce their own emissions somewhat, however small the increase in emissions by the poor, if the global total of emissions is to come down while the contribution of the poor nations to that total is rising. The smaller the increase in emissions necessary for the poor nations to rise out of poverty, the smaller the reduction in emissions necessary for the rich nations—environmentally sound development by the poor is in the interest of all.

Consequently, two complementary challenges must be met—and paid for—which is where the less obvious issues of justice come in.[2] First, the economic development of the poor nations must be as 'clean' as possible, minimizing any additional CO_2 emissions. Second, the CO_2 emissions of the wealthy nations must be reduced by more than the amount by which the emissions of the poor nations increase. The bills for both must be paid: someone must pay to make the economic development of the poor as clean as possible, and someone must pay to reduce the emissions of the wealthy. These are the two components of the first issue of justice: allocating the costs of prevention.

B. Allocating the Costs of Coping

No matter what we do for the sake of prevention from this moment forward, it is highly unlikely that all global warming can be prevented, for the following reason. What the atmospheric scientists call a commitment to warming is already in place simply because of all the additional greenhouse gases that have already been thrust into the atmosphere by human activities since around 1860. Today is already the morning after. We have done whatever we have done, and now its consequences, both those we understand and those we do not understand, will play themselves out, if not this month, some later month. Even a good-faith transition to sustainable levels of CO_2 emissions would make the problem of warming worse for quite a few years before it could begin to allow it to become better; years of fiddling while our commitment to the warming of future generations expands, will make the problem considerably worse still than it already has to be.

The second issue of justice, then, is: how should the costs of coping with the social consequences of global warming be allocated? The two thoughts that immediately spring to mind are, I believe, profoundly misguided; they are, crudely put, to-each-his-own and wait-and-see. The first thought is: let each nation that suffers negative consequences deal with 'its own' problems, since this is how the world generally operates. The second is: since we cannot be sure what negative consequences will occur anyway, it is only reasonable to wait and see what they are before becoming embroiled in arguments about who should pay for dealing with which of them. However sensible these two strategic suggestions may seem, I believe that they are quite wrong and that this issue of paying for coping is both far more immediate and much more complex than it seems. This brief overview is not the place to pursue the arguments in any depth, but I would like to telegraph why I think these two obvious-seeming solutions need at the very least to be debated.

i. To-Each-His-Own

Instantly adopting this solution depends upon assuming without question a highly debatable description of the nature of the problem, namely, as it was put just above: "let each nation that suffers negative consequences deal with its own problems." The fateful and contentious assumption here is that whatever problems arise within one nation's territory are *its own* in some sense that entails that it can and ought to deal with them on its own, with (only) its own resources. This assumption depends, in turn, upon both of two implicit and dubious premises.

First, it is taken for granted that every nation now has all its own resources under its control. Stating the same point negatively, one can say that it is assumed that no significant proportion of any nation's own resources are physically, legally, or in any other way outside its own control. This assumes, in effect, that the international distribution of wealth is perfectly just, requiring no adjustments whatsoever across national boundaries! To put it mildly, that the world is perfectly just as it is, is not entirely clear without further discussion. In a world in which, in fact, major portions of the natural resources of many of the poorer nations are under the control of multinational firms operated from elsewhere and, in fact, many Third World states are crippled by burdens of international debt contracted for them, and then wasted, by illegitimate authoritarian governments, the assumption that the international distribution of wealth is entirely as it should be is hard to swallow.

Second is an entirely independent question that is also too quickly assumed to be closed: it is taken for granted that no responsibility for problems resulting within one nation's territory could fall upon another nation or upon other actors or institutions outside the territory. Tackling this question seriously would mean attempting to wrestle with slippery issues about the causation of global warming and about the connection, if any, between causal responsibility and moral responsibility. Once the issues are raised, however, it is certainly not a foregone conclusion, for instance, that coastal flooding in Bangladesh (or the total submersion of, for example, the Maldives and Vanuatu) would be entirely the responsibility of, in effect, its victims and not at least partly the responsibility of those who produced, or profited from, the greenhouse gases that led to the warming that made the ocean water expand and advance inland. On quite a few readings of, for example, the widely accepted principle, the polluter pays, those who caused the change in natural processes that resulted in the human harm would be expected to bear the costs of making the victims whole. Once again, I am not trying to settle the question here, but merely to establish that it is indeed open until the various arguments are heard and considered.

ii. Wait-and-See

The other tactic that is supposed to be readily apparent and eminently sensible is: stay out of messy arguments about the allocation of responsibility for potential problems until we see which problems actually arise—we can then restrict our arguments to real problems and avoid imagined ones. Unfortunately, this too is less commonsensical than it may sound. To see why, one must step back and look at the whole picture.

The potential costs of any initiative to deal comprehensively with global warming can be divided into two separate accounts, corresponding to two possible components of the initiative. The first component, introduced in the previous section of this chapter, is the attempted prevention of as much warming as possible, the costs of which can be thought of as falling into the prevention account. The second component, briefly sketched in this section, is the attempted correction of, or adjustment to—what I have generally called "coping with"—the damage done by the warming that for whatever reasons goes unprevented.

It may seem that if costs can be separated into prevention costs and coping costs, the two kinds of costs could then be allocated separately, and perhaps even according to unrelated principles. Indeed, the advice to wait-and-see about any coping problems assumes that precisely such independent handling is acceptable, because it assumes, in effect, that prevention costs can be allocated—or anyhow that the principles according to which they will be allocated, once they are known, can be agreed upon—and prevention efforts put in motion, before the possibly unrelated principles for allocating coping costs need to be agreed upon. What is wrong with this picture of two basically independent operations is, that what is either a reasonable or a fair allocation of the one set of costs depends upon how the other set of costs are allocated. The respective principles for the two allocations must not merely be related, but be complementary.

In particular, the allocation of the costs of prevention will directly affect the ability to cope later of those who abide by their agreed-upon allocation. To take an extreme case, suppose that what a nation was being asked to do for the sake of prevention could be expected to leave it much less able to cope with "its own" unprevented problems on its own than it would be if it refused to contribute to the prevention efforts—or refused to contribute on the specific terms proposed—and instead invested all or some of whatever it might have contributed to prevention in its own preparations for its own coping. For example, suppose that in the end more of Shanghai could be saved from the actual eventual rise in sea level due to global warming if China simply began work immediately on an elaborate and massive, Dutch-style system of sea walls, dikes, canals, and sophisticated flood-gates—a kind of Great Sea Wall of China—rather than spending its severely constrained resources on, say, reducing fossil fuel demand or other prevention measures. From a strictly Chinese point of view, the Great Sea Wall might be preferable even if China's refusal to contribute to the prevention efforts resulted in a higher sea level at Shanghai than would result if the Chinese did cooperate with prevention (but then did not have time or resources to build the Sea Wall fast enough or high enough).

This fact, that the same resources that might be contributed to a multilateral effort at prevention might alternatively be invested in a unilateral effort at coping raises two different questions, one primarily ethical and one primarily non-ethical (although these two questions are related). First, would it be fair to expect cooperation with a multilateral initiative on prevention, given one particular allocation of those costs, if the costs of coping are to be allocated in a specific other way (which may or may not be cooperative)? Second, would it be reasonable for a nation to agree to the one set of terms, given the other set of terms—or, most relevantly, given that the other set of terms remained unspecified? Doing your part under one set now while the other set is up for grabs later leaves you vulnerable to the possibility of the second set's being stacked against you in spite of, or because of, your cooperation with the first set. It is because the fairness and the reasonableness of any way of allocating the costs of prevention depends partly upon the way of allocating the costs of coping that it is both unfair and unreasonable to propose that concerning prevention binding agreement should be reached now, while concerning coping we should wait-and-see.

C. Background Allocations and Fair Bargaining

This last point about potential vulnerability in bargaining about the coping terms, for those who have already complied with the prevention terms, is a specific instance of a general problem so fundamental that it lies beneath the surface of the more obvious questions and constitutes a third issue of justice. The outcome of bargaining among two or more parties, such as various nations, can be binding upon those parties that would have preferred a different outcome only if the bargaining situation satisfies minimal standards of fairness. An unfair process does not yield an outcome that anyone ought to feel bound to abide by, if one can in fact do better. A process of bargaining about coping in which the positions of some parties were too weak precisely because they had invested so much of their resources in prevention would be unfair in the precise sense that those parties that had already benefitted from the invested resources of the consequently weakened parties were exploiting that very weakness for further advantage in the terms on which coping would be handled.

In general, of course, if several parties (individuals, groups, or institutions) are in contact with each other and have conflicting preferences, they obviously would do well to talk with each other and simply work out some mutually acceptable arrangement. They do not need to have and apply a complete theory of justice before they can arrive at a limited plan of action. If parties are more or less equally situated, the obvious method by which they should explore the various terms on which various combinations of them could agree upon a division of resources or sacrifices (or a process for allocating the resources or sacrifices)—and thereby settle which of them will in fact cooperate with the others for some purpose like preventing or coping with global warming—is actual direct bargaining with each other. Other things being equal, it may be best if parties can simply work out, among themselves, the terms of any dealings they will have with each other.

Even lawyers, however, have the concept of an unconscionable agreement; and ordinary non-lawyers have no difficulty seeing that voluntarily entered agreements can have objectionable terms if some parties were subject, through excessive weakness, to undue influence by other parties. Parties can be unacceptably vulnerable to other parties in more than one way, naturally, but perhaps the clearest case is extreme inequality in initial positions. This means that morally acceptable bargains depend upon initial holdings that are morally acceptable—not, for one thing, so outrageously unequal that some parties are at the mercy of others.

Obviously, this entails, in turn, that the recognition of acceptable bargaining presupposes knowledge of standards for fair shares, which are one kind of standards of justice. If we do not know whether the actual shares that parties currently hold are fair, we do not know whether any actual agreement they might reach would be morally unconscionable. The simple fact that they all agreed is never enough. The judgment that an outcome ought to be binding presupposes a judgment that the process that produced it was minimally fair. While this may not mean that they must have "a complete theory of justice" before they can agree upon practical plans, it does mean that they need to know the relevant criteria for minimally fair shares of holdings before they can be confident that any plan they actually work out should in any way constrain those who might have preferred different plans.

If bargaining among nations about the terms on which they will cooperate to prevent global warming is to yield any outcome that can be morally binding on the nations who do not like it, the "initial" holdings at the time of the bargaining must be fair. Similarly, the "initial" holdings at the time of the bargaining about the terms on which they will cooperate to cope with the unprevented damage from global warming depends, once again, upon minimally fair shares at that point; holdings at the point of bargaining over the arrangements for coping will have been influenced by the terms of the cooperation on prevention. Consequently, one requirement upon the terms for prevention is that they should not result in shares that would be unfair at the time that the terms of coping are to be negotiated. The best way to prevent unfair terms of coping would appear to be to negotiate both sets of terms at the same time and to design them to be complementary and fair taken together.

D. Allocating Emissions: Transition & Goal

The third kind of standard of justice is general but minimal: general in that it concerns all the resources and wealth that contribute to the distribution of bargaining strength and weakness, and minimal in that it specifies, not thoroughly fair distributions, but distributions not so unfair as to undermine the bargaining process. The fourth kind of standard is neither so general nor so minimal. It is far less general because its subject is not the international distribution of all wealth and resources, but the international distribution only of greenhouse gas emissions in particular. And rather than identifying a minimal standard, it identifies an ultimate goal: what distribution of emissions should we be trying to end up with? How should shares of the limited global total of emissions of a greenhouse gas like CO_2 be allocated among nations and among individual humans? Once the efforts at prevention of avoidable warming are complete, and once the tasks of coping with unprevented harms are dealt with, how should the scarce capacity of the globe to recycle the net emissions be divided?

So far, of course, nations and firms have behaved as if each of them had an unlimited and unshakable entitlement to discharge any amount of greenhouse gases that it was convenient to release. Everyone has simply thrust greenhouse gases into the atmosphere at will. The danger of global warming requires that a ceiling be placed upon total net emissions. This total must somehow be shared among the nations and individuals of the world. By what process and according to what standards should the allocation be made?

I noted above the contrast between the minimal and general third kind of standard and this fourth challenge of specifying a particular (to greenhouse emissions) final goal. I should also indicate a contrast between this fourth issue and the first two. Both of the first two issues are about the allocation of costs: who pays for various undertakings (preventing warming and coping with unprevented warming)? The fourth issue is about the allocation of the emissions themselves: of the total emissions of CO_2 compatible with preventing global warming, what percentage may, say, China and India use—and, more fundamentally, by what standard do we decide? Crudely put, issues one and two are about money, and issue four is about CO_2. We need separate answers to, who pays? and to, who emits? because of the distinct possibility that one nation should, for any of a number of reasons, pay so that another nation can emit more. The right answer about emissions will not simply fall out of the right answer about costs, or vice versa.[3]

We will be trying to delineate a goal: a just pattern of allocation of something scarce and valuable, namely greenhouse-gas emissions capacities. However, a transition period during which the pattern of allocation does not satisfy the ultimate standard may well be necessary because of political or economic obstacles to an immediate switch away from the status quo. For instance, current emissions of CO_2 are very nearly as unequal as they could possibly be: a few rich countries with small populations are generating the vast bulk of the emissions, while the majority of humanity, living in poor countries with large populations, produce less altogether than the rich minority. It seems reasonable to assume that, whatever exactly will be the content of the standard of justice for allocating emissions, the emissions should be divided more equally than they currently are. Especially if the total cannot be allowed to keep rising, or must even be reduced, the *per capita* emissions of the rich few will have to decline so that the *per capita* emissions of the poor majority can rise.

Nevertheless, members of the rich minority who do not care about justice will almost certainly veto any change they consider too great an infringement upon their comfort and convenience, and they may well have the power and wealth to enforce their veto. The choice at that point for people who are committed to justice might be between vainly trying to resist an almost certainly irresistible veto and temporarily acquiescing in a far-from ideal but significant improvement over the status quo. In short, the question would be: which compromises, if any, are ethically tolerable? To answer this question responsibly, one needs guidelines for transitions as well as ultimate goals: not, however, guidelines for transitions instead of ultimate goals, but guidelines for transitions in addition to ultimate goals. For, one central consideration in judging what is presented as a transitional move in the direction of a certain goal is the extent of distance travelled toward the goal. The goal must have been specified in order for this assessment to be made.[4]

II. TWO KINDS OF ANSWERS

A principle of justice may specify to whom an allocation should go, from whom the allocation should come, or, most usefully, both. The distinction between the questions, from whom and to whom, would seem too obvious to be worth comment except that "theories" of justice actually tend in this regard to be only half-theories. They tend, that is, to devote almost all their attention to answering the question, to whom, and to fail to tackle the challenges to the firm specification of the sources for the recommended transfers. This is one legitimate complaint practical people tend to have against such "theories": "you have shown me it would be nice if so-and-so received more, but you have not told me who is to keep less for that purpose—I cannot assess your proposal until I have heard the other half."

Unfortunately, the answer to, from whom? does not flow automatically from all answers to, to whom? Often a given specification of the recipients of transfers leaves open a wide variety of possible allocations of the responsibility for making the transfers. For instance, if the principle governing the allocation of certain transfers were, "to those who had been severely injured by the pollution from the process," the potential sources of the transfers would include: those who were operating the process, the owners of the firm that authorized the process, the insurance company for the firm, the agency that was supposed to be regulating the process, society in general, only the direct beneficiaries of the process and no one else, and so on. Quite often proposals about justice are not so much wrong as too incomplete to be judged either right or wrong.

I have phrased the first four kinds of issues about justice, which arise from different aspects of the challenge of global warming, as, in effect, from-whom questions, precisely because this is the neglected side of the discussion of justice. What we are now noticing is simply that there is, in addition, always the question, to whom? It is more likely that, to whom? will have an obvious answer than it is that, from whom? will, but it is always necessary to check. If we are discussing the costs of coping, for example, it might seem obvious that from whomsoever the transfers should come, they should go to those having the most difficulty coping. However, if the specification of the sources of the transfers is, "those who caused the problem being coped with," then country A, which did in fact cause the problem in country X, might be expected to assist country X, and not country Y, even though country Y was having much more difficulty coping (but with problems that were not A's responsibility).

One vital point that this abstract example of countries A, X, and Y illustrates is: answers to, from whom? and answers to, to whom? are interconnected. Once one has an answer to one question or the other, certain answers to the remaining question are inappropriate and, sometimes, another answer to the remaining question becomes the only one that really makes any sense. Often, these logical connections are very helpful.

III. TWO KINDS OF GROUNDS

We saw, in Section I, that if one thinks hard enough about how the international community should respond to global warming, questions about justice arise unavoidably at four points, and in Section II we have just now observed that besides answering these more difficult questions about the bearers of responsibility who should be the sources of any necessary transfers, there is always in principle, and often in practice, a further question to answer in each case about the appropriate recipients of any transfers. It is helpful, finally, to notice that individual principles of justice for the assignment of responsibility fall into one or the other of two general kinds, which I will call fault-based principles and no-fault principles.

A well-known fault-based principle is: "the polluter pays"; and a widely accepted no-fault principle is "payment according to ability to pay." The principle of payment according to ability is pay is no-fault in the sense that alleged fault, putative guilt, and past misbehavior in general are all completely irrelevant to the assignment of responsibility to pay. Those with the most should pay at the highest rate, but this is not because they have done wrong in acquiring what they own, even if they have in fact done wrong. The basis for the assignment of progressive rates of contribution, which are the kind of rates that follow from the principle of payment according to ability to pay, is not how wealth was acquired but simply how much is held.

In contrast, the polluter-pays principle is based precisely upon fault or, anyhow, causal responsibility. "Why should I pay for the clean up?" "Because you created the problem that has to be cleaned up." The kind of fault invoked here need not be a moralized kind—the fault need not be construed as moral guilt so much as simply a useful barometer or symptom to be used to assign the burden of payment to the source of the need for the payment. That is, one need not, in order to rely upon this principle, believe that polluters are wicked or even unethical in some milder sense. The rationale for relying upon polluter-pays could, in particular, be an entirely amoral argument about incentives: the polluter should pay because this assignment of clean-up burdens creates the strongest disincentive to pollute. Even so, this would be a fault-based principle in my sense of "fault-based," which simply means that the inquiry into who should pay depends upon a factual inquiry into the origins of the problem. The moral responsibility for contributing to the solution of the problem is proportional to the causal responsibility for creating the problem. The pursuit of this proportionality can itself, in turn, have a moral basis (guilty parties deserve to pay) or an amoral basis (the best incentive structure makes polluters pay). The label "fault-based" has the disadvantage that it may sound as if it must have a moral basis, which it may or may not have, as well as having a moral implication about who ought to pay, which it definitely does have.

An alternative label, which avoids this possible moralistic misunderstanding of "fault-based," would have been to call this category of principles, not "fault-based," but "causal" or "historical," since such principles make the assignment of responsibility for payment depend upon an accurate understanding of how the problem in question arose. This, however, has the greater disadvantage of suggesting as the natural label for what I call "no-fault" principles, "acausal" or "ahistorical" principles. That would, I think, be more misleading still, because it would make the no-fault principles sound much more ethereal and oblivious to the facts than they are. "Payment according to ability to pay" does not call for an inquiry into the origins of the problem, but neither is it ahistorical or acausal. An historical analysis or a view about the dynamics of political economy might be a part of the rationale for an ability-to-pay principle, so it would be seriously misleading to label this principle "ahistorical" or "acausal" just because it does not depend upon a search for the villain in the not necessarily moralistic sense in which "fault-based" principles do depend upon identifying the villain, that is, in the sense of who produced the problem. So, I use "fault-based" for principles according to which the answer to, from whom? depends upon an inquiry into the question, by whom was this problem caused? and to "no-fault" for principles according to which, from whom? can be answered on grounds other than an analysis of the production of the problem.

Principles for arriving at the second kind of answer noted in section II, to whom transfers should be made, also fall into the general categories of fault-based and no-fault. The principle, “make the victims whole,” is ultimately fault-based in that the rightful recipients of required transfers are identified as specifically those who suffered from the faulty behavior on the basis of which it will be decided from whom the transfers should come: on this principle, the transfers should come from those who caused the injury or harm and go to those who suffered the injury or harm. Indeed, one of the great advantages of fault-based principles is precisely that their cause/effect structure provides complementary answers to both questions: transfers go to those negatively affected, from those who negatively affected them. This specific principle, “make the victims whole,” embodies a perfectly ordinary view—and an especially clear one, since it also partly answers yet another question, how much should be transferred, by indicating that the transfer should be at least enough to restore the victims to their condition prior to the infliction of the harm. The victims (to whom) are to be “made whole” (how much—minimum amount, anyway) by those who left them less than whole (from whom). This principle does not completely answer the question of how much because it leaves open the option that the victims are entitled to more than enough merely to restore them to their condition *ex ante*, that is, it leaves open the possibility of additional compensation.

An ordinary example of a kind of no-fault principle for answering to whom an allocation should go is: “maintain an adequate minimum.” Naturally, the level of what was claimed to be the minimum would have to be specified and defended for this to be a usable concrete version of this kind of principle. It has the general advantage of all no-fault principles, however, in that no inquiry needs to be conducted into who was in fact injured, who injured them, how much they were injured, and to what extent their problems had other sources, and so forth. Transfers go to those below the minimum until they reach the minimum; then something else happens (for example, they are retrained for available jobs). Quite a bit of information is still needed to use such a no-fault principle, both to justify the original specification of the minimum level and to select those who are in fact below it. Yet this information is of different types from the information needed to apply a fault-based principle: one does not need an understanding of possibly highly complex systems of causal interactions and positive and negative feedbacks and/or lengthy chains of historical connections among potentially vast numbers of agents and multiple levels of analysis. The information needed to apply no-fault principles tends to be contemporaneous information about current functioning, which is often easier to obtain than the convincing analysis of fault needed for the use of a fault-based principle.

The evident disadvantage of a no-fault principle for specifying to whom transfers go is that it lacks the kind of naturally complementary identification of from whom the transfers should come that flows from the cause/effect structure of fault-based principles. In particular, it does not imply that the transfers should come from whomever caused those who are below the minimum to be below the minimum; in fact, it does not even assume that there is any clear answer or, for that matter, any meaningful question of the form, who caused those below the minimum to be there? The consequence of the absence of the convenient complementary answers implied by fault-based principles is that with no-fault principles the answers to the to-whom question and the from-whom question must be argued for separately, not by a single argument like arguments about fault. It might be, for example, that if the answer to the question, to whom? is, those below the minimum, the answer to the question, from whom, may be: those with the greatest ability to pay. The point, however, is that the argument for using ability-to-pay to answer the one question and the argument for using maintenance-of-a-minimum to answer the other question have to be two separate arguments.

So, this is a framework for facing unavoidable questions about justice. All that remains are the specific answers to the questions and the arguments to justify them!

* This chapter is excerpted from a longer analysis that first appeared in *Law and Policy*.

NOTES

[1] For a fuller discussion, see Shue (1992).

[2] Less obvious, that is, than the issue whether the poor should have to sacrifice their own economic development so that the rich can maintain all their accustomed affluence. As already indicated, if someone honestly thought this demand could be fair, we would belong to such different worlds that I do not know what I could appeal to that we might have in common.

[3] For a provocative proposal, see Agarwal and Narain (1991).

[4] For a serious attempt to grapple with transitional formulae, see Grubb and Sebenius (1992).

LITERATURE CITED

Agarwal, A. and S. Narain. 1991. *Global Warming in an Unequal World.* Centre for Science and Environment: New Delhi.

Grubb, M. and J.K. Sebenius. 1992. Participation, allocation and adaptability in international tradeable emission permit systems for greenhouse gas control. In *Climate Change: Designing A Tradeable Permit System.* Chap. 9. Organisation for Economic Co-operation and Development: Paris.

Intergovernmental Panel on Climate Change (IPCC). 1990. *Climate Change: The IPCC Scientific Assessment.* eds. J.T. Houghton, G.J. Jenkins, and J.J. Ephraums. Cambridge University Press: Cambridge.

International Environment Reporter. 1992. Top Environmental Official Welcomes Summit Aid Pledges from Developed Nations–China. *Current Reports.* 15(13):444, July 1, 1992.

Shue, H. 1992. The unavoidability of justice. In *The International Politics of the Environment.* eds. A. Hurrell and B. Kingsbury, Chap. 14. Oxford University Press: New York.

12

STATUS and ISSUES CONCERNING AGRICULTURAL EMISSIONS of GREENHOUSE GASES

John M. Duxbury
Dept. of Soil Crop and Atmospheric Sciences, Cornell University

Arvin R. Mosier
USDA – ARS, Fort Collins, Colorado

In this chapter we evaluate the contribution of agriculture to radiative forcing of climate through emissions of greenhouse gases associated with land clearing and agricultural production. We choose to combine both activities under the aegis of "agriculture," although it could be argued that some land clearing is for purposes of human settlement rather than for agriculture *per se.*

For the most part, we limit our discussion to three greenhouse gases, namely carbon dioxide (CO_2), methane (CH_4), and nitrous oxide (N_2O). We only briefly mention other greenhouse gases produced or affected by agriculture because of uncertainties in emission estimates and/or in evaluation of their radiative forcing effects. We first discuss global budgets of the three gases and agricultural contributions to them, then we make comparisons of the relative effects of energy and agriculture to radiative forcing at both global and country scales. Finally we present our views on selected "issues" related to the assessment and mitigation of greenhouse gas emissions.

I. GLOBAL ANNUAL BUDGETS OF CO_2, CH_4, AND N_2O

A. Carbon Dioxide

The global concentration of CO_2 in the atmosphere is currently about 357 parts per million by volume (ppmv) and is increasing at a rate of 1.8 ppmv yr^{-1} (Watson et al., 1992). The increasing atmospheric accumulation of CO_2 is the balance between anthropogenic emissions, primarily caused by fossil fuel combustion and deforestation in the tropics, and uptake by terrestrial and oceanic sinks. The present rate of atmospheric CO_2 increase is, however, about half that expected from estimates of source and sink strengths. This is perhaps not surprising since natural terrestrial and oceanic processes dominate the CO_2 budget (Figure 12-1), accounting for 97% or more of all sources and sinks. Relatively small changes in natural CO_2 emission or uptake rates can therefore have a significant effect on atmospheric loading. Several researchers agree that there is an unidentified biospheric sink for CO_2 in the northern temperate zone (Keeling et al., 1989a, 1989b; Tans et al., 1990; Enting and Mansbridge, 1991) but disagree on its allocation between terrestrial and oceanic biospheres. Enhanced productivity in the terrestrial biosphere, due to the increased level of atmospheric CO_2 (often referred to as CO_2 fertilization) and/or increased atmospheric nitrogen (N) deposition, and the strong carbon storage capacity of immature forests are the most likely driving forces behind increased sink strength (Watson et al., 1992).

There is a longitudinal gradient in CO_2 concentration of about 3 ppmv from the South Pole to the North Pole, indicating a continuous flux of CO_2 from the Northern Hemisphere where about 90% of fossil fuel emissions occur. Isotopic analysis of atmospheric CO_2 (on a carbon content basis) shows that the terrestrial biosphere is largely responsible for the seasonal changes of up to 10 ppmv observed in the Northern Hemisphere.

B. Methane

The current global average atmospheric concentration of CH_4 is 1720 parts per billion by volume (ppbv), more than double its pre-industrial value of 800 ppbv. The concentration of CH_4 in the Northern Hemisphere is about 100 ppbv more than in the Southern Hemisphere, indicating either greater source or lower sink strength in the Northern Hemisphere (Watson et al., 1992). The rate of increase in atmospheric CH_4 concentration has slowed from about 20 ppbv yr^{-1} in the 1970s (Blake and Rowland, 1988) to about 10 ppbv yr^{-1} currently (Steele et al., 1992), but the reason for this is uncertain (Watson et al., 1992).

About 70% of CH_4 production arises from anthropogenic sources and about 30% from natural sources (Figure 12-1). Agriculture is considered to be responsible for about two-thirds of the anthropogenic sources (Duxbury et al., 1993). Biological generation in anaerobic environments (natural and man made wetlands, enteric fermentation, and anaerobic waste processing) is the major source of CH_4, although losses associated with coal and natural gas industries are also significant. The primary sink for CH_4 is reaction with hydroxyl (OH) radicals in the troposphere (Crutzen, 1991; Fung et al., 1991), but small soil (Steudler et al., 1989; Whalen and Reeburgh, 1990; Mosier et

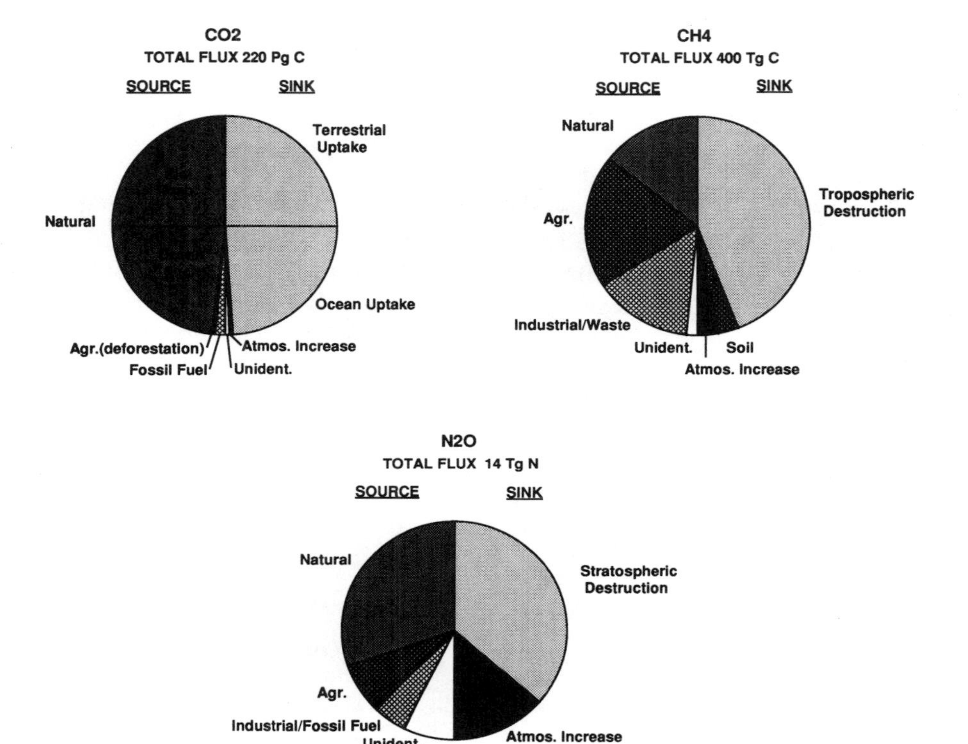

Figure 12-1. Global annual source-sink budgets for carbon dioxide, methane, and nitrous oxide.

al., 1991) and stratospheric (Crutzen, 1991) sinks , equivalent to about 10 and 2.5%, respectively, of the tropospheric sink, have also been identified.

C. Nitrous Oxide

The present global average atmospheric concentration of N_2O is about 310 ppbv and it is increasing at a rate of 0.6–0.9 ppbv yr^{-1} (Prinn et al., 1990; Watson et al., 1992). The concentration of N_2O is about 0.75 ppbv higher in the Northern Hemisphere than in the Southern Hemisphere (Prinn et al., 1990), indicating greater source strength in the former. It is generally agreed that soils are the major source of N_2O but global N_2O budgeting exercises (Figure 12-1) suggest that the strength of known sources is underestimated or that unidentified sources exist (Duxbury et al., 1993; Robertson, 1993).

Analysis of the latitudinal distribution of atmospheric N_2O suggests that emissions of N_2O between 90–30N, 30N–equator, equator–30S, and 30–90S are 22–24%, 32–29%, 20–29%, and 11–15%, respectively, of the global total, and that there is a large tropical source (Prinn et al., 1990). The somewhat different estimate of Bouwman (1992) that these same latitudes contribute 32%, 31%, 29%, and 9%, respectively, to global N_2O production, reinforces the conclusion that our knowledge of N_2O sources is incomplete.

The only known significant sink for N_2O is photolysis in the stratosphere (Watson et al., 1990). Anaerobic soils have large potentials for reduction of N_2O to N_2 (Erich et al., 1984) and, in fact, the major product of denitrification in soils is usually N_2 rather than N_2O. However, slow rates of dissolution of atmospheric N_2O and its transport in wet and/or flooded soils prevent this process from being a significant regulator of atmospheric N_2O.

II. SIGNIFICANCE OF GREENHOUSE GAS EMISSIONS FROM AGRICULTURE

A. Contributions to Anthropogenic Fluxes of CO_2, CH_4, N_2O and Other Trace Gases

Agriculture presently contributes about 21–25%, 60%, and 65–80% of total anthropogenic emissions of CO_2, CH_4, and N_2O, respectively (Duxbury et al., 1993; Isermann, 1992; Watson et al., 1992). Agriculture is also thought to be responsible for over 95% of the ammonia (NH_3), 50% of the carbon monoxide (CO), and 35% of the nitrogen oxides (NO_X) released into the atmosphere as a result of human activities (Isermann, 1992).

i. Carbon Dioxide

Although fossil fuel consumption is presently the major source of anthropogenic CO_2 (generating 6 Gt CO_2-C yr^{-1}), cumulative releases of CO_2 over the last 160 years are higher from land conversion to agriculture (264 Gt) than from fossil fuel sources (168 Gt) (Figure 12-2). It was not until the early 1950s that annual CO_2 emissions from fossil fuels exceeded those from land clearing activities. Prior to 1900, conversion of land to agriculture occurred primarily in Western Europe, the former USSR, and North America. However, some reforestation of land in the Northeast U.S. began in the late 1800s, coincident with the shift from horses to the internal combustion engine as the main source of power for farming and transportation and hence reduced need of land for pasture and production of feed. The dynamic of forest cover in Connecticut (Figure 12-3) is illustrative of these changes. Other factors that contributed to reforestation in the Northeast U.S. include

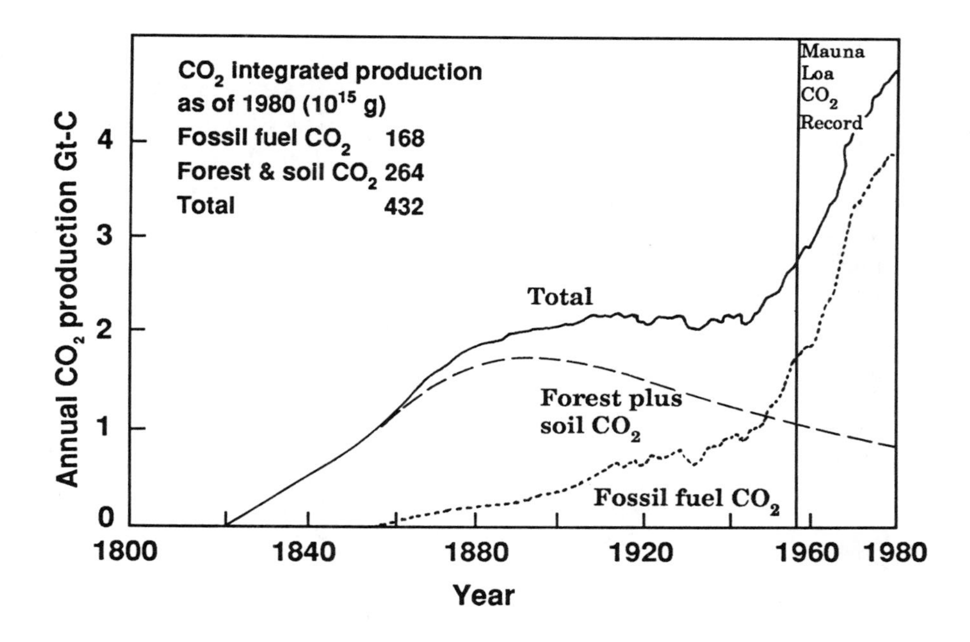

Figure 12-2. Carbon dioxide emissions associated with land clearing activities and fossil fuel combustion between 1820 and 1980.

utilization of lands better suited for agriculture in the Midwest and, beginning in the 1950s, dramatic increases in crop yields associated with the development of high yielding cultivars of the major grain crops and widespread use of fertilizers and pesticides.

Large scale land clearing is now occurring only in the tropics and associated CO_2 emissions are estimated to be between 1.1 to 3.6 Gt C yr^{-1} (Houghton, 1991). This rather wide range results from uncertainties in rates of forest clearing, the fate of deforested lands (permanent or temporary clearing), and forest carbon stocks. For example, recent studies have shown that deforestation in the Brazilian Amazon in the 1980s was overestimated by a factor of at least 2 (see Watson et al., 1992). Although several aspects of the deforestation story have recently been modified, the IPCC (Watson et al., 1992) believes that its 1990 estimate of 1.6 Gt C yr^{-1} for CO_2-C release from deforestation (Watson et al., 1990) is still valid; this value is about 21% of present anthropogenic CO_2 emissions.

Carbon dioxide emissions associated with the manufacture of agricultural equipment and agrochemicals and with fuel use on farms is less than 3% of that generated by other uses of fossil fuel (CAST, 1992; Isermann, 1992).

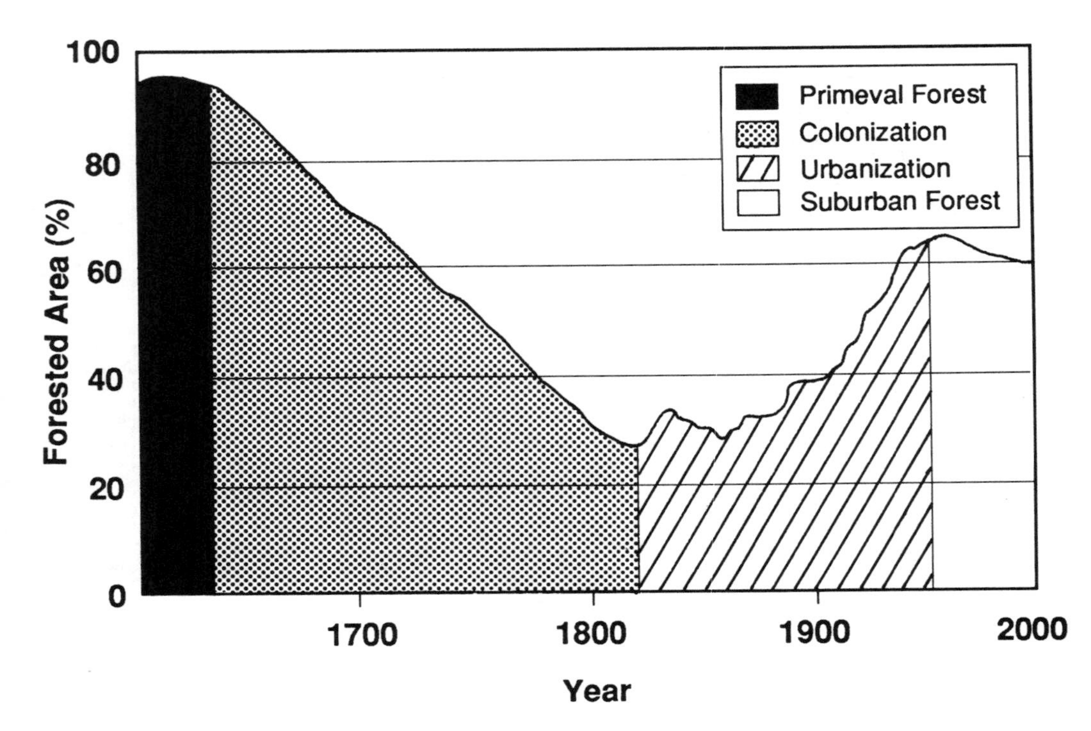

Figure 12-3. Changes in forest cover in Connecticut, U.S. (from CAST, 1992).

ii. Methane

The release of an estimated 205–245 Tg CH_4 yr^{-1} from agricultural sources is derived from enteric fermentation in ruminant animals (80 Tg), paddy rice production (60–100 Tg), biomass burning (40 Tg), and animal wastes (25 Tg) (Watson et al., 1992). Recent research in India (Parashar et al., 1991, 1992) caused the IPCC to lower its estimate of CH_4 emissions from paddy rice from 100 to 60 Tg CH_4 yr^{-1} (Watson et al., 1992), but such a change is speculative given the limited number of measurements made in these studies. Further measurements under actual production conditions in Asia are needed to improve estimates of CH_4 emissions from flooded rice.

The soil sink strength for CH_4 appears to have been reduced by changes in land use, chronic deposition of N from the atmosphere, and alterations in N dynamics of agricultural soils (Steudler et al., 1989; Keller et al., 1990; Scharffe et al., 1990; Mosier et al., 1991). Ojima et al. (1993) estimate that the consumption of atmospheric CH_4 by soils of temperate forest and grassland ecosystems has been reduced by 30%. On the other hand, Summerfeld et al.

(1993) recently showed that snow covered soil continued to oxidize CH_4 and suggested that the importance of soil consumption may have been underestimated. Without the temperate soil sink for CH_4, the atmospheric concentration of CH_4 would be increasing at about 1.5 times the current rate; consequently it would seem prudent to investigate the global significance of this effect.

iii. Nitrous Oxide

The budget for N_2O presented by IPCC (Watson et al., 1992) is poorly constructed and incomplete. It does not include any value for grasslands although the global area of this biome is almost as great as that of forest land (3.1×10^9 ha vs 4.0×10^9 ha). Using this area and N_2O flux data from Parton et al. (1988), significant emissions of N_2O may be ascribed to grasslands (Duxbury et al., 1993). Additionally, no contribution was included for increased N mineralization rates and biological N fixation (BNF) in agricultural systems; by analogy to experience in the tropical environment (Matson and Vitousek, 1987), increased N cycling should lead to higher N_2O emissions in agricultural systems compared to the natural ecosystems they replaced.

It is also likely that N_2O production resulting from fertilizer and increased use of BNF is underestimated because the effect of a nitrogen input is usually only partially traced through the environment. Figure 12-4, taken from Duxbury et al. (1993), illustrates some of the flows of N following application of 100 kg ha^{-1} of fertilizer N to a field on a typical dairy farm in the U.S. Primary and secondary flows of N are shown by dashed and solid lines, respectively. In this example, 50 of the 100 kg are harvested in the crop and 50 are lost by the combination of leaching (25), surface run-off (5), and volatilization (20, primarily denitrification). If N_2O comprises 10% of the volatilized N, 2 kg N_2O-N would be generated in the primary cycle. Assessments of fertilizer effects on N_2O emissions usually stop at this point even though only 20 of the 100 kg N have been returned to the atmosphere and it has been estimated that 80% of the N would be returned within a decade (McElroy et al., 1977).

Secondary flows shown by the solid lines include feeding of the 50 kg of harvested N to animals, which generate 45 kg of manure N. The manure is returned to cropland to fertilize a second crop, however about half of this N is volatilized as NH_3 prior to or during manure application. Volatilized NH_3 is aerially dispersed and subsequently returned to and cycled through both natural ecosystems and cropland. Ammonia volatilization from agricultural systems is globally important (Isermann, 1992) but its impact on N_2O emissions has not been explicitly addressed. To provide some perspective, it should be noted that the quantities of fertilizer N used and animal manure N generated by U.S. agriculture are equal (Bouldin et al., 1984). On a global scale, we estimate that 30 of the 80 Tg fertilizer N used each year are volatilized as ammonia (NH_3).

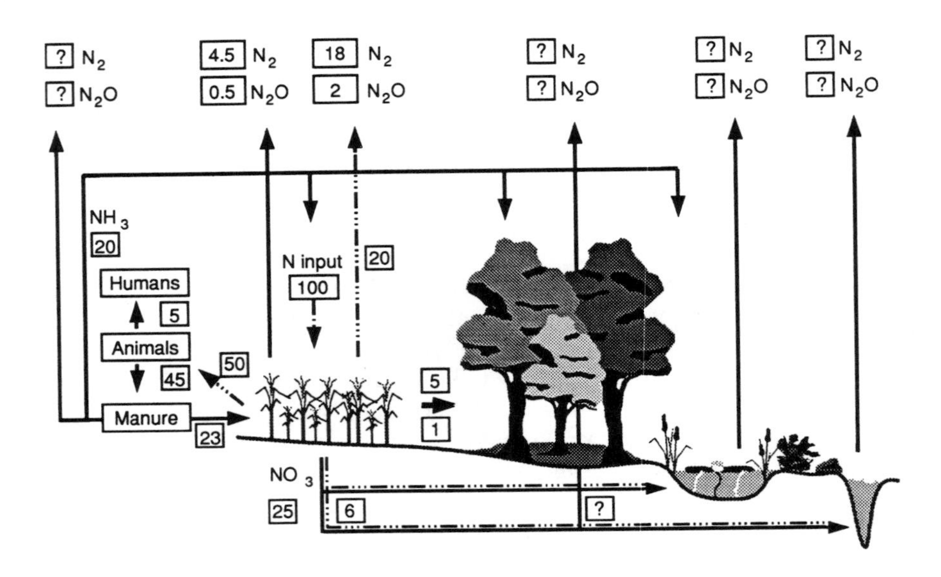

Figure 12-4. Fate of fertilizer N applied to a dairy farm in the U.S. (from Duxbury et al., 1993).

Similarly, the amount of N_2O arising from leached nitrate (NO_3^-), which may average 20–25% of applied N (Hauck, 1984), is not known but much may be denitrified in riparian zones or cycled through wetland or aquatic vegetation. A complete accounting of fertilizer N, biologically fixed N, and N mineralized from soil organic matter is difficult to achieve, but necessary if we are to accurately assess the impact of increased use of N in agricultural ecosystems on terrestrial N_2O emissions (Duxbury et al., 1993).

iv. Other Trace Gases and Aerosols

While the contributions of industrial and agricultural activities to rising atmospheric levels of CO_2, CH_4, and N_2O are documented, the quantitative impacts of other nitrogen gases (including NO_x and NH_3), sulfur gases (including SO_2, COS, and CS_2), and ozone (O_3) are less certain. The extent and impact of changing O_3 levels in both the stratosphere and the troposphere are difficult to evaluate. Reductions in stratospheric O_3 now occur routinely over the Antarctic every spring and may be becoming more widespread. These reductions change processes that have opposing effects on global climate.

Tropospheric O_3 levels are controlled by a complex set of reactions and feedbacks that are also temporally and spatially variable. Ozone is generated in the troposphere by a series of reactions involving photo-oxidation of CO,

CH_4, and non-methane hydrocarbons in the presence of NO_x, a common anthropogenic pollutant. Reactant concentrations and reaction outcomes are variable; at NO_x concentrations less than about 10 parts per trillion by volume (pptv), the net result is a decrease in tropospheric O_3 and OH radicals, whereas production of O_3 and OH radicals occurs at higher NO_x levels. Additionally, feedbacks between O_3, its precursor gases, and OH radicals introduce self-regulation of both O_3 and CH_4. Elevated levels of O_3 and NO_x enhance OH radical production which in turn oxidizes CO and CH_4 with consequent reduction in O_3 generation (Duxbury et al., 1993). Current evidence indicates that NO_x emissions, primarily from nitrification in soils, are large enough to influence local and regional atmospheric chemistry (Williams et al., 1992). Nitrogen fertilized soils seem to be a major source for NO_x, and emissions from agricultural fields may dominate NO_x production in agricultural regions during the cropping season (Williams et al., 1992).

B. Contributions of Agriculture to the Anthropogenic Greenhouse Effect

i. Global Warming Potential Indices

Evaluation of the contributions of different greenhouse gases to the overall greenhouse effect is generally done using their *global warming potential* (GWP) indices. The GWP index combines the capacity of a gas to absorb infrared radiation and its residence time in the atmosphere with a time frame of analysis, then expresses the result relative to CO_2. A GWP index should also consider any indirect effects of a gas that arise due to its atmospheric chemistry, including effects upon and conversion to other greenhouse gases. The most recent IPCC analysis (Isaken et al., 1992) has, however, withdrawn earlier quantitative estimates of indirect GWPs (Shine et al., 1990) citing insufficient understanding of atmospheric chemical processes.

Other factors that need to be considered when formulating GWPs include a reduced, non-linear greenhouse response to future increases in atmospheric CO_2 (which will increase the importance of other greenhouse gases relative to CO_2) (Isaken et al., 1992), overlapping absorption bands of different greenhouse gases, and possible changes in the residence time of gases as atmospheric chemistry continues to be altered (Shine et al., 1990).

Many of the features that complicate assignment of a GWP value apply to CH_4; for example, CH_4 is oxidized to CO and then to CO_2 through atmospheric chemical processes that may also lead to the formation of tropospheric O_3 (also a greenhouse gas) and OH radicals (which react with CH_4). Furthermore, continuing increases in CH_4 emissions may cause its atmospheric residence time to lengthen because the present annual production of tropospheric OH radicals is almost all consumed by reaction with CH_4 and CO (Watson et al.,

Table 12-1. Global warming potential (GWP) of greenhouse gases.

Gas	Lifetime	Relative radiative	Direct	GWP	Indirect effects
	yr	Forcing/kg	20 yr	100 yr	
CO_2	120	1	1	1	None
CH_4	10.5	58	35	11	Positive, magnitude similar to direct GWP?
N_20	132	206	260	270	Uncertain
CFC 12	116	5750	7100	7100	Negative

Source: Watson et al. (1990, 1992)

1990).

Direct GWP values for selected greenhouse gases for 20 and 100 year time frames are given in Table 12-1. The relative GWPs of CO_2, N_2O, and CFC 12 change little with time because they have similar atmospheric residence times. In contrast, the 100 yr GWP of CH_4 is one-third of its 20 year GWP because of its shorter atmospheric residence time.

ii. Global GWPs Created by Agricultural and Energy Sectors

A comparison of GWPs created by anthropogenic emissions of CO_2, CH_4, and N_2O from global agricultural and energy sectors in 1990 is shown in Table 12-2. Several conclusions can be drawn from this data:

- agriculture and energy sectors dominate both anthropogenic emissions and GWPs of these gases, together accounting for essentially all of the CO_2, 74% of the CH_4, and 79% of the N_2O derived from anthropogenic sources, and 92% and 95% of the total anthropogenic GWPs for these gases for 20 and 100 year time frames of analysis, respectively.
- agriculture contributes between one-quarter and one-third of the anthropogenic GWP attributable to these sources, depending on the time frame of the analysis.
- CO_2 emissions are the major contributor to the total GWP, accounting for 67% and 84% of the 20 and 100 year GWPs, respectively.
- N_2O emissions contribute only 3 to 4% to the calculated GWPs.

If it is assumed that the indirect effects of CH_4 are equal to its direct effect and that there are no indirect effects of N_2O, then total GWP will increase and the contributions of agriculture and energy sectors change to 35 and 54%, respectively, of the new 20 year GWP and to 29 and 65%, respectively, of the new 100 year GWP. This increases the contribution of agriculture from 26–31% to 29–35%.

Table 12-2. Comparison of global anthropogenic GWP created by agriculture (A) and energy sectors (E) for 1990 estimated emissions of greenhouse gases.

Gas	Anthropogenic emissions[a]			Share of GWP[b]			
				20 yr		100 yr	
	Total	A	E	A	E	A	E
	Tg	— % of total anthropogenic —					
CO_2	28600	23	77	15	52	19	65
CH_4	360	46	28	14	8	5	3
N_2O	5	61	18	2	<1	2	<1
			Total	31	61	26	69

[a] From Watson et al. (1992)
[b] Using direct GWP only

Several interesting comparisons have been made between the GWP created by a dairy cow and familiar examples of energy use by humans. Isermann (1992) estimated that the GWP of 3 dairy cows is equivalent to that of 2 automobiles. However, Drennen and Chapman (1992) calculated a much lower effect. They found that the total greenhouse effect of a dairy cow is equivalent to one 75 W incandescent light bulb operated continuously. These differences are due to the wide range of values for CH_4 in different GWP indices. Further, Drennen and Chapman made an allowance for the carbon content of the feed consumed by the dairy cow.

iii. Agricultural Contributions to Anthropogenic GWPs In The U.S., India, and Brazil

Agriculture contributes less than 5% of the present annual anthropogenic GWP created by the U.S. (CAST, 1992). Similar analyses for India and Brazil (and other countries less developed than the U.S.) give different results. Present annual emissions of CO_2 and CH_4 associated with energy and agricultural sectors in the U.S., India, and Brazil are shown in Table 12-3, and comparisons of the direct GWP created in these countries are shown in Figure 12-5. Nitrous oxide is not included because it does not contribute significantly to the GWPs.

Depending on the time frame of analysis (20 or 100 years), the total anthropogenic GWP created by the U.S. is three to five times that created by India and six to seven times that created by Brazil (Figure 12-5a). In contrast to the U.S., the agricultural sectors in India and Brazil are major contributors to country total GWPs, accounting for 40–70% and 70–80% of the GWP generated in India and Brazil, respectively (Figure 12-5a). Furthermore, the agricultural sector in India generates two to four times the GWP created by

Table 12-3. Annual emissions of carbon dioxide and methane associated with agriculture (A) and energy (E) sectors in the U.S., India, and Brazil.

	CO_2		CH_4			
Country	A[a]	E[b]	Livestock[c]	Manure[d]	Rice[e]	E[f]
	Tg yr^{-1}					
U.S.	120	4800	7.0	3.9	0.4	8.6
India	NA	600	10.3	1.6	29.0	NA
Brazil	450	200	7.5	1.1	0.0	NA

[a] U.S. data from CAST (1992); direct energy consumption in agriculture is not available for India and Brazil. Brazil value is for deforestation (Houghton, 1991)

[b] From National Academy of Sciences (1991)

[c] From Lerner et al. (1988)

[d] From Safley et al. (1992)

[e] U.S. data from Hogan (1992) and value for India is based on flux being proportional to production, and a global flux of 100 Tg yr^{-1}

[f] From Hogan (1992)

agriculture in the U.S., while agriculture in Brazil generates about twice the GWP of US agriculture (Figure 12-5a). Doubling the direct GWP value for CH_4 to include a likely maximum contribution from its indirect effects has little impact on this result. The relative contributions of US and Indian agriculture are proportional to population in these countries, whereas Brazilian agriculture generates almost 4 times that of US agriculture on a *per capita* basis. The disproportionately large *per capita* Brazilian contribution is due to current land conversion activities. Similar land conversion effects occurred at an earlier time in both the U.S. and India and are not assessed against current agriculture in our analysis.

The contributions of CO_2 and CH_4 to the GWPs created by energy and agriculture in the U.S., India, and Brazil are shown if Figure 12-5b. Carbon dioxide from fossil fuel combustion dominates the U.S. GWP and agricultural and energy sectors contribute about equally to that associated with CH_4. The 20 year GWP in India is dominated by CH_4 from rice paddies (70%) and ruminant animals (30%), but energy and agriculture are about equal contributors to the 100 year GWP. Carbon dioxide is the major source of Brazil's GWP, and two-thirds of this is associated with deforestation for agricultural development. It should be noted, however, that Brazilian agriculture could be given a CO_2 credit for its production of ethanol for fuel. In 1990, ethanol production was the equivalent of 18 $x10^6$ t of petroleum or 55 Tg CO_2, which would offset 12% of the CO_2 estimated to be generated by deforestation or 28% of that from fossil fuel combustion.

This analysis shows that reduction in energy use in the U.S. would be the most effective strategy for mitigation of GWP created by these countries. Reduction in emissions of CH_4 from rice paddies and ruminant animals in India

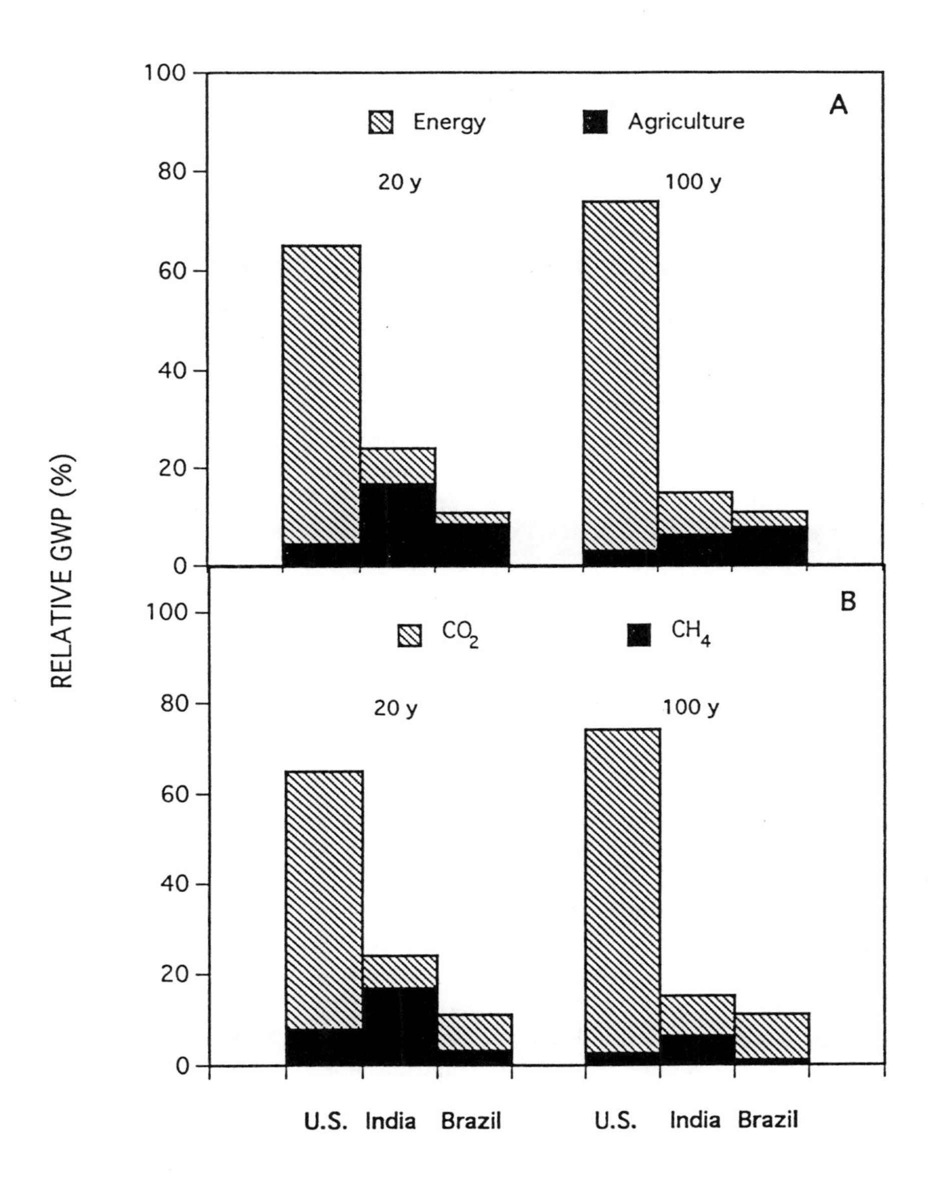

Figure 12-5. Global warming potentials created by a) agriculture and energy sectors and b) carbon dioxide and methane in the U.S., India, and Brazil.

would yield small benefits, but reducing deforestation in Brazil would have little impact. This is not to advocate deforestation in the tropics; it is still a significant contributor to the global anthropogenic GWP. Moreover, current deforestation in the tropics is generally associated with settlement by poor people which leads to an extensive, low productivity agriculture that is not the best option for either food production or conservation of biodiversity. It also often leads to land degradation and environmental pollution.

iv. Projected Future Greenhouse Gas Emissions

Projected future global and agricultural emissions of CO_2, CH_4, and N_2O according to the IPCC IS92a scenario (Leggett et al., 1992) are shown in Figure 12-6a and b, respectively. The IPCC IS92a scenario, which is a low to moderate economic growth scenario (2.3–2.9% yr^{-1}), includes an increase in the world's population from 5.2 billion in 1990 to 11.3 billion in 2100, a phase out of CFCs, and controls on SO_x, NO_x, and non-methane volatile hydrocarbons. Under this scenario, CO_2 and CH_4 emissions are expected to increase more or less linearly with time and to parallel population growth after 2025, whereas N_2O is expected to increase until 2025 then stabilize (Figure 12-6a).

Agriculturally related emissions of CO_2, CH_4, and N_2O (Figure 12-6b) are projected to follow different patterns than do total emissions. Emissions of CO_2 are expected to decline as rates of deforestation and land conversion to agriculture decrease, whereas CH_4 and N_2O emissions continue to increase over the next 60 years before stabilizing. The relative contribution of agriculture to future emissions of CH_4 and N_2O is expected to remain constant, whereas agriculture is expected to be a progressively less important source of CO_2 (Figure 12-6b). Increases in agricultural sources of CH_4 are mostly due to the projection that there will be a rapid increase in consumption of meat and dairy products, and therefore much greater contributions from enteric fermentation and associated animal wastes than there is at present. This projection assumes constant emissions at the present rate per unit of production, ignoring probable improvements in production methodologies and potential reductions in CH_4 due to altered animal diets (Leng, 1991; CAST, 1992). Moreover, land and feed constraints not reflected in these scenarios may prevent the suggested growth in animal production from being achieved.

Increases in emissions of N_2O follow fertilizer use, which is expected to more than double by 2100. Fertilizer N use in developed countries has essentially been level since the early 1980s while use in developing countries is increasing (FAO, 1992) and will likely continue to do so. The impact of changing fertilization practices and the dependency of N_2O emissions on local soil types, moisture, and agricultural practices was not considered in the IPCC analysis.

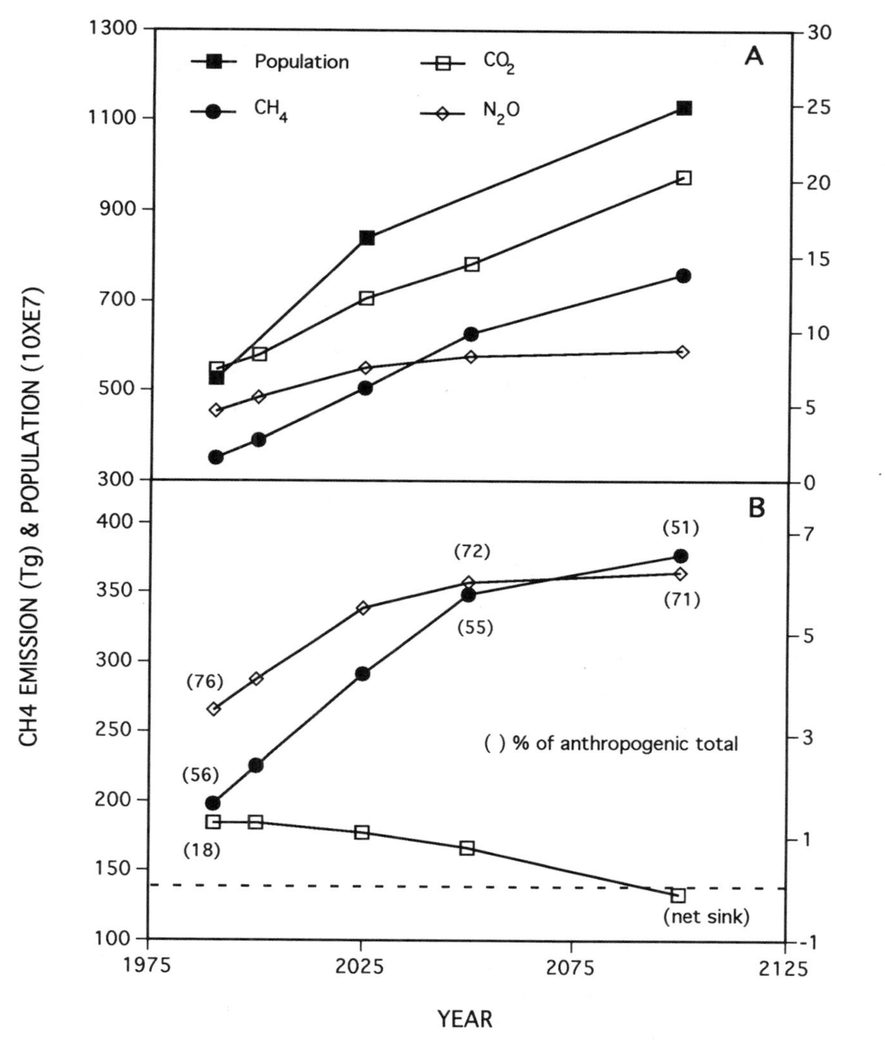

Figure 12-6. Projected future global growth in population and anthropogenic emissions of carbon dioxide, methane, and nitrous oxide: a) total and b) agricultural for IPCC scenario IS92a (constructed from data in Leggett et al., 1992).

The IPCC scenarios (Leggett et al., 1992) do not consider any impact of climate change on the demographics of agriculture. If, for example, climate change were to cause agriculture to move north into zones which have large forest biomass and soil carbon reserves, a whole new cycle of CO_2 release would occur. It should also be recognized that without climate change the potential for future land development is limited. In Asia, an estimated 82% of potential cropland is already under cultivation. There are large reserves of potential cropland in Latin America and sub-Sahara Africa, but the soil is marginal on much of this land, or rainfall is unreliable. Oxisols are the best soils for agricultural development in the tropics. These soils have chemical restrictions, but they can easily be overcome. However, the present development of land in the tropics is frequently driven by political decisions related to poor people, rather than by sound agricultural policy. An unfortunate consequence is the deforestation of marginal lands that are ecologically fragile. Clearly, the development of a land extensive, low productivity agriculture is illogical in terms of food production, storage of carbon, biodiversity, and other environmental issues. Instead, an intensive, highly productive agriculture using as little land as possible should be encouraged.

III. SELECTED ISSUES REGARDING GREENHOUSE GAS EMISSIONS AND AGRICULTURE

In this section we discuss a few key aspects concerning greenhouse gas emissions from agriculture. These are

- reliability of methods of measuring trace gas fluxes with reference to estimates of global fluxes.
- assessment of baseline contributions and interactions between greenhouse gases.
- feasibility of sequestering carbon in forests and soils.
- choices of mitigation options; energy or agriculture?

A. Accuracy of Measurement of Agricultural Sources of Greenhouse Gases

i. Methods of Flux Measurement

It would be preferable to measure fluxes of trace gases to and from soil surfaces without perturbation of natural concentration gradients. Until recently, this has proved to be impossible for many trace gases, including CH_4 and N_2O, because analytical methods were either not sufficiently sensitive or detector response times were too slow. Consequently, soil cover or chamber techniques, which concentrate gases but disturb the soil-plant-atmosphere continuum, have routinely been used.

Numerous studies have shown that CH_4 and N_2O fluxes are extremely variable in both space and time at the scale of measurement of chamber methods, which is generally up to $1m^2$ in space and 1 hour in time. Coefficients of variation for simultaneous measurements from replicate chambers range from about 30% to as high as 500% of the mean value (for examples see Mosier, 1990). While variability may be reduced by comparing cumulative values from chambers at fixed sites (Duxbury et al., 1986), it appears that flux estimates using chamber methods will always have a wide range.

Analytical methodology has now progressed to the point that field scale measurement of CH_4 and N_2O fluxes can be made without disturbing the system. Thurtell et al. (personal communication) have developed a fast response gas monitor, based on tunable diode laser (TDL) infrared absorption, for measuring fluxes of both gases using micrometeorological techniques. They have extensively field tested the TDL gas monitors and commercial systems will soon be available. However, these are expensive and their use is limited to sites that meet specifications for the micrometeorological method. The advantage of the micrometeorological approach is that it integrates fluxes over a distance of 100 m or so, but continuous monitoring may still be necessary to include temporal variability. Moreover, since a wide range of field variables control fluxes, improved field measurements may only provide a small gain in prediction capability.

ii. Scaling and Extrapolation to Global Flux Estimates

Figure 12-7, modified from Stewart et al. (1989), illustrates the time and space scales that can be addressed by various methods of measuring trace gas fluxes. Most regional and global trace gas budgets of CH_4 and N_2O have been extrapolated from chamber measurements (Matson et al., 1989). This is usually done by simple linear methods, coupling representative or stratified flux data with the amount of surface to which the data apply. Unfortunately, extrapolation of flux data has generally been done without considering how well spatial and temporal variability have been represented in individual studies. Such an approach is highly susceptible to bias and could lead to significant errors (Hicks, 1989).

A major problem, then, in constructing global budgets is how to estimate large-scale emission rates when the available data and measurement methods are limited in their applicability. One possibility for systematic extrapolation would be to make flux measurements by chamber and/or tower techniques according to some stratification of landscapes, such as across vegetation or soil types, for which geographic data bases exist or can be generated (preferably

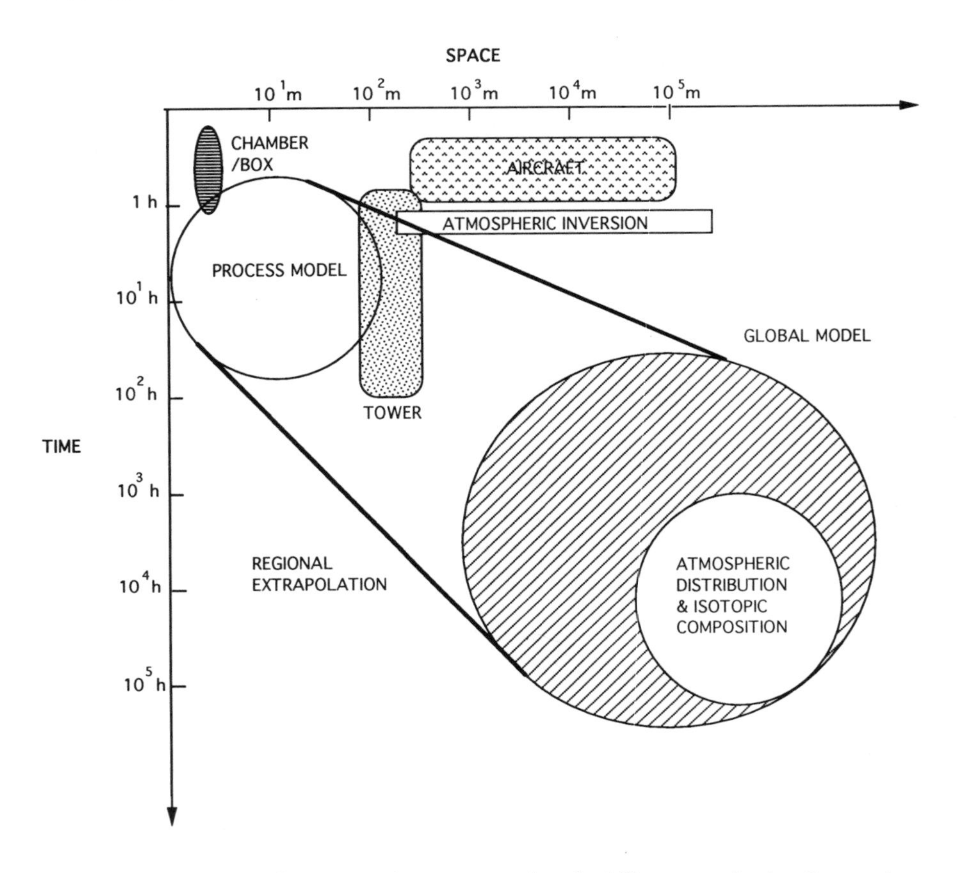

Figure 12-7. Time and space scales associated with different methods of assessing greenhouse gas fluxes from soil surfaces (modified from Stewart et al., 1989).

by remote sensing). An alternative approach would be to model trace gas fluxes over a landscape using data bases for the controlling environmental and edaphic variables (Stewart et al., 1989). Such models are evolving from a number of research locations around the world. (See Chapter 13 for more detail on measuring agricultural sources of GHGs.)

B. Baseline Emission Data and Tradeoffs

Assessment of agricultural effects on emissions of greenhouse gases and GWPs requires that baseline values be subtracted from the measured emissions and associated GWPs of agricultural systems. Thus, for example, if N_2O emissions from native and improved rangelands are 0.5 and 1.0 kg N ha^{-1} $yr^{-1,}$ respectively, the net effect of agriculture is to increase the flux of N_20-

N by 0.5 kg ha^{-1} yr^{-1}. While this type of adjustment is obvious, it is not always made; examples are ignoring changes in populations of native ruminant animals when appraising emissions associated with livestock, and reductions in CH_4 emissions as wetlands in agricultural landscapes are drained. Care must also be taken to use appropriate baseline emission data. For example, where drainage of agricultural lands has been improved, baseline emissions should be from undrained lands.

Tradeoffs between greenhouse gases should also be considered when comparing management alternatives that could mitigate agricultural emissions. For example, N_2O is not evolved from flooded rice soils because it is reduced to N_2, but water management practices that could reduce CH_4 emissions would likely increase N_2O emissions. Tradeoffs may also exist between greenhouse gas emissions and other environmental issues. For example, the use of legume green manures to supply N for flooded rice promotes CH_4 production (Lauren and Duxbury, 1993), although this practice can be considered desirable from a sustainable agriculture perspective. Moreover, organic sources of N may be needed to sustain high yields of flooded rice. A similar tradeoff with enhanced CH_4 emissions likely occurs with summer flooding of organic soils in the Florida Everglades agricultural area, which is used as an alternative to the use of pesticides for weed and pest control.

C. Opportunities for Sequestration of Carbon in Forests and Soils

i. Reforestation

The large historical release of CO_2 from conversion of forested lands to agriculture (Figure 12-2) shows the potential to counteract increasing atmospheric CO_2 levels by reforestation programs. Estimates of the mean annual rate of biomass accumulation by trees growing for periods from 30–100 years in natural forests range from 1.7 t C ha^{-1} for somewhat degraded lands in the tropics (Houghton, 1991) to 3 t C ha^{-1} for productive forests in temperate climates (Moulton and Richards, 1990; Sedjo, 1989). The dynamics of tree growth vary with tree species and tree age so that actual carbon accretion rates depend on forest type and the time frame of analysis. Although reforestation to counteract releases of CO_2 is an intuitively appealing idea, a few simple scenarios illustrate some of the difficulties faced by this approach:

- release of tree biomass to CO_2 when tropical forests are cleared occurs within 10 years; consequently, the mean annual rate of CO_2 release from the average tropical forest containing 170 t biomass C ha^{-1} is 17 t C ha yr^{-1}, which is 10 times the rate suggested by Houghton (1990) for regrowth of tropical forests.

- at an annual accretion rate of 3 t C ha^{-1}, 10^9 ha of land, or an area the size of the continental U.S., would need to be reforested in order to balance the present rate of increase of CO_2 in the atmosphere (3 Gt C yr^{-1}).
- there are a combined 30 $x10^6$ ha of idle and marginal cropland in the U.S. that could be reforested. At an accretion rate of 3 t C ha-1 yr^{-1}, reforestation of all of this land would sequester the equivalent of 7% of current CO_2 release from fossil fuel combustion in the U.S..

Utilizing trees as a renewable energy source could lead to greater short term carbon benefits compared to simply growing forests. Production of tree biomass in temperate climates can be increased to 5–6 t C ha^{-1} yr^{-1} by using short term rotations of trees that grow quickly when they are young, including species such as hybrid poplar and some hardwoods (Cannell and Smith, 1980; Riha, 1988). Achievement of high yields usually requires site improvement, e.g. nutrient inputs, and dense planting. The 5–6 t C ha^{-1} yr^{-1} mean annual yield can be realized in 5 years at a density of 100,000 plants ha^{-1}, while it may take 10 years to get this yield at a density of 2000 plants ha^{-1}. Management choices can therefore be made depending on how quickly a usable product is desired.

Mean annual biomass accumulation rates up to 20 t C ha^{-1} have been achieved with tropical legume trees such as *Leucaena Leucocephala* (Riha, 1988) and plantations of such trees would appear to be a viable approach to providing fuel wood for domestic purposes in the tropics. Furthermore, taking advantage of biological N fixation could help overcome some of the nutrient constraints that are sure to arise when trees are grown like agricultural crops. Crop plants can also rival trees in terms of biomass accumulation; alfalfa-grass or clover-grass mixtures will yield 4.5 t C ha^{-1} yr^{-1} on reasonably drained soils in the northern U.S. (Pfeifer et al., 1990), napier grass will yield 13–14 t C ha^{-1} yr^{-1} in Puerto Rico (M.J. Wright, personal communication), and sugarcane can yield as much as 50 t C ha^{-1} yr^{-1} (Irvine, 1983). Several options therefore exist when selecting plants for biomass production and decisions can be made depending on the type of biomass desired.

ii. Soil Carbon

Interest in sequestration of carbon in soils stems from the knowledge that this is the largest terrestrial reservoir of biospheric carbon and that agricultural use of land almost always leads to loss of soil organic carbon (SOC). Global SOC stocks are estimated to be 1400–1500 Gt C (Schlesinger, 1977; Post et al., 1982), compared to 720 and 560 Gt C in the atmosphere and terrestrial vegetation, respectively. Estimates of global loss of SOC in the period from the mid- to late-1800s to the present vary from 10 to 100 Gt C (Bolin, 1977; Revelle and Monk, 1977; Houghton et al., 1983; Schlesinger, 1986) or from 0.7 to 7% of the global pool. These values translate to a mean annual flux of between 0.1 and 1 Gt C. More recent studies of C loss, which may be more accurate, give values at the high end of this range or one-third of the annual increase in atmospheric CO_2-C.

Cultivation of soils lowers their SOC content and many studies have shown SOC losses in the range of 30–50% for surface soils in temperate climates. Estimates of SOC loss are generally made using paired plots of cultivated and uncultivated soils. Examples from the North American Great Plains are given by Haas and Evans (1957), Hobbs and Brown (1965), Voroney et al. (1981), Campbell and Souster (1982), and Tiessen et al. (1982). Loss of SOC occurs via enhanced mineralization of organic matter and by erosion of surface soils. Assessment of the effects of agriculture on C storage in soils, therefore, requires apportionment of changes in SOC content between these two processes and evaluation of the fate of SOC in eroded materials. Unfortunately, this type of analysis has not been done and most of the literature on SOC loss does not contain information that would allow soil erosion to be estimated. Most studies report % C in surface soils only, rather than total C in a soil profile (this requires measurement of soil bulk density) and only a few consider changes in horizonation due to tillage and/or erosion.

There is, however, adequate documentation to illustrate the importance of erosion on assessment of soil C storage. Two examples of the effects of wind erosion in the Great Plains region are given here; first, measurements of erosion made by the ^{137}Cs technique after 14 years of a tillage experiment in Nebraska showed that losses of soil from the "plow layer" were between 15–40, 15–20, and 5% for plow, subtillage, and no-tillage treatments, respectively (O'Halloran et al., 1987). Second, in a study along paired rangeland/cropland toposequences on sandstone, siltstone, and shale parent materials in North Dakota, Aguilar et al. (1988) found that substantial amounts of soil had been lost from the cropland after 44 years of a wheat-fallow rotation. Solum thickness and carbon storage in cropland soils was 5–60% less than that in control rangeland, with only occasional evidence of accumulation of soil either on the shoulder or at the bottom of a slope (e.g., Figure 12-8 for the sandstone toposequence). Regressions between soil loss and SOC loss had r^2 values of 0.52, 0.75, and 0.96 for sandstone, siltstone, and shale toposequences, respectively, consistent with a causal relationship.

Because erosion generally transfers soil to lower and wetter landscape positions, it probably leads more to preservation of SOC than it does to its destruction. Moreover, severe erosion events that occur infrequently are a major landscape forming factor and these can lead to buried soil horizons where SOC is at least partially preserved. For example, Yonker et al. (1988) found that 20% of the SOC in the top meter of soils of the Central Plains Experimental Range in northeastern Colorado was in buried horizons.

Tillage of soils gradually destroys macroaggregates and promotes the mineralization of their associated organic matter (see review by Duxbury et

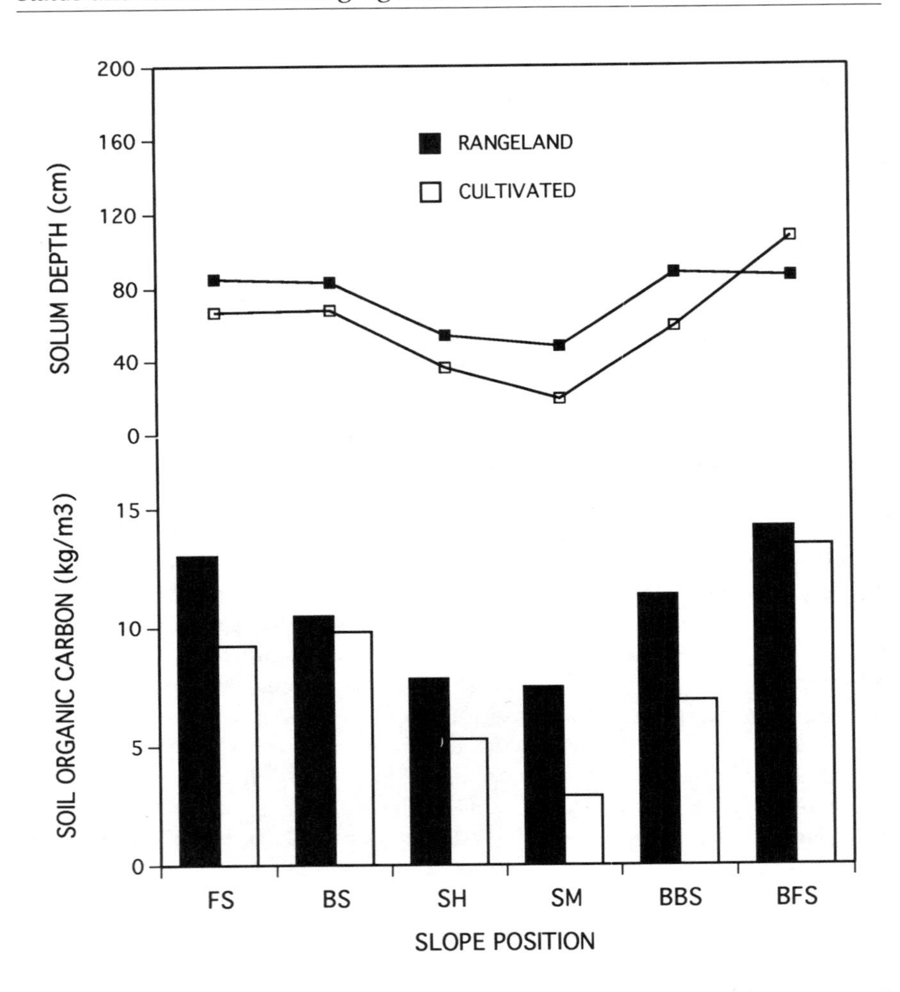

Figure 12-8. Solum depth and soil organic carbon storage to a depth of 1m in paired rangeland/cultivated agricultural plots across a toposequence in North Dakota, U.S. (constructed from data in Aguilar et al., 1988).

al., 1989). Consequently, reducing tillage on cropland that has been cultivated for many years will lead to increases in their SOC content. There are, however, few documented examples of SOC recovery upon adoption of reduced/no-tillage practices, mostly because of a lack of baseline data (most of these experiments are carried out from an agronomic rather than a C storage perspective). Long term (25–80 years) mean rates of SOC gain between 0.22–0.5 t C ha^{-1} yr^{-1} (22–50 g C $m^2 yr^{-1}$) have been found when cultivated cropland is converted to no-till agriculture, pasture, or simply abandoned to naturally revegetate (Jenkinson, 1981; Lugo et al., 1986). Carbon sequestration rates of this magnitude would not significantly offset C release from fossil fuel combustion. For example, Kern and Johnson (1991) calculated that increasing no-till production of the major grain crops in the U.S. from the present 27% to 76% of the cropland over the next 30 years would lead to a gain of 0.42 Gt C in SOC and a saving of 0.01 Gt C in fossil fuel use. The combined savings of 0.43 Gt C would amount to 1.1% of projected fossil fuel use in the U.S. during this same time period.

Although sequestration of C in soils via adoption of no-till agriculture is unlikely to be significant, it is important to note that tillage induced mineralization of SOC occurs much more quickly than its recovery upon ceasing tillage. No-till practices should therefore be adopted on newly developed agricultural lands as a means of reducing or preventing SOC mineralization.

D. Mitigation Options and Choices

i. General Options

Isermann (1992) suggests a number of general ways in which losses of nitrogen, phosphorous, sulfur, and carbon from the agricultural sector may be decreased. These include:

- a reduction of food/nutrient surpluses in human, animal, and plant nutrition in the developed countries, and
- more efficient external and internal nutrient recycling within the agricultural industry, especially in the developed countries. The spatial division or segregation of country and city inhabitants (i.e., of food producers and food consumers) and the frequent separation of animal and plant production in intensive agriculture encourage nutrient losses. Matching production to need, integration between society and agriculture in developed countries, and integration within agriculture would all lead to better use of resources and would provide multiple environmental benefits, including reduced emissions of greenhouse gases.

ii. Specific Options for Mitigating Agricultural Sources of CO_2, CH_4, and N_2O

We have previously concluded that forestry and agricultural options for sequestration of C in trees and soils are likely to be of limited value, especially in the near term. The best approach from a scientific viewpoint would be to use as small a land area as possible for agriculture, thereby preserving as much land as possible in forests. This could be achieved by encouraging use of resource evaluation technologies and sound agricultural development and production policies, including wise use of resources and inputs to achieve and sustain high levels of productivity. The most appropriate soil management goal would be to maintain or enhance soil fertility by using minimum tillage and other conservation techniques that reduce soil erosion and maintain high SOC levels.

The major agricultural sources of CH_4 are flooded rice, the ruminant digestive tract, and animal wastes. Decreasing CH_4 emissions from these sources by 10 to 15% would stabilize atmospheric methane at its present level and is a realistic objective. Improvements in ruminant nutrition through use of feed additives can alter fermentation in the rumen and improve efficiencies of protein production (Gibbs et al., 1989; Leng, 1991; CAST, 1992). These measures can easily be applied to intensively managed, confined animals but free ranging animals will most probably continue to yield more CH_4 per unit of feed energy and grow more slowly than those on improved feeds. Opportunities to reduce CH_4 emissions from flooded rice are still at the research stage but may include cultivar selection and development, and changes in nutrient and water management (Braatz and Hogan, 1991).

Nitrous oxide emissions from agricultural soils can be minimized through improved N management. CAST (1992) suggests a number of N management strategies:

- use soil testing to determine fertilizer N requirement; this will project and adjust for N mineralization from soil, legumes, manures, organic wastes, and any mineral N added by irrigation water;
- dispense with the "maintenance" concept;
- adjust the rate of N to a reasonable yield goal for specific fields;
- place N deep enough in the soil to lower the N_2O/N_2 ratio when denitrification occurs; and
- time N application to when it is needed by the crop.

The amount of fertilizer N used and the timing of application should have a goal of leaving as little residual N as possible in the soil during the non-cropped periods of the year. Additionally, agricultural systems that provide continuous plant cover should be utilized wherever feasible in order to minimize leaching and denitrification of NO_3^- associated with bare soils and to enhance nutrient recycling. In irrigated systems, better water and fertilizer management can be used to limit denitrification. Additives such as nitrification inhibitors (Braatz and Hogan, 1991), slow release fertilizer formulations, and multiple fertilizer applications can also be used to control the supply of nitrate and limit denitrification.

iii. Mitigation Choices

It would seem logical to ask three simple questions when investigating options for mitigation of radiative forcing of climate by anthropogenic emissions of greenhouse gases. These are:

Table 12-4. Cost-benefit comparison of mitigation options for the U.S.

Option	Potential emission reduction	Net Cost[a]
	10^8 t CO_2 equiv. yr^{-1}	
Energy efficiency of buildings	9	B
Automobile gas mileage	3	B
Industrial energy management	5	B → L
Landfill gas collection	2	L
Halocarbon-CFC reduction	14	L
Agriculture	2	L
Reforestation	2	L → M
Electricity supply	10	L → M
Total of B&L options	36 or one-third of 1990 US emissions	

[a] B = net benefit
L = low cost, \$1-9 per t CO_2 equivalent
M = moderate cost, \$10-99 per to CO_2 equivalent
Source: National Academy of Sciences (1991).

- does the activity contribute significantly to radiative forcing of climate;
- how essential is the activity; and
- what is the cost/benefit of the proposed mitigation practice?

In this paper we have shown that agriculture is a relatively minor contributor to anthropogenic emissions of CO_2 and a major contributor to anthropogenic emissions of both CH_4 and N_2O. However, emissions of CO_2 dominate anthropogenic radiative forcing of climate and this is predicted to continue into the foreseeable future. Clearly then, anthropogenic emissions of CO_2 should be reduced and this essentially means reducing use of fossil fuel in the developed countries. An impartial cost-benefit analysis of mitigation options for the U.S. carried out by the National Academy of Sciences (NAS, 1991) showed financial benefits or low costs associated with more efficient use of fossil fuels, but low impacts of targeting agriculture (Table 12-4). At the global level, efforts should also be made to reduce deforestation in the tropics as associated CO_2 emissions presently contribute about 20% of the anthropogenic CO_2 total. The major mitigation effort must, however, be directed at fossil fuel consumption because its use is expected to increase in the future while the rate of deforestation is expected to decline (Figure 12-2).

We also argue that agriculture is an essential activity of mankind, whereas this is not always the case with fossil fuel use. Nevertheless, reducing emissions of greenhouse gases should be a consideration in developing best management practices for agriculture as well as for industry and people. At the same time, we must question whether growth economics and environmental sustainability of our planet are compatible objectives.

LITERATURE CITED

Aguilar, R., E.F. Kelly, and R.D. Heil. 1988. Effects of cultivation on soils in Northern Great Plains rangeland. *Soil Sci. Soc. Amer. J.* 52:1081–85.

Blake, D.R. and F.S. Rowland. 1988. Continuing worldwide increase in tropospheric methane, 1978–1987. *Science.* 239:1129–31.

Bolin, B. 1977. Changes of land biota and their importance for the carbon cycle. *Science.* 196:613–15.

Bouldin, D.R., S.D. Klausner, and W.S. Reid. 1984. Use of nitrogen from manure. In *Nitrogen in Crop Production.* ed. R.D. Hauck, 221–45. Amer. Soc. Agron.: Madison, WI.

Bouwmann, A.F. 1992. Latitudinal distribution of atmospheric emissions of nitrous oxide from natural and agricultural soils. In *Proceedings of CH_4 and N_2O Workshop: CH_4 and N_2O emissions from Natural and Anthropogenic Sources and their Reduction Research Plan.* ed. K. Minami, 128–34. Nat. Inst. of Agro-Environ. Sci.: Tsukuba, Japan.

Braatz, B.V. and K.B. Hogan. 1991. *Sustainable rice productivity and methane reduction research plan.* US EPA: Washington, D.C.

Campbell, C.A. and W. Souster. 1982. Loss of organic matter and potentially mineralizable nitrogen from Saskatchewan soils due to cropping. *Can. J. Soil Sci.* 62:651–56.

Cannell, M.G.R. and R.I. Smith. 1980. Yields of minirotation closely spaced hardwoods in temperate regions; Review and appraisal. *Forest Sci.* 3:415–28.

CAST. 1992. *Preparing U.S. Agriculture for Global Climate Change.* Report no.119. Council for Agric. Sci. and Technol.: Ames, IA.

Crutzen, P.J. 1991. Methane's sinks and sources. *Nature.* 350:380–81.

Drennen, T.E. and D. Chapman. 1992. Negotiating a response to climate change: Role of biological emissions. *Contemp. Policy Issues.* 10:49–58.

Duxbury, J.M., D.R. Bouldin, R.E. Terry, and R.L. Tate, III. 1986. Emissions of nitrous oxide from soils. *Nature.* 298:462–64.

Duxbury, J.M., L.A. Harper, and A.R. Mosier. 1993. Contributions of agroecosystems to global climate change. In *Agroecosystem Effects on Radiatively Important Trace Gases and Global Climate Change.* eds. L.A. Harper, A.R. Mosier, J.M. Duxbury, and D.E. Rolston, 1–18. ASA Special Publication no. 55. Amer. Soc. Agron.: Madison WI.

Duxbury, J.M., M.S. Smith, and J.W. Doran. 1989. Soil organic matter as a source and sink of plant nutrients. In *Dynamics of Soil Organic Matter in Tropical Ecosystems.* eds. D.C. Coleman, J.M. Oades, and G. Uehara, 33–67. Univ. of Hawaii Press: Honolulu.

Eichner, M.J. 1990. Nitrous oxide emissions from fertilized soils: Summary of available data. *J. Env. Qual.* 19:272–80.

Enting, I.G. and J.V. Mansbridge. 1991. Latitudinal distribution of sources and sinks of CO_2: Results of an inversion study. *Tellus.* 43B:156–70.

Erich, M.S., A. Bekerie, and J.M. Duxbury. 1984. Activities of denitrifying enzymes in freshly sampled soils. *Soil Sci.* 138: 25–32.

FAO. 1992. *FAO Yearbook: Fertilizer.* Vol. 40. Food and Agriculture Organization of the U.N: Rome.

Fung, I., J. John, J. Lerner, E. Matthews, M. Prather, L.P. Steele, and P.J. Fraser. 1991. Three-dimensional model synthesis of the global methane cycle. *J. Geophys. Res.* D7:13033–65.

Gibbs, M.J., L. Lewis, and J.S. Hoffman. 1989. *Reducing methane emissions from livestock: Opportunities and issues.* EPA 400/1-89/002. Office of Air and Radiation, US EPA: Washington, D.C.

Haas, H.J. and C.E. Evans. 1957. *Nitrogen and carbon changes in Great Plains soils as influenced by cropping and soil treatments.* Tech. Bull. 1164. USDA: Washington, D.C.

Hauck, R.D. 1984. Epilogue. In *Nitrogen in Crop Production.* ed. R.D. Hauck, 781–86. Amer. Soc. Agron.: Madison, WI.

Hicks, B.B. 1989. Regional extrapolation: Vegetation-atmosphere approach. In *Exchange of Trace Gases between Terrestrial Ecosystems and the Atmosphere.* eds. M.O. Andreae and D.S. Schimel, 109–18. John Wiley & Sons: New York.

Hobbs, J.A. and P.L. Brown. 1965. *Effects of cropping history and management on nitrogen and organic carbon contents of a Western Kansas soil.* Kansas Ag. Expt. Sta. Tech. Bull. no.144. Kansas State Univ.: Manhattan, KS.

Houghton, R.A. 1991. Tropical deforestation and atmospheric carbon dioxide. *Climatic Change.* 19:99–118.

Houghton, R.A., J.E. Hobbie, and J.M. Melillo. 1983. Changes in the carbon content of terrestrial biota and soils between 1860 and 1980: A net release of CO2 to the atmosphere. *Ecol. Monographs.* 53:235–62.

Irvine, J.E. 1983. Sugarcane. In *Potential productivity of field crops under different environments.* 361–81. IRRI: Los Banos, Phillipines.

Isaksen, I.S.A., V. Ramaswamy, H. Rodhe, and T.M.L. Wigley. 1992. In *Climate Change 1992.* eds. J.T. Houghton, B.A. Callander, and S.K Varney, 51–67. Cambridge Univ. Press: Cambridge.

Isermann, K. 1992. Territorial, continental and global aspects of C, N, P and S emissions from agricultural ecosystems. In *NATO Advanced Research Workshop (ARW) on Interactions of C,N,P and S Biochemical Cycles.* Springer-Verlag: Heidelberg. In Press.

Jenkinson, D.S. 1981. The fate of plant and animal residues in soil. In *The Chemistry of Soil Processes.* eds. D.J. Greenland and M.H. B. Hayes, 505–61. John Wiley & Sons: New York.

Keeling, C.D., R.B. Bacastow, A.F. Carter, S.C. Piper, T.P. Whorf, M. Heiman, W.G. Mook, and H. Roeloffzen. 1989a. A three-dimensional model of atmospheric CO_2 based on observed winds: 1. Analysis of observational data. In *Aspects of Climate Variability in the Pacific and the Western Americas.* ed. D.H. Peterson, 165–236. Geophys. Monograph no. 55. Amer. Geophys. Union: Washington, D.C.

Keeling, C.D., S.C. Piper, and M. Heiman. 1989b. A three-dimensional model of atmospheric CO_2 based on observed winds: 4. Mean annual gradients and interannual variations. In *Aspects of Climate Variability in the Pacific and the Western Americas.* ed. D.H. Peterson, 305–63. Geophys. Monograph no. 55. Amer. Geophys. Union: Washington, D.C.

Keller, M., M.E. Mitre, and R.F. Stallard. 1990. Consumption of atmospheric methane in soils of central Panama: Effects of agricultural development. *Global Biogeochem. Cycles.* 4:21–27.

Kern, J.S. and M.G. Johnson. 1991. *The impact of conservation tillage use on soil and atmospheric carbon in the contiguous United States.* EPA/600/3–91/056. US EPA, Environ Res. Lab: Corvallis, OR.

Lauren, J.G. and J.M. Duxbury. 1993. Methane emissions from flooded rice amended with a green manure. In *Agroecosystem Effects on Radiatively Important Trace Gases and Global Climate Change.* eds. L.A. Harper, A.R. Mosier, J.M. Duxbury, and D.E. Rolston, 183–92. ASA Special Publication no. 55. Amer. Soc. Agron.: Madison, WI.

Leggett, J., W.J. Pepper, and R.J. Swart. 1992. Emission scenarios for IPCC: An update. In *Climate Change 1992.* eds. J.T. Houghton, B.A. Callander, and S.K Varney, 69–96. Cambridge Univ. Press: Cambridge.

Leng, R.A. 1991. *Improving ruminant production and reducing methane emissions from ruminants by strategic supplementation.* EPA/400/1-91/004. US EPA, Office of Air and Radiation: Washington, D.C.

Lerner, J., E. Matthews, and I. Fung. 1988. Methane emissions from animals: A global resolution data base. *Global Geochem. Cycles*. 2:139–56.

Lugo, A.E., M.J. Sanchez, and S. Brown. 1986. Land use and organic carbon content of some subtropical soils. *Plant and Soil*. 96:185–96.

Matson, P.A. and P.M. Vitousek. 1987. Cross system comparisons of soil nitrogen transformations and nitrous oxide flux in tropical forest ecosystems. *Global Biogeochem. Cycles.* 1:163–70.

Matson, P.A., P.M. Vitousek, and D.S. Schimel. 1989. Regional extrapolation of trace gas flux based on soils and ecosystems. In *Exchange of Trace Gases between Terrestrial Ecosystems and the Atmosphere*. eds. M.O. Andreae and D.S. Schimel, 97–108. John Wiley & Sons: New York.

McElroy, M.P., S.C. Wofsy, and Y.L. Yung. 1977. The nitrogen cycle: Perturbations due to man and their impact on atmospheric N20 and O_3. *Philosophical Trans. of the Royal Soc.* (London) B. 277:159–81.

Mosier, A.R. 1990. Gas flux measurement techniques with special reference to techniques suitable for measurements over large ecologically uniform areas. In *Soils and the Greenhouse Effect*. ed. A.F. Bouwman, 289–301. John Wiley & Sons: New York.

Mosier, A.R., D. Schimel, D. Valentine, K. Bronson, and W.J. Parton. 1991. Methane and nitrous oxide fluxes in native, fertilized, and cultivated grasslands. *Nature*. 350:330–32.

Moulton, R.J. and K.R. Richards. 1990. *Cost of sequestering carbon through tree planting and forest management in the U.S.* Tech. Report WO-58. USDA Forest Service: Washington, D.C.

NAS. 1991. *Policy implications of greenhouse warming*. Nat. Acad. of Sci. Press: Washington, D.C.

O'Halloran, I.P., J.W.B. Stewart, and E. De Jong. 1987. Changes in P forms and availability as influenced by management practices. *Plant and Soil*. 100:113–26.

Ojima, D.S., D.W. Valentine, A.R. Mosier, W.J. Parton, and D.S. Schimel. 1993. Effect of land use change on methane oxidation in temperate forest and grassland soils. *Chemosphere*. 26:675–85.

Parashar, D.C., J. Rai, P.K. Gupta, and N. Singh. 1991. Parameters affecting methane emissions from paddy fields. *Indian J. of Radio and Space Physics*. 20:12–17.

Parashar, D.C., A.P. Mitra, and S.K. Sinha. 1992. Methane budget from Indian paddy fields. In *Proc. of CH_4 and N_2O Workshop: CH_4 and N_2O emissions from Natural and Anthropogenic Sources and their Reduction Research Plan*. ed. K. Minami , 57–69. Nat. Inst. of Agro-Environ. Sci.: Tsukuba, Japan.

Parton, W.J., A.R. Mosier, and D.S. Schimel. 1988. Rates and pathways of nitrous oxide production in a shortgrass steppe. *Biogeochemistry*. 6:45–58.

Pfeifer, R.A., G.W. Fick, D.J. Lathwell, and C. Maybee. 1990. *Screening and selection of herbaceous species for biomass production in the Midwest/Lake states*. ORNL/Sub/85-27410/5. U.S. Dept. of Commerce: Springfield, VA.

Post, W.M., W.R. Emanuel, P.J. Zinke, and A.G. Stangenberger. 1982. Soil carbon pools and world life zones. *Nature*. 298:157–59.

Prinn, R., D. Cunnold, R. Rasmussen, P. Simmonds, F. Alyea, A. Crawford, P. Fraser, and R. Rosen. 1990. Atmospheric emissions and trends of nitrous oxide deduced from 10 years of ALE-GAGE data. *J. Geophys. Res.* 95:18369–85.

Revelle, R. and W. Monk. 1977. The carbon dioxide cycle and the biosphere. In *Energy and Climate*. 243–80. Nat Acad Sci.: Washington, D.C.

Riha, S. 1988. Understanding yield in short rotations of multipurpose tree species. In *Modeling Growth and Yield of Multipurpose Trees*. eds. N.R. Adams and F.B. Cady, 26–33. Winrock International: Arlington, VA.

Robertson, G.P. 1993. Fluxes of nitrous oxide and other nitrogen trace gases from intensively managed landscapes. In *Agroecosystem Effects on Radiatively Important Trace Gases and Global Climate Change*. eds. L.A. Harper, A.R. Mosier, J.M. Duxbury, and D.E. Rolston, 95–108. ASA Special Publication no. 55. Amer. Soc. Agron.: Madison, WI.

Safley, L.M., M.E. Casada, J.W. Woodbury, and K.F. Roos. 1992. *Global methane emissions from livestock and poultry manure*. EPA/400/1-91/048. Office of Air and Radiation, US EPA: Washington, D.C.

Scharffe, D., W.M. Hao, L. Donoso, P.J. Crutzen, and E. Sanhueza. 1990. Soil fluxes and atmospheric concentrations of CO and CH_4 in the northern part of the Guayana shield, Venezuela. *J. Geophys. Res.* 95:22475–80.

Schlesinger, W.H. 1977. Carbon balance in terrestrial detritus. *Ann. Rev. Ecol. and Systematics*. 8:51–81.

Schlesinger, W.H. 1986. Changes in soil carbon and associated properties with disturbance and recovery. In *The Changing Carbon Cycle: A Global Analysis*. eds. J.R. Trabalka and D.E. Reichle, 194–220. Springer Verlag: New York.

Sedjo, R.A. 1989. Forests to offset the greenhouse effect. *J. Forestry*. 87:12–15.

Shine, K.P., R.G. Derwent, D.J. Wuebbles, and J-J. Morcrette. 1990. Radiative forcing of climate. In *Climate Change: The IPCC Scientific Assessment*. eds. J.T. Houghton, G.J. Jenkins, and J.J. Ephraums, 47–68. Cambridge Univ. Press: Cambridge.

Steele, L.P., E. J. Dlugokencky, P.M. Lang, P.P. Tans, R.C. Martin, and K.A. Masarie. 1992. Slowing down of the global accumulation of atmospheric methane during the 1980s. *Nature*. 358: 313–16.

Steudler, P.A., R.D. Bowden, J.M. Melillo, and J.D. Aber. 1989. Influence of nitrogen fertilization on methane uptake in temperate soils. *Nature*. 341:314–16.

Stewart, J.W., I. Aselmann, A.F. Bouwman, R.L. Desjardins, B.B. Hicks, P.A. Matson, H. Rhode, D.S. Schimel, B.H. Svensson, R. Wassmann, M.J. Whiticar, and W-X.Yang. 1989. Extrapolation of flux measurements to regional and global scales. In *Exchange of Trace Gases Between Terrestrial Ecosystems and the Atmosphere*. eds. M.O. Andreae and D. S. Schimel, 155–74. John Wiley & Sons: NY.

Summerfeld, R.A., A.R. Mosier, and R.C. Musselman. 1993. CO2, CH_4, and N_2O flux through a Wyoming snowpack and implications for global budgets. *Nature*. 361:140–42.

Tans, P.P., I.Y. Fung, and T. Takahashi. 1990. Observational constraints on the global atmospheric carbon dioxide budget. *Science*. 247: 1431–38.

Tiessen, H., J.W.B. Stewart, and J.R. Bettany. 1982. Cultivation effects on the amounts and concentration of carbon, nitrogen, and phosphorus in grassland soils. *Agron. J.* 74: 831–35.

USEPA 1992. Anthropogenic methane emissions in United States: Estimates for 1990. Draft Report to Congress. Office of Air and Radiation, Washington, D.C., 20460.

Voroney, R.P., J.A. Van Veen, and E.A. Paul. 1981. Organic C dynamics in grassland soils. 2. Model validation and simulation of the long term effects of cultivation and rainfall erosion. *Can. J. Soil Sci.* 61:211–24.

Watson, R.T., H. Rhode, H. Oeschger, and U. Siegenthaler. 1990. Greenhouse gases and aerosols. In *Climate Change: The IPCC Scientific Assessment*. eds. J.T. Houghton, G.J. Jenkins, and J.J. Ephraums, 7–40. Cambridge Univ. Press: Cambridge.

Watson, R.T., L.G. Meira Filho, E. Sanhueza, and T. Janetos. 1992. Sources and sinks. In *Climate Change 1992*. eds. J.T. Houghton, B.A. Callander, and S.K Varney, 25–46. Cambridge Univ. Press: Cambridge.

Whalen, M. and W. Reeburg. 1990. Consumption of atmospheric methane by tundra soils. *Nature*. 346:160–62.

Williams, E.J., G.L. Hutchinson, and F.C. Fehsenfeld. 1992. NO_x and N_2O emissions from soil. *Global Biogeochem. Cycles*. 6:351–88.

Yonker, C.M., D.S. Schimel, E. Paroussis, and R.D. Heil. 1988. Patterns of organic carbon accumulation in a semiarid shortgrass steppe, Colorado. *Soil Sci. Soc. Amer. J.* 52:478–83.

13

AGRICULTURAL EMISSIONS of GREENHOUSE GASES

MONITORING and VERIFICATION

Owen Greene

Julian E. Salt
Department of Peace Studies, Bradford University

I. INTRODUCTION

According to the best available scientific assessments, without urgent and substantial international action, average temperatures at the Earth's surface are likely to rise by 0.15–0.45 degrees Celsius per decade during the next century and sea levels could rise by 3–10 cm per decade (IPCC, 1990, 1992). This magnitude of climatic change could cause massive social, economic, and environmental damage. Estimates of the importance of agriculturally emitted greenhouse gases (GHGs) vary widely. The U.S. EPA (1989) estimates agriculture's role at approximately 10%; the U.S. Office of Technology Assessment (1991) estimates the contribution of "food-related" sources in the 1980s to be about 16%, which included CFCs for refrigeration and assumed that 50% of biomass burning was agricultural related. (See Chapter 12 by Duxbury and Mosier for further discussion of this subject.) Thus, the direct and indirect monitoring of greenhouse gas emissions from this sector of human activities is of importance in the development of international measures to limit climate change.

Monitoring of agricultural emissions is important for two reasons. First, it improves scientific understanding of the processes of climate change. Second, it may be used for assessment, or verification, of commitments made by countries to limit their contribution to climate change. The monitoring requirements for each of these tasks overlap significantly, but they are nevertheless distinct.

For this chapter, it is important not only to review monitoring capabilities but also to clarify the ways in which the monitoring of greenhouse gas emissions (or removal through sinks) relate to the international policy making process. The next section introduces both the scientific and implementation review processes and highlights the distinctly different monitoring requirements that they each imply. The following section puts the task of monitoring emissions arising from agricultural activities into context by outlining the requirements and policy debates relating to greenhouse gas emissions from other sectors of human activity. Subsequent sections examine the prospects for monitoring emissions and related activities for each of the agricultural practices that contribute to climate change. They examine relevant monitoring techniques and the prospects for their use in establishing reliable estimates relating to baseline data and future emissions. The chapter thus aims both to review technical monitoring capabilities and to discuss their implications for policy.

II. THE RELEVANCE OF MONITORING: VERIFICATION AND SCIENTIFIC REVIEW

At the United Nations Conference on the Environment and Development (UNCED) in June 1992, representatives of 153 countries signed a Framework Convention on Climate Change. As a "framework" convention, it establishes a number of basic principles and objectives. (See Chapter 10 for a discussion of the basic obligations of the Framework Convention.) It establishes institutions and procedures for a review process, and also sets up structures for negotiating future protocols detailing more substantial commitments.

The review process is central to the development of the Climate Convention. As part of the broader scientific review process, the Conference of the Parties will regularly receive reports on the state of current scientific and technological knowledge from international scientific bodies. At the same time there will be regular reviews of the implementation of countries' existing commitments. Both the scientific and implementation reviews will be used as a basis for decisions by the Conference of the Parties on whether to adopt amendments or new protocols to the Convention. More broadly, such review processes will inform policy making by all the actors involved, be they international organizations, nations, regional groupings, multilateral companies, or pressure groups.

The Framework Convention creates a number of institutions relevant to the review processes. The Conference of the Parties is the supreme body of the convention, and will normally meet about once a year. A permanent secretariat will service and support the Conference of the Parties and manage the convention on a day-to-day basis. In addition, a Subsidiary Body for Scientific and Technological Advice will be established to provide timely information and advice on scientific and technological matters relating to the

Convention. This will be multidisciplinary and will be open to all relevant experts as nominated by their governments (Article 9). A similar Subsidiary Body for the Implementation will also be established to: provide assessments of the overall aggregated effect of the measures taken to limit climate change; review implementation of countries' commitments (focusing mostly on developed countries at first); and assist the Conference of the Parties in the preparation and implementation of its decisions (Article 10).

The scientific committee will in many ways be a continuation of the Intergovernmental Panel on Climate Change (IPCC), widely regarded as having carried out its tasks effectively since it was established in 1988. To facilitate its work, all parties to the Framework Convention are obliged to promote and cooperate in research, systematic observation, and the development of data archives related to the climate system and potential measures to limit climate change. In addition, all parties are required to exchange relevant information, and are specifically obliged to "develop, periodically update, publish and make available to the Conference of the Parties... national inventories of anthropogenic emissions by sources and removals by sinks of all greenhouse gases not controlled by the Montreal Protocol, using comparable methodologies to be agreed by the Conference of the Parties" (Article 4).

Further, all parties to the Framework Convention are required to formulate, publish, implement, and regularly update national or regional programs to limit greenhouse gas emissions; and cooperate in and promote a range of beneficial measures (such as technology transfer, changes in forest management, education, and training). Until further protocols are negotiated, countries are free to choose which measures they adopt. However, they must announce the programs or commitments they have adopted, and report on progress towards implementation. Developed countries must regularly provide a particularly detailed description of the measures they have taken and a specific estimate of the effects that these will have on their greenhouse gas emissions (or removals). Together with the data collected for scientific review, these reports will provide the main basis for the implementation committee's assessments.

In view of the enormous complexity and uncertainty associated with assessments of climate change, and with measures to limit or adapt to it, the tasks of the scientific and implementation committees (and of the broader scientific community) are manifestly Herculean. However, the importance of good scientific and technical advice for the development of an effective international regime is clear, as is the importance of good assessments of the aggregate effect of measures already taken. Less widely appreciated is the central importance of effectively reviewing each country's implementation progress.

In fact, implementation review is at the core of the Framework Convention's approach to promoting international action to limit climate change. Countries are invited to commit themselves to whatever measures they deem appropriate, but the substance and implementation of these measures will be reviewed internationally. The hope is that the international review process will increase political pressures on governments to make substantial commitments and to honor them. Unless commitments are verifiable and appropriate international procedures are established to monitor implementation, governments will be under less pressure to implement their commitments fully, and suspicions of non-compliance will tend to undermine international cooperation on this issue. If this is true for the Framework Convention, it will be even more true for future protocols imposing substantial limitations on countries' greenhouse gas emissions (Greene, 1991; Lanchbery et al., 1992; Greene and Salt, 1992, 1993).

The monitoring requirements for effective verification and implementation review can therefore be distinct from those of scientific review. For verification, it is important to specifically monitor national measures or greenhouse gas inventories, and to distinguish one state's contribution from another's. One needs to monitor implementation of specific commitments: for example, an ability to estimate methane emissions to an accuracy of +/- 50% may be adequate for some scientific purposes but it is not adequate to verify a state's compliance with a commitment to reduce its emissions by 10%. Moreover, data used for verification has more immediate political significance: countries have greater incentives to mislead or to challenge its validity. As far as possible then, commitments should be designed so as to be amenable to verification.

III. AGRICULTURAL GREENHOUSE GAS EMISSIONS IN CONTEXT

Agriculture is only one of several sectors of human activity that increase atmospheric concentrations of greenhouse gases and thereby contribute to climate change. As shown in Table 13-1, greenhouse gas emissions from the production and use of fuels for energy are by far the most important, and emissions from deforestation and waste are estimated by the IPCC to contribute to climate change on roughly the same scale as agriculture.

Proposals to limit greenhouse gas emissions have so far focused mostly on limiting carbon dioxide emissions from burning fossil fuels and on regulating (increasing) forest areas. This is not only because these two factors are particularly significant in contributing to climate change, but also because they seem most amenable to international regulation and change. Not only could fossil fuel-derived carbon dioxide emissions from developed countries be

Table 13-1. Instantaneous radiative forcing components of GHGs by Sector.

Sector	Breakdown	Gases					% Global
		CO_2	CH_4	N_2O	O_3	CFC	GHGs
Energy	Power generation	X		X	X		
	Industry (metals/chemicals, etc.)	X		X	X		
	Domestic (heating/lighting)	X		X	X		54
	Transport (road/rail/sea/air)	X		X	X		
	Biomass burning (wood fuel)	X		X	X		3
	Others (oil/gas/coal manufacture)	X	X	X	X		[a]
Forests	Selective	X		X			8
	Clearing	X	X	X	X		
	Flooded (reservoirs)		X	X			
	Biomass decay	X		X			
Agriculture	Rice paddies		X				4.5
	Ruminants	X	X				3
	Fertilizers	X		X			1.5
	Land conversion	X		X			
	Animal waste	X	X				
Waste Management	Landfill waste sites		X		X	X	5[a]
	Incineration	X		X	X		
	Recycling schemes	X		X			
Others	Cement manufacture	X		X			1
	CFC production/use	X				X	11.5
	Sub-totals %	55	15	6	[b]	24	Others: 8.5

[a] Figure includes methane released from mining/drilling activities

[b] Difficult to determine directly

Source: IPCC (1990).

reduced significantly, but each state's emissions could be indirectly monitored with some accuracy using energy statistics; thus commitments in this area promise to be verifiable (Fischer et al., 1990). Similarly, forest areas could be reliably monitored, and measures to halt and reverse deforestation can be justified on many grounds (habitats, biodiversity, climate change, long term economic self-interest), making international commitments seem relatively negotiable.

In contrast, international agreements to regulate emissions from rice paddies, ruminants, use of fertilizers, or agricultural waste are widely perceived to be less negotiable, and more difficult to monitor and implement. Nevertheless, monitoring is important for the scientific review process. Moreover, it is worthwhile examining whether any commitments relating to agricultural greenhouse gas emissions can be identified that are potentially verifiable. These are the concerns of the following sections.

IV. MONITORING AGRICULTURAL SECTOR GHG EMISSIONS

The aim of this section is to describe the methodologies required to prepare proper inventories of greenhouse gas emissions from the agricultural sector. Each methodology will be described and assessed for its ability to perform the task at hand and an indication of the reliability of the collected data will be provided. This process is important as the data will form the core of the database from which all future decisions are made concerning the viability of monitoring (and verification) regimes. The following section will then analyze the monitoring regimes required for each sub-sector of the agricultural sector and make an assessment of the appropriateness and consequent reliability of such methods for the monitoring of each sub-sector. The final section will look at the consequences of both inventory methodologies and monitoring regimes in terms of the ability to verify GHG emissions from the agricultural sector in general and the consequences this will impose on any commitments to reduce or stabilize GHG emissions in both this sector and in the climate change convention as a whole.

A. Background

The agricultural sector is probably the most diverse of all the major GHG emitting sectors on a global basis. This in itself poses a challenge as a plethora of societies and cultures practice many varieties of agriculture, the variations being determined by local climatic conditions and terrain as well as nutritional needs and desires.

Four major sources of agricultural greenhouse gas emissions exist: ruminants; animal waste; rice paddies; and soils and fertilizer use. Each of these sources will be examined in turn. Besides these four main sub-sectors within the agricultural sector, there are other smaller sub-sectors that are responsible for GHG emissions. However, by their nature they do not entirely fall under the label of "Agriculture" and as such shall not be treated here. A good example of this is "Land-use Change," which covers many areas and is not considered here. (Note that in Chapter 12, Duxbury and Mosier *attribute* the emissions resulting from land-use change to the agricultural sector.)

Monitoring of GHGs from the agricultural sector can be treated in one of two ways. They can either be monitored *directly* from the emission source or they can be calculated/estimated *indirectly* from the monitoring of a sector function. In fact, both types of monitoring are required, albeit at different levels in the monitoring regime.

B. Four-Tier Regime for Monitoring

It is possible to have a system of monitoring which is composed of four levels that reflect varying degrees of resolution. At the highest level of resolution, monitoring can take place by *ground-based* techniques, comprised of teams of people taking measurements with accurate instruments designed, usually, to do a very specific job. Examples of this approach would include ground-based air measurements with sensors specific to particular GHGs (e.g., N_2O or CH_4). By definition these monitoring techniques will yield highly resolved specific data covering a small area. The measurements will also be *direct* measurements in that they measure the greenhouse gas in question directly.

At the intermediate level, use is made of *low-level aircraft* as platforms to carry sensors that can either monitor *directly* such things as air pollution or *indirectly* monitor changes in land-use patterns by remote-sensing techniques. Due to the nature of the techniques, they tend to be of intermediate resolution but able to cover larger areas.

At the opposite end from ground-based monitoring there are *satellite-based* remote sensing techniques that monitor *indirectly* the greenhouse gas emissions from the sector in question. By virtue of the high altitudes, these techniques give the lowest resolution but cover the largest areas.

Finally, there is the technique of using *statistics* to indirectly determine GHG emissions from primary data pertaining to the sub-sector in question. A good example of this is represented by the data on fertilizer use for N_2O emissions. This type of monitoring can generate high resolution data—providing the appropriate data has been collected. Usually, the data will have been collected for other reasons, perhaps connected with tax estimation on commercial products (by government departments). This data will possibly be collated by a central agency (e.g., FAO) for global databases. Gaps in the data may occur for a number of reasons, including commercial sensitivities.

Thus, for each component of the agricultural sector under consideration a combination of these techniques can be used to gain an estimate of each state's GHG emissions. The skill of designing an effective monitoring system lies in the ability to apply the most relevant monitoring technique to each individual sub-sector. A summary of this approach is shown in Table 13-2.

One priority in monitoring emissions is to establish reliable "*baseline*" data. Without this baseline, any changes that occur in emission patterns in response to policy changes or otherwise become relatively unmonitorable. In addition, the concept of a baseline (year) also helps negotiators when discussing potential stabilization or reduction (of GHG) scenarios. The following sub-sections examine the prospects for establishing baseline data as well as for monitoring and verifying subsequent emissions.

Table 13-2. Monitoring regime for agricultural emissions.

Sector	Monitoring		
	Ground	Remote-Sensing	Statistics
Methane-emitting			
Natural wetlands	Spot-check	Area-method	
Rice paddies	Model studies	Area-method	Rice sales Area coverage Water-use
Ruminants	Model studies		Livestock numbers Feed sales
Termites	Spot-check		
Nitrous oxide-emitting			
Grasslands	Model studies	Area-method	
Fertilizers	Model studies		Production figures Use figures
Contaminated aquifer (manure)	Spot-check		Waste production Figures/storage
Soils (natural)	Model studies	Area-method	
Land clearance	Model studies	Area-method	Land-use figures

i. Enteric Fermentation – CH_4 Emissions

This process involves the breakdown of carbohydrates to produce methane and occurs in both ruminant (cattle, sheep) and non-ruminant (pigs, horses) animals, although ruminants produce the majority of the methane due to their ability to digest cellulose. The quantity of methane produced depends on such factors as the type, weight, and age of the animal as well as the quality and quantity of feed and energy expenditure of the animal in question.

To date, the best global inventories of methane emissions from animals have arrived at a total figure of 78 Tg CH_4/yr for farm animals (Crutzen et al., 1986), while wild animals collectively generate 5 Tg CH_4, and humans a mere 1 Tg CH_4. The bulk of these emissions is derived from dairy cows and cattle principally located in India, Russia, Brazil, U.S., and China (Lerner et al., 1988).

Termites are another potential methane source. These are not strictly in the agricultural sector, but in practice many are closely linked with animal husbandry practices. Estimates of global methane emissions from this source are highly varied, ranging from 0-200 Tg CH_4 (Cicerone and Oremland, 1988).

Estimates of methane emissions from animals are calculated as follows:

1. Estimate the numbers of a given type of animal (species/variety).
2. Estimate the average emissions from each type of animal.
3. Global total methane emissions are arrived at by summing the product of (1) and (2) for all relevant animal types.

For certain varieties and species of animals the estimation of total numbers in each state is relatively easy, as such animals are kept commercially and as such will be documented for tax or regulatory purposes. This is the case in developed countries and also to an extent in some lesser developed countries due to the fact that animal husbandry may represent a principal industry for exports (e.g., Brazil or Argentina). When the figures of a particular animal are not well known, a less direct method can be applied whereby the ratio of animal/human is calculated on model areas and extrapolated to the total human population. Obviously this method is prone to significant errors concerned with poor sampling.

The FAO Production Yearbook (FAO, 1985) probably holds the best global database of animal types and numbers. Global numbers are approximately (in millions): cattle (1270), sheep (1140), goats (460), and buffalo (126).

Once the total number of each type of animal is known, estimates of emissions per animal need to be made for each species and variety. One of the great problems presented by this task is the vast variation in animal living conditions, even within a given species. Feed type and regularity will vary, as will the climate in which the animal lives. For example, a cow living in the Netherlands will consume a different feed and experience radically different climatic conditions than a cow in India. This is exemplified by the emission level of 94 kg CH_4/animal/yr for dairy cattle from the Netherlands compared to 35 kg CH_4/animal/yr for cows in India (OECD, 1991).

Ideally, model studies have to be carried out for all realistic situations experienced by all animal types. However, this is probably unrealistic and averaging must be carried out.

In conclusion, the use of this approach and its accuracy depends upon the extent to which the nation/state in question has documented its livestock. Obviously, for the more developed nations this will be an easier task than for lesser developed nations and account (initially) should be made of this. With farm animals in general it would be reasonable to expect an error of +/- 10–20% on total livestock numbers for developed countries and somewhat greater for developing nations. However, data concerning feeding patterns will be less developed and will increase the overall error of the total methodology to estimate GHG emissions to a level of around 30–60%. This is assuming that the systems to monitor the livestock and feeding patterns are in place and running. Obviously inaccuracies will be even greater where this is not the case.

ii. Animal Waste – CH_4 Emissions

Waste from animals is composed mainly of organic matter, and if allowed to decompose in an anaerobic environment it will form methane. Aerobic

Table 13-3. Waste characteristics and methane emission potentials.

Animal type	Developed countries			Developing countries		
	Waste production[a]	Volatile solids	Emissions potential (B_0)	Waste production[a]	Volatile solids	Emissions potential (B_0)
	(kg/d/1000kg)	(%)	(m^3/kg VS)[b]	(kg/d/TAM)[c]	(%)	(m^3/kg VS)[b]
Feed cattle	86	11.6	0.33	NA[e]	15	0.10
Grazing cattle[d]	58	12.4	0.17	12.5	15	0.10
Dairy cattle	86	11.6	0.24	15.6	15	0.14
Swine	84	10.1	0.45	4.1	10	0.29
Sheep	40	23.0	0.19	1.6	23	0.13
Goats	41	26.6	0.17	1.8	27	0.13
Chickens/ducks	85	19.4	0.32	0.12	19	0.26
Turkeys	47	19.4	0.30	0.26	19	0.26
Horses, mules	51	19.6	0.33	18.4	20	0.26
Camels	NA[e]	NA[e]	NA[e]	18.4	16	0.21
Donkeys	51	19.6	0.33	12.2	20	0.26

[a] Waste production in kilograms per day per 1000 kilograms of live weight
[b] m^3/kg VS = methane production in cubic meters per kilogram of volatile solids
[c] TAM - Typical Animal Mass
[d] Includes buffalo in developing countries
[e] Not applicable
Source: Casada and Safley (1990).

decomposition, by contrast, produces carbon dioxide. As very large quantities of organic waste are produced annually in connection with animal husbandry, this sector is responsible for considerable methane emissions to the environment. Producing an estimate for methane emissions from animal waste depends critically on the manner in which the waste is stored. If the majority of waste is allowed to decompose in the open, then methane emissions are relatively low. Conversely, if the waste is stored in a closed system, then methane emissions are typically high. The estimation of methane emissions is also quite dependent upon detailed knowledge concerning the animal population in question.

The best approach for estimating emissions from agricultural waste is to probably to use the following procedure:

1. Define categories of waste type and storage methods, assigning typical methane emission coefficients to each.
2. Estimate the amount of waste from each category.
3. Sum over all types for the product of 1) and 2).

The most precise method of determining the emission potential of each waste type is to actually carry out measurements of the volatile solids (VS) under controlled situations. Again, the distinction between the average for

Table 13-4. Percentage of potential methane production (MCF) for various waste management systems.

System	MCF
	% of B_0
Pasture	2[b]
Liquid/slurry storage (less than one month)	10
Liquid/slurry storage (more than one month)	60[b]
Solid storage	10
Anaerobic lagoon	90
Drylot	10
Burned for fuel (arid/semi-arid regions)	5
Burned for fuel (moist regions)	10
Daily spread	5
Paddock	10
Deep pit stacking	5
Litter	10
Anaerobic digester[a]	14

[a] The MCF for anaerobic digesters refers to the amount of methane that is released into the atmostphere. The amount of methane produced from the waste and captured for use will be much larger.

[b] These estimates have been revised since the Casad and Safley (1990) estimates. Pasture was initially at 5%; liquid/slurry storage (more than one month) was at 20%.

Source: Casada and Safley (1990); U.S. EPA (1993).

animals from developed and developing countries becomes important, as is shown in Table 13-3. Thus, to achieve an accurate estimate of animal waste emissions a total-breakdown of the sector (animal and waste storage types) has to be achieved. A typical range of methane emissions is shown for a list of storage systems in Table 13-4 (as a percentage of methane emissions potential (B_0); m^3 CH_4/ kg-VS). It soon becomes apparent that a good terminology (universally accepted) is required for all the storage types, otherwise great confusion will exist amongst different countries' practices, leading to an erroneous inventory of storage systems. This will ultimately lead to an incorrect estimate of potential methane emissions from such systems.

It is clear that the estimation of methane emissions from agricultural waste is closely linked with the methodology that is required for the estimation of methane emissions from ruminants in general as, by and large, the same animals are central to both inventories.

As a sector, the emissions from agricultural waste are very segmented and will require a specific methodology for each possible emission type. Present estimates of methane emissions from this sector vary widely and span the range of 15–152 Tg/yr (OECD, 1991). It is clear that a systematic approach is required to tabulate all potential emissions from this sub-sector. As it stands, it is believed that methane emissions from agricultural waste stored in various storage systems can probably be estimated with associated errors of +/- 20%

for each type of storage system. However, the waste for which there is no direct storage data (which probably represents the bulk globally) can only be estimated from animal numbers and as such may have an error of +/- 50% or greater. A divergence in estimation accuracies will become most noticeable in the poorer developing countries, where reporting methods for animal waste are probably non-existent.

It is clear from the above analysis that states' greenhouse gas emissions from animal wastes cannot be reliably monitored, making national commitments relating specifically to such emissions virtually impossible to verify adequately. In some respects, this situation may gradually improve. Animal waste is increasingly regarded as a major source of pollution, particularly in regions of intensive animal rearing such as the Netherlands (NEPP, 1989). To the extent that this leads to increased regulation and requirements for storage, it will increase the proportion of animal waste for which reliable national data exists. However, this is bound to be a slow and incomplete process in the developed world, to say nothing of developing states.

However, the situation may be more promising if national commitments lead to the implementation of "good (climate friendly) agricultural practice" within states. In this case, the monitoring task for implementation review and verification purposes would be to check that declared programs designed to encourage waste management techniques that reduced methane emissions were being implemented—a potentially more realistic task.

iii. Rice Paddies – H_4 Emissions

Methane is emitted from rice paddies only under certain conditions. Conditions that give rise to the greatest methane emissions are caused by flooding of the rice paddy (so that the rice plant is in an anaerobic environment), in combination with high temperatures. The methane is produced by methanogenic bacteria that reduce carbon compounds (including CO_2) and the produced methane is transported to the surface by migration within the plant's tubular structure. Thus, the emission rate from a given paddy depends upon a combination of factors that will vary from region to region and even on a smaller scale (field to field).

Most of the world's wetland rice production occurs in Asia (90%) covering an area of 123 million hectares. Estimates of the methane emissions from laboratory based studies generate a global figure ranging between 50 to 280 Tg CH_4/yr (OECD, 1991). The large range in estimates arises partly from extrapolations based on different rice varieties (typically only a few varieties are taken into account, although there are over 60,000 varieties known) and partly from the fact that the CH_4 emissions are predominantly determined by local conditions that vary tremendously.

The flux of methane gas from the site under question is also difficult to monitor due to its high variability; methane emissions tend to be at their highest in the afternoon when temperatures are the highest. Account also has to be made of the number of crops of rice that are produced on a given site. In some parts of India, China, and Southeast Asia, two or three crops can be grown per year, increasing the 'effective' area of rice paddy under cultivation. Finally, complications may arise due to the use of inorganic fertilizers (ammonium sulfate) as they introduce an additional oxidant to the soil, tending to reduce the emissions of methane. Organic fertilizers, conversely, have been shown to increase CH_4 emission due to the presence of extra carbonaceous materials.

Large seasonal variations of methane emissions within a given site are known to occur, partly due to water management practices. If a paddy is drained for a period of time, the surface soil's chemistry changes in such a way as to reduce methane production. Other seasonal effects are caused by the actual growing cycle of the rice plant, tending to peak during the reproductive stages.

The current methodology used to estimate the emissions from rice paddies attempts to estimate an average daily emission rate (g $CH_4/m^2/day$) and to sum this over the growing season, which can vary from 80–180 days. This approach is used in the hope that the natural variability of emissions will be minimized. However, the daily emission rate is seasonally averaged to take account of any seasonal variations that occur. A recommended daily rate of 0.19–0.69 g $CH_4/m^2/day$ has been used (Schütz et al., 1989), based on measurements in a field trial in Hangzhou in China. The low side of the range is derived from daily emissions monitoring of the early crop, while the high side of the range pertains to the late crop on the same site. Both rates are averaged over the whole season. An average daily flux of 0.5 g $CH_4/m^2/day$ is assumed to be a realistic figure (Mathews et al., 1991).

The season's emissions should ideally be based on a three-year rolling schedule as fluctuations from one year to the next will also introduce some variability. A particularly good harvest, for instance, could set a false baseline if chosen as a representative year.

More elaborate methodologies exist for deriving methane inventories for rice paddies but they rely upon more detailed local information (e.g., soil type, agricultural practice, climate etc.). Such detailed information is available for some regions, but certainly not for most relevant areas in the world. A huge increase in the amount of ground-monitoring would have to be supplied to facilitate such an enhanced approach.

Data concerning the growing of rice is collected by the FAO and reported in the FAO Yearbook. This contains details of rice-crop cultivation in the 103 rice-growing nations (annually harvested area, number of crops/year, length of each cycle, area used by each cycle). A summation of the area used per crop and the length of growing season over the whole of the globe will give a first estimate of global methane emissions due to rice production. Further refinement can probably be incorporated by supplementing the national data with more regional or local data on soil properties (e.g., soil maps, etc.), however, this will have to be done at the regional level and supplied to the FAO.

The accuracy of the methodology used to estimate rice-derived GHG emissions is dependent primarily on data concerning the land-coverage by rice paddies. This is reported as relatively accurate, with an associated error as small as +/-10% for the main rice-growing areas of the world. However, the greatest uncertainty concerns the variation of methane emissions from each individual paddy site and rice variety. This can probably range over an order of magnitude from site to site and from species to species and, as such, will affect the accuracy of the total estimate accordingly. It is estimated that GHG emissions from the whole sub-sector, in light of the many compounding uncertainties associated with the local growing conditions (soil, etc.), may have errors that could be +/- 50% or worse.

Another related potential methane source that has to be considered concerns the sub-sector of wetlands in general. Wetlands (bogs and marshes) are naturally occurring ecosystems that generate methane anyway and have been doing so for a long time. As such, their emissions should be considered as biogenic in nature and not included in a nation's inventory. However, if an area of marsh is converted to a rice paddy the difference in methane emission should really be included in an inventory as it now represents an anthropogenic emission. However, this would imply a need for data pertaining to the original emission rates of the wetlands prior to cultivation, which is typically unavailable. A similar argument could also be applied to any area of land that due to man-made interference is changed in such a manner as to either increase or decrease its methane emissions. Strictly, these emissions should be categorized under another sub-group, that of "Land-Change."

Rice paddies cover extensive areas and as such lend themselves quite well to indirect monitoring by satellite-based remote sensing techniques. Thus, high flying satellites could be used to photograph paddies, which would allow a rough estimate of the total area of land disposed to rice cultivation. In addition, flooded rice paddies show up clearly (and distinctly from dry paddies) with certain microwave techniques. This data could be used to determine the percentage of paddies that are "wet" and correlate this with their exact location. This would form a good baseline of data upon which further detailed layers of data could be added. The more detailed data could be gained from a combination of government statistics on crop types/ volumes produced (e.g., FAO) which could be backed up by in-the-field measurements of actual GHG emissions from geo-referenced paddies. Model studies on particular varieties' emissions could also be added to the study to help corroborate the validity of in-field measurements. However, simulating a variety of growing conditions for 60,000 types of rice plant seems unrealistic, and variations in local conditions imply that estimates of national methane emissions from rice paddies will remain sufficiently inaccurate to render national commitments on such emissions unverifiable. Commitments relative to paddy *areas,* or flooding practices, might however be more amenable to verification.

iv. Fertilizer Use – N_2O Emissions

Nitrous oxide (N_2O) is produced naturally in soils by two mechanisms: first, by the reduction of naturally occurring nitrite or nitrate (denitrification) and second, by the oxidation of ammonium to nitrate (nitrification). The addition of commercial inorganic fertilizers (containing either ammonium and/or nitrate) is another source. Estimates of N_2O emissions due to the addition of inorganic fertilizers vary widely, covering the range 9–31.4 Tg N_2O/yr (Hahn and Junge, 1977). Estimates of the natural emission (biogenic) of N_2O from temperate and sub-tropical soils have been put at 4.5 Tg N/yr (Slemr et al., 1984), with an additional figure of 6 Tg N/yr (Seiler and Conrad, 1987) attributed to tropical soils. This latter sub-sector is notoriously difficult to estimate as the rate of nitrous oxide release is proportional to rate of nutrient cycling within the ecosystems under question.

Besides the direct emission of nitrous oxide from soils, it is thought that considerable N_2O is released from contaminated surface and groundwater due to nutrient run off. Estimates of the percentage of applied fertilizer reaching aquifer layers vary from 5–30% of applied fertilizer, resulting in a global flux of nitrous oxide from aquifers in the range of 1.3–2.7 Tg N_2O/yr (Breitenbeck, 1990). Additional nitrous oxide emission is also possible from animal wastes containing N-compounds, although accurate estimates are unavailable.

Factors that appear to affect N_2O release from soils include soil aeration, drying and wetting cycles (oxygen pressure in soil), soil temperature, and pH. In terms of management practices, various factors will combine to determine overall N_2O emissions, including: fertilizer type, application rate and timing, application technique, crop type, use of other chemicals, irrigation and tillage practices, and residual N and C from previous crops.

Studies (OECD, 1991) have shown that there probably exists a non-linear relationship between fertilizer application and N_2O emissions and that a wide range of emission coefficients exist for the various fertilizer types, depending upon their chemical constituents. Table 13-5 lists the N_2O emission

Table 13-5. Fertilizer derived N_20 emissions by fertilizer type.

Fertilizer type	N_2O-N produced (Median) %	N_20-N produced (Range) %
Anhydrous ammonia and aqua ammonia	1.63	0.86–6.84
Ammonium nitrate	0.26	0.04–1.71
Ammonium sulfate nitrate		
Calcium ammonium nitrate		
Ammonium type	0.12	0.02–1.5
Ammonium sulfate		
Ammonium phosphate		
Urea	0.11	0.07–1.5
Nitrate	0.03	0.001–0.5
Calcium nitrate		
Potassium nitrate		
Sodium nitrate		
Other nitrogen fertilizers	0.11	0.001–6.84
Other complex fertilizers	0.11	0.001–6.84

Source: OECD (1991).

factors of certain fertilizer types in common use. World Bank estimates have shown that use of inorganic fertilizers is increasing by 1.3%/yr in the developed world and 4.0% in the developing world (World Bank, 1988).

There are two methodologies presently used to estimate N_2O emissions from fertilizer use. The first method simply involves summing all fertilizer usage (aggregated by fertilizer type) in combination with the respective emission coefficient for each fertilizer type. A three-year rolling data record is used to average fertilizer use so that a more consistent estimate is obtained. It is evident, due to the uncertain nature of nitrous oxide emissions (caused by a combination of external factors), that the ranges of N_2O emissions are quite wide, and so the median of the range typically is taken as the representative figure.

The second methodology takes this calculation one step further by including the crop type to which the fertilizer is being applied. Then total N_2O emissions are calculated by summing over all fertilizer types (and emission coefficients) as well as crop types. Table 13-6 shows the typical range of emission values obtained from various crop types. This approach is quite sophisticated and requires extensive data input, and as a consequence the simpler first method is generally used. However, it is recognized that to gain further insight on N_2O emissions, more research into these areas is required.

Overall, the accuracy of related GHG emissions from the fertilizer sector is quite poor. Data concerning the production of fertilizers will probably be

Table 13-6. Fertilizer-derived emissions from soil systems and fertilizer types.

Fertilizer type	Crop system	N_2O-N produced
		%
Anhydrous ammonia	Soil[a]	0.860–6.84
	Corn	0.000–1.80
Ammonium type	Grass	0.030–0.70
	Soil	0.040–0.18
	Plant	0.090–0.90
Ammonium nitrate	Grass	0.040–1.71
	Plant	0.05
	Grains	0.040–0.70
Ca, K, Na nitrate	Grass	0.001–0.50
	Soil	0.010–0.04
	Plant	0.007–0.10
Urea	Grass	0.18
	Soil	0.017–0.14

[a] Soil as used here refers to the experimental conditions under which fertilizer was applied to soil with no crops planted.

Source: Eichner (1990).

very accurate; however, there will be greater uncertainty of the use of such fertilizers (i.e., used or stored). If and when the fertilizer is used there is yet another layer of uncertainty associated with details of exactly where and in what amount these fertilizers are used. This will inevitably culminate in serious inaccuracies, resulting in estimates that have associated errors of some +/- 100%. In effect, the uncertainty associated with soil properties undermines the reliability of other data in this sector.

C. Summary

In summary, estimates of national greenhouse gas emissions from each source in the agricultural sector are quite inaccurate. Some of the relevant data, such as fertilizer use and paddy areas, are amenable to reasonably accurate monitoring. Improved record keeping at the national level could lead to substantial improvements in present data. For example, reporting systems relating to animal numbers, feed and fertilizer inputs, and waste generators and management, could usefully and realistically be developed. In addition to improved accounting systems, a combination of remote sensing and ground-based monitoring systems can be used to improve the overall reliability of data. In most agricultural sectors, the UN Food and Agriculture Organization (FAO) is probably the best placed organization to verify and collate national reports, and to help states improve their own national reporting and management systems. Nevertheless, although industrial progress can be made to improve the reliability of much relevant data, the prospects for using this to calculate accurate national inventories for greenhouse gas emissions from agricultural activities typically remain poor. In most cases, the variability due to local conditions and other factors undermines improvements in reliability of other data.

V. COMMITMENTS AND VERIFICATION

The implication of the above analysis is that national commitments to specific limits on greenhouse gas emissions from each of the agricultural sectors considered are likely to be unverifiable for the foreseeable future. If estimates of certain emissions are accurate to +/- 50%, compliance with commitments, for example, to reduce national emissions by 10% cannot adequately be monitored. However, verification of other types of commitment may be more practical. For example, in principle it is relatively practical to monitor commitments relating to *areas* of rice-paddy production, quantities of fertilizer used, or numbers of animals kept.

In any case, it is not clear that direct commitments to limit greenhouse gas emissions due to agricultural practices would often be negotiable. Even if we were able to monitor Indonesia's national methane emissions due to wet-rice production, it is unlikely that they would agree to direct constraints in this area. In most of the world, agricultural activities are seen as essential to well-being or survival, and there would be great resistance to international limits. Only in OECD countries would such direct limits seem at all realistic.

In our view, the most promising way forward is to negotiate commitments relating not to actual GHG emitting sources, but to management practices that through their implementation bring about associated direct or indirect limits in GHG emissions without directly limiting agricultural production. In relation to rice, one could, for instance, negotiate commitments to implement certain water-management practices to be followed which in themselves could be monitored and verified by the techniques described. It may be that the implementation of these management practices could be monitored and verified with greater reliability than other types of commitment, thus creating a situation whereby greater influence in changing the GHG emissions from agriculture was achieved while avoiding specific limits on agricultural production. This would be very positive in that good agricultural practices would be promoted and the feeling of a police-style monitoring of very specific emission sources would be reduced, again aiding the whole international regime-building process. The climate-friendly management practices would probably, due to their nature, be less sub-sector specific and as a consequence have far greater impact on overall GHG management as a whole.

VI. CONCLUSION

This chapter has described the current approach used by such bodies as the IPCC and OECD to estimate GHG emissions from the agricultural sector. It has been shown that baseline conditions need to be established and methodologies set-up to estimate emissions from various sub-sectors. Monitoring techniques required to establish the validity of these emission scenarios are then discussed and assessed. Next, the data has to be collated by an appropriate authority (first at national and then international levels). Finally, this data, in the form of a report, can be used by bodies such as the IPCC in its work with the Framework Convention on Climate Change.

This chapter has highlighted the fact that, of all the sectors, agriculture is probably the sector where reliable estimates of GHG emissions are hardest to achieve. Certain parameters could be monitored relatively accurately, but overall the accuracy in this sector will never be able to match that of the energy sector.

The monitoring activities in the agricultural sector are best viewed as taking place on two distinct levels. The direct and indirect monitoring of emissions from sources (via a four-tier system) must be viewed as primarily for scientific review purposes, allowing each state to compile inventories of its own emissions so that a national report can be prepared and forwarded to the Secretariat of the Framework Convention. This database can also be used to form the basis for the scientific work carried out by the IPCC.

Secondly, for verification purposes concerning the implementation of commitments made in the context of the Framework Convention, monitoring of agricultural practices or improved "management systems" as opposed to actual emissions is most appropriate in general. Not only is it less taxing methodologically and technically, but such commitments may actually be more negotiable.

Acknowledgements

The authors gratefully acknowledge the support of a research grant from the UK Economic and Social Research Council, Global Environmental Change Programme.

LITERATURE CITED

Breitenbeck, G.A. 1990. *Proceedings of the Workshop on Greenhouse Gas Emissions for Agricultural Systems.* US-EPA Summary report. VII-6-VII-10. Washington, D.C.

Casada, M.E. and L.M. Safley. 1990. *Global methane emissions from livestock and poultry manure.* Report to the Global Change Division. US EPA: Washington, D.C.

Cicerone, R.J., and R.S. Oremland. 1988. Biogeological aspects of atmospheric methane. *Global Biogeochemical Cycles.* 2(4):299–327.

Crutzen, P.J., I. Aselmann, and W. Seiler. 1986. Methane production by domestic animals, wild animals, other herbivorous fauna, and humans. *Tellus.* 38B:271–84.

Eichner, M.J. 1990. Nitrous oxide emissions from fertilized soils: Summary of available data. *J. Env. Qual.* 19:272–80.

FAO. 1985. FAO Yearbook. Vo. 39. FAO: Rome, Italy.

Fischer, W., J. Di Primio, and G. Stein. 1990. *A Convention on Greenhouse Gases: Towards the Design of a Verification System.* KFA Julich Report JUL-2390. KFA Julich: Germany.

Framework Convention on Climate Change. 1992. UN, A/AC.237/18.

Greene, O. 1991. Building a global warming convention: Lessons from the arms control experience. In *Pledge and Review Processes: Possible Components of a Climate Convention.* eds. M. Grubb and N. Sateen, xxi-xxxiii. Royal Institute for International Affairs: London.

Greene, O. and J. Salt. 1992. Limiting Climate Change: Verifying National Commitments. *Ecodecision.* 7:9–13.

Greene, O. and Salt, J. (1993) After UNCED: Building an effective climate change convention. submitted to: *Environmental Values*, Feb. 1993.

Hahn, J. and C. Junge. 1977. Atmospheric nitrous oxide: A critical review. *Z. Naturforsch.* 32a:190–214.

Houghton, J.T., G.J. Jenkins, and J.J. Ephraums. eds. 1990. *Climate Change: The IPCC Scientific Assessment.* Cambridge University Press: New York.

Houghton, J.T., B.A. Callander, and S.K. Varney. eds. 1992. *Climate Change 1992: The supplementary report to the IPCC Scientific Assessment.* UNEP/WMO. Cambridge University Press, New York.

Kruse, M., H. A. Simon, and J.N.B. Bell. 1989. Validity and uncertainty in the calculation of an emission inventory for ammonia arising from agriculture in Great Britain. *Environmental Pollution.* 56:237–57.

Lanchbery, J., O. Greene, and J. Salt. 1992. *Verification and the Framework Convention on Climate Change.* A briefing document for UNCED, Rio de Janeiro, June 1992. (Document May, 1992.) VERTIC: London.

Lerner, J., E. Mathews, and I. Fung. 1988. A global high-resolution database. *Global Biogeochemical Cycles.* 2:139–56.

Mathews, E., I. Fung, and J. Lerner. 1991. Methane emissions from rice cultivation: Geographic and seasonal distribution of cultivated areas and emissions. *Global Biogeochemical Cycles.* 5:3–24.

National Environmental Policy Plan (NEPP). 1989. Ministry of Housing, Physical Planning and Environment. The Hague, Netherlands.

OECD. 1991. *Estimation of Greenhouse Gas Emissions and Sinks.* Final report from the OECD Experts Meeting, Feb. 1991. Prepared for IPCC. Revised Aug. 1991.

Office of Technology Assessment. 1991. *Changing by Degrees: Steps to Reduce Greenhouse Gases.* U.S. Congress: Washington, DC.

Schütz, H., W. Seiler, and H. Rennenberg. 1989. Presentation. International Conference on Soils and the Greenhouse Effect. August 1989. Wageningen, Netherlands.

Seiler, F. and R. Conrad. 1987. Contribution of tropical ecosystems to the global budget of trace gases, especially CH_4, H_2, CO and N_2O. In *Geophysiology of Amazonia: Vegetation and Climate Interactions.* ed. R. Dickenson, 132–62. John Wiley & Sons: New York.

Slemr, F., R. Conrad, and W. Seiler. 1984. Nitrous oxide emissions from fertilised and unfertilised soils in a subtropical region (Andalusia, Spain). *Journal of Atmospheric Chemistry.* 1:159–169.

Smith, K.A. and J.R.M. Arah. 1990. *Losses of Nitrogen by Denitrification and Emissions of Nitrogen Oxides from Soils.* The Fertiliser Society, No.299. Peterborough, U.K.

United States Environmental Protection Agency. 1989. *Policy Options for Stabilizing Global Climate.* Draft Report to Congress. US EPA: Washington, D.C.

United States Environmental Protection Agency. Kurt Roos, personal communication.

World Bank. 1988. *World Development Report, 1988.* Oxford University Press: New York.

14

BRIDGING the GAP between POLICYMAKERS and ACADEMICS

Susan Offutt
Executive Director, Board on Agriculture
National Research Council/National Academy of Sciences

What is this gap between global change policymakers and academics that the paper's title presumes? What might be its cause? And why, if a gap indeed exists, should it matter whether or not it is bridged? What would be involved in an attempt?

I. WHAT DEFINES A GAP?

This supposed gap in the arena of climate change debate and analysis can be seen as one manifestation of the general phenomenon that occurs when public policy appears (to scientists, anyway) to be made without full regard for the results of scientific analysis. As a consequence, the policy choice set may include options that science would advise against and exclude others that science might indicate merit consideration. From the point of view of the academic scientist, then, a "gap" may be defined to exist when the range of public policy actions or options is affected by policymakers' lack of knowledge or appreciation of the relevant science base. A good case in point is the debate over the need for public regulation of the deliberate release of genetically-engineered organisms in the environment, wherein the science community generally minimizes potential risks and thus the need for oversight, while policymakers have typically disagreed (more on the source of this disagreement in a moment).

The gap may arise out of poor communication of science-based information by academics and/or the inability of the policymaker to understand it. While acknowledging the validity of the concern about the scientific literacy of society, of which policymakers are a reflection, it is worth pointing out that scientists do not always make salient points clearly to the public. As an example, consider the use of ionizing radiation in food preservation, a process that scientific consensus has held safe for 40 years but which still lacks widespread public acceptance. Is it possible that the science community has done all it can to make itself understood?

Of course, there is more to the story than the inability of policymakers to comprehend science. Many are, in fact, quite capable but simply have other, nonscience-based reasons for rejecting academic advice. Policymakers' decision criteria include concerns about economic impacts and political fallout that take an option's technological feasibility as a necessary but not sufficient condition for selection. When academics discount the importance (or the validity) of these concerns, their advice is likely to be ignored. The solution, though, is not to create a master race of scientist-politicians but to close the gap through improving scientists' appreciation of the role and responsibilities of the policymakers.

An assessment of the current state of play in the agricultural dimensions of the global climate change debate and analysis follows. Then, observations on potentially constructive ways to close the gap are offered, along with a discussion of problems peculiar to working in agricultural science and policy.

II. HOW BIG IS THE GAP?

A gap probably does indeed exist on the topic of agriculture and global climate change, although its size has narrowed considerably over the past three or four years. The unprecedented emergence of the Intergovernmental Panel on Climate Change (IPCC) as an influential source of scientific opinion and advice has greatly enhanced the communication between academics and policymakers. The official participation of the US government in IPCC has been supported by the institutionalization of the Federal global change research program under the auspices of the Office of Science and Technology Policy. This cross-cutting evaluation of Federal global change research itself set a precedent and surely helped to expedite consensus about scientific priorities. And now, the cross-cut has spawned an economics research initiative. While this initiative is carefully distinguished from more short term policy analysis, it represents the necessary intermediate step in laying the foundation for such analyses.

Compared to other contemporary science policy issues, global climate change has been the target of extraordinary coordinated effort within the Federal science establishment and with academia. As a result, the development of the science base to support policymaking is relatively far along. Contrast this situation to the comparatively little that is known about AIDS' treatment, an issue that also emerged in the mid-part of the eighties. The attention afforded global climate change is obviously directly related to its hold on the public through its apparent affirmation of the "end of nature" and of the need for concern about environmental quality. For these reasons, the scientists' audience is considerably more receptive than usual.

What remains to be done in closing the gap between academics and policymakers concerns the specifics of possibilities for mitigation and adaptation as developed against the background of ongoing efforts to better understand global physical processes. For agriculture and climate change, it is indeed true that the devil is in the details.

III. CLOSING THE GAP: THE LESSONS OF EXPERIENCE

In my time at the Office of Management and Budget (OMB), I had the opportunity to participate in IPCC deliberations, in the initiation of the global change research cross-cut, and in the development and analysis of Bush Administration policy options. I will briefly recount two episodes from that period that are intended to be instructive in considering how to narrow the gap in coming years.

In the spring of 1989, after travel to Moscow to participate in IPCC meetings on the impacts of global climate change, I began to work with John Reilly of USDA's Economic Research Service and Garth Paltridge of the National Oceanic and Atmospheric Administration (NOAA) on methodology for assessing those impacts. Our initial work was on the development of the conceptual framework of an "expert system" with which the US government could define the worth of any mix of preventative or adaptive actions by itself and by any particular set of countries. Besides the expert system architecture, information was required on "impact response functions" for climate-sensitive activities and on "cost functions" of possible preventative and adaptive actions. We were groping toward a way to translate large quantities of scientific information into a format that would be "policymaker-friendly." At that point, there was little activity underway in the Federal program to address questions about the costs and benefits of alternative policy actions in response to global climate change.

To support our efforts, we were able to convince OMB policy officials to sanction a workshop on impact response functions to be put together by NOAA's National Climate Program Office (NCPO). With about two dozen scientists in attendance, the workshop was held in early September 1989 at the Coolfont resort in West Virginia. Its findings on climate impact response functions were subsequently summarized and published by NCPO. What is interesting in the current context about the Coolfont workshop is not the estimation of the response functions themselves but what the Coolfont group attempted to do and how it was received in the science and policy communities.

Our expert system, crude as it was, approached the problem of climate change from the perspective of a policymaker. Necessarily, then, it operated at a high degree of aggregation and embodied the notion of multiple choices of combinations of policy options. The input it required on climate response functions was aggregated to a corresponding degree in order to establish at least the qualitative directions of trade-offs across natural resource and economic sectors and countries under alternative policy configurations. Policymakers could apply their own subjective weights to alternative outcomes. The intent of the expert system was to identify which policy actions were technologically feasible, to estimate associated costs and benefits, and to let policymakers decide what was tolerable.

Although our instincts were sound, the Coolfont exercise was not warmly embraced by either the science or the policy community. In the case of the scientists, Coolfont appears to have been widely criticized according to the standard criteria: not peer-reviewed, insufficiently rigorous and empirical, etc. Policymakers in the Administration, on the other hand, were already looking the other way, having decided that the uncertainties associated with predictions of climate change justified only further research, not policy action. There was, therefore, little interest in discussion of even hypothetical options. As it turned out, our expert system experiment was too policy-oriented for the scientists and too academic for the policymakers.

Undaunted by (or perhaps ignorant of) the reaction to Coolfont, I continued to pursue within OMB the idea of comparison of alternative policy actions across sectors and countries. Because the OMB environment required just about instantaneous response to any policy question, it was clear that even our embryonic expert system was too complex. For the purposes of promoting strategic thinking, though, the exercise could be adapted to qualitative dimensions that still gave a feel for the alignment of winners and losers across the globe.

Both of these efforts were aimed at servicing policymakers by defining the problem and characterizing its solutions from their perspective. In taking this approach, the inadequacies of the existing science knowledge base became apparent, in particular how little could be said about subcontinental regional impacts. Because of the very different work environments they inhabit, it is too much to expect scientists and policymakers to ask the same questions about climate change coincidentally. While policymakers can hardly be expected to "do" science, it would not seem unreasonable to have scientists "do" policymaking at least by asking how their findings can be structured to be compatible with a public decision process.

The idea of a matrix or a game based on strategic tradeoffs has yet to be resurrected, although it still seems to me, anyway, a potentially useful tool. Could it be that scientists consider global climate change to be too complicated and important to be susceptible to anything as simple as the characterizations I have described? Yet, ultimately, critical decisions will be made using a conceptual framework no more complicated (and probably much simpler and less consistent) than that Reilly, Partridge, and I described.

IV. THE PECULIARITIES OF AGRICULTURE

The agricultural community is widely and justifiably perceived as insular, resistant to ideas that originate outside its traditionally prescribed boundaries. As an example, consider that it is only in the past few years that any non-farm interests have even been allowed to participate (albeit with uncertain effect) in shaping government policy on agriculture and its impacts on the environment. And I believe it is fair to say that, by and large, the agricultural science community suffers from a comparable insularity, born out of its traditional sequestration in colleges of agriculture. When confronted with issues that dictate interaction with non-agricultural interests and disciplines, the agricultural community is slow to respond, let alone take the initiative.

In the case of climate change, this insularity was reflected in the initial structure of the Federal cross-cut, which did not explicitly include participation by USDA. Although agriculture and energy are the two major sectors that both affect and are affected by the processes of climate change, the Department of Energy was a player from the start while USDA was not. What little there was to do with agriculture was housed in EPA. This omission of USDA reflected the bias of the Federal science agencies, which quite often overlook its significant role in the funding and conduct of research in life sciences and natural resources. After intervention by OMB, USDA was included in the cross-cut, when, ironically enough, it became evident that USDA itself was unsure of its desire to be included. Despite the support of the Office of Science and Education, non-science policy officials in USDA were hesitant to divert resources to climate change research. The basis for compromise was found by agreeing (with justification) that large portions of ongoing USDA programs were directly supportive of the goals of the climate change initiative. USDA went on to contribute to the cross-cut effort in important ways, not the least of which was providing to its management strong staff, capable of assuring agriculture be considered in setting climate change research and program priorities.

The problem of the insularity of agriculture is not limited to the Federal family but is detectable in the academic community's approach to climate change, as well. There, the isolation of agriculture probably arises partly out of the organization of academic science (with agriculture in its own unit) and also out of the sense that agriculture, as a sector, will be capable of adaptation to climate change without extraordinary effort. There is, of course, validity in the idea of the adaptability of agriculture, but unfortunately it can be employed as a rationalization for excluding agricultural priorities in research design. Since arriving at the National Research Council, I have experienced "*deja vu* all over again" in having to argue the position that agriculture is not only a logical but a significant component of climate change research.

Happily enough, conferences such as this act to counter insulationist tendencies, and there is every reason to expect them ultimately to be overcome. One of the forces working against continued isolation is generational change in the agricultural science community itself. Younger researchers are less likely to have been educated and spent their entire careers in colleges of agriculture, so they may have acquired "outside" contacts and views along the way. Integration of agricultural priorities into research plans might therefore become less problematic. However, to the extent that agriculture does indeed exhibit characteristics that distinguish it from other industries (e.g., extreme sensitivity to weather and climate), there will be a continuing need for both agricultural researchers and policymakers to take the initiative in explaining these circumstances to those with less familiarity (that is, to the 98 percent of the US population not living on farms).

V. BRIDGING THE GAP

To finish, I will offer some opinions as to what could be done to bridge the gap between researchers and policymakers. Some of these efforts ought to be aimed at providing a better science base for anticipated policy decisions and some should help academics better appreciate the circumstances under which their research results are interpreted and applied.

With respect to building a better science base, I have three suggestions. First, agricultural researchers ought to press the modelers to consider their needs in refining the specifications and regional resolution of general circulation models (GCMs). I am told that there are many aspects of the GCMs that require further attention, and perhaps these needs should be addressed first in order to lay the foundation for better regional analysis down the line. Nonetheless, agricultural aspects of climate change need to be appreciated now, and they might not be, in the absence of affirmative action by agricultural researchers. Second, the specifics of any mitigation strategy for agricultural emissions need to be considered carefully. To what extent would such strategies require regulations of practices or input use on the farmstead? Here, the US has little experience and is only beginning to come to grips with regulation governing non-point source water pollution originating in farming practices. The traditional solution, voluntary (or subsidized) compliance on the part of farmers, may or may not work. Third, increased attention should be given by US researchers to the implications of climate change for developing countries, whose interests have not been well represented in the global research effort to date. Particularly with respect to agriculture, developing countries' vulnerabilities to climate change coupled with the pressing need to increase food supplies makes their prospects for adaptation considerably less sanguine than extrapolation of analyses of US agriculture would suggest. The science community should anticipate that, ultimately, agreement on transfer of enabling technologies and resources from the developed to the developing world will be conditioned on the donors' understanding exactly what measures need to be financed.

Finally, I see a need for investment in human capital. Specifically, scientists who want to increase the probability that science contributes positively in responding to the challenges of climate change ought to get smart about the public policy process. The response to climate change will largely be formulated in the executive branch agencies. US commitments to international agreements on climate change mitigation and adaptation will be negotiated as treaties by the President. The example of the North American Free Trade Agreement is instructive, wherein the Congress is put in the position of reacting to an already-negotiated agreement. To get in on the ground level, scientists must gain access to the agencies that support the executive's negotiations.

Most usually, academics interact with Federal agency science career staff, not the policy officials who ultimately make decisions. But it is not necessary to meet the Secretary of a cabinet agency in order to affect her policy decisions. Instead, agency staff should be cultivated for knowledge about how decisions are made: in what form is information presented? what questions are usually asked? what are the particular priorities of that official? If regular contacts on the agency staff cannot be helpful, get the names of individuals who play a more direct role in policy advising. Go to see them: I believe you will find they are surprisingly accessible, especially if you are there to learn about how they do their jobs and not to push for more funding for your particular research project. And, learn about the different roles of different agencies in forming policy. Establishing new relationships with non-traditional agencies can have a big payoff.

If you conclude from these suggestions that I think scientists should spend more time inside the Beltway, you are correct. In the case of climate change, and particularly with respect to its agricultural dimensions, "ivory tower" academics will have a difficult time making an impact. Learning about government is not such a daunting prospect; at any rate, if one has even just the slightest interest in human behavior, it is never boring. And, organizations such as the Board on Agriculture exist to facilitate the kind of interchange I have described, and I would welcome your willingness to work with us.

15

COMMUNICATIONS between DECISION MAKERS and RESEARCHERS:

An OVERVIEW of the JOINT CLIMATE PROJECT

J. Christopher Bernabo
Science and Policy Associates, Inc.

Peter D. Eglinton
Science and Policy Associates, Inc.

Chester L. Cooper
Battelle Pacific Northwest Laboratories

I. INTRODUCTION

Climate research efforts over the last few decades have been driven primarily by a desire to understand and reduce the scientific uncertainties related to predicting global climate change. This fundamental, long term research is critical for developing a thorough understanding of the climate system and a foundation for more detailed and reliable predictions. However, the fundamental questions of interest to scientists are not necessarily the questions of most relevance to decision makers.

There has been a need for a broad assessment of what questions and uncertainties are most useful to address from the perspective of decision makers. A systematic and iterative dialogue between researchers and decision makers is required to define what scientific information can be provided over various time frames, in order to address pending policy questions on global climate change.

The *Joint Climate Project to Address Decision Makers' Uncertainties*[1] was designed to achieve more explicit communication between decision makers and scientists on the informational requirements for policy development and the capabilities of research to meet them. The project's findings will help to manage decision makers' expectations about the kinds of assistance they can

Table 15-1. Groups that participate in identifying decision makers' concerns.

Government	Non-Government
U.S. Congress	Coal
Office of Technology Assessment	Gas
Environmental Protection Agency	Oil
Department of Energy	Chemical
Department of Agriculture	Manufacturing
Department of the Interior	Forest Products
Department of State	Transportation
Office of Management and Budget	Electric Utility
Council of Economic Advisors	Environmental Groups
Council on Environmental Quality	
National Governors' Association	

anticipate from research, and also will assist research planners in identifying policy-relevant priorities.

II. UNIQUE APPROACH

There is a widely recognized gap in the understanding between the science and policy communities. Bridging this gap is crucial not only for developing sound policies, but also for formulating research relevant to decision making.

The *Joint Climate Project* approached the problem of identifying policy-relevant research using two distinct phases: decision makers first defined their information needs; then, scientists determined the research required to meet those policy-relevant needs. This two-step process avoided the common fallacy of implicitly assuming that scientists somehow know what is relevant to policy or that policy makers should know what is good science.

In the first phase of this project dozens of U.S. government and private sector officials, ranging from working level experts to Congressmen, Administration officials, and industry CEOs, were convened to define their key questions about global climate change. The iterative process lasted six months and included a series of interviews, workshops, and focus groups that resulted in a consensus set of broad, policy-relevant questions for researchers to address. Examples of groups represented in defining the decision makers' questions are listed in Table 15-1.

During the second phase of the project, leading experts in climate-related disciplines were assembled to discuss the decision makers' questions developed during the preceding months. The workshop participants examined (i) the research needed to address the questions and (ii) the expectations for providing better information over the next two, five, ten years, and beyond.

In addition, the project examined the needs and opportunities for improving the dialogue between decision makers and researchers.

III. FINDINGS

The consensus-identifying approach of this project yielded several key findings that reflect the general concerns of decision makers and the responses of the research community. In discussions with these two communities, several common themes emerged for enhancing communication and increasing the value of research results.

A. The Concerns Of Decision Makers

This section describes several of the guiding principles, suggested by the participants, from which the policy-relevant questions were derived. In particular, they stressed the importance of augmenting atmospheric research with the economic, social, and ecological sciences, in order to inform policy decisions.

- International Perspectives Drive Policy: The project was conducted during the year before the United Nations Conference on Environment and Development (UNCED). Talks were well underway to craft a Framework Convention on Climate Change. Therefore, many government policy makers focused on these and other ongoing international negotiations and conferences. The officials specifically asked for information to support follow-up actions to UNCED and preparations for future events. For their part, non-government decision makers expressed concern with the possible regulatory implications of proposed actions.
- Climate Change Impacts and Responses Key to Decision Making: Decision making is driven by concerns about the potential impacts of changing climate at the regional level, rather than predictions of changing global mean values of climate variables. More information is needed from the economic, social, and ecological sciences on the potential regional impacts of climate change, and the consequences of possible response strategies. Any response to the threat of climate change must be measured against what is at stake. Therefore, more information is needed on the ecosystems, regions, and human populations that are most at risk from potential climate changes—even if atmospheric research is still unable to provide reliable predictions of the specific changes that will drive the effects.
- Interim Information Needed: Researchers need to provide interim information and iterative assessments while developing long term answers. These interim reports should be as explicit as possible about the inherent uncertainties and avoid compromising scientific objectivity. This need for interim information is driven by the decisions made during relatively short time periods, depending on cycles in politics, budgets, and public concern.

- Implications Of Uncertainties Need Clarification: Researchers need to clarify the sources and implications of policy-relevant scientific uncertainties, and estimate time frames for reducing them. Many uncertainties, although scientifically profound, may be relatively insignificant for developing policies.
- Certainty Not A Prerequisite For Action: Resolution of all scientific uncertainties is not a prerequisite for policy action. Decision makers will apply their constituents' values to determine how much certainty they judge is enough to take a given policy action. Decision makers regularly deal with uncertainties in complex issues and do not often demand scientific proof as a prerequisite for action.

Decision makers developed a list of priority questions regarding the climate system, potential impacts of climate change, and human responses, Table 15-2. In general, the questions focused on the major sources of uncertainty, the importance of regional information, and on ways to evaluate the full range of mitigation and adaptation options.[2]

B. The Response of Researchers

In the next phase of the project, a diverse group of U.S. experts in climate-related fields were convened to examine how research could best address these questions posed by decision makers. Specifically, the scientists examined what types of research are needed to reduce the uncertainties in these policy relevant-questions and estimated the time frames for possible results.

The researchers were sometimes reluctant to focus on the questions framed by decision makers, and expressed concern about the logic behind some of the questions. For example, some of the scientists questioned why decision makers needed information on climate change impacts and response options before an understanding of the full extent of the problem was available. Although this provided an example of the communications gap between scientists and decision makers, the participants did ultimately provide all the necessary information to respond to the key questions. The major conclusions of this session are listed below.

- Timely Results: Some of the key questions decision makers have about climate change can be addressed within a short time frame on the basis of the analysis and interpretation of already available scientific information. Although more complete scientific understanding of climate change may be decades away, much of the information needed to begin addressing decision makers' questions can be provided within two to five years. This could include a comprehensive evaluation of indicators of global climate change, a preliminary vulnerability analysis for systems and regions most sensitive to climate change, and an assessment of the sources and levels of greenhouse gas emissions for use in identifying potential mitigation and adaptation options.
- Parallel Approach To Climate, Impacts, And Responses Research: Scientists need not wait for accurate climate predictions before beginning their research on

Table 15-2. Questions of decision makers.

Generic questions

- What are the major sources of uncertainties regarding the climate system, the impacts of climate change, and potential responses?
- What uncertainties can be quantified, and with what degree of reliability?
- When will key uncertainties be significantly reduced and to what extent?
- How could or should scientists provide information to decision makers on the reliability of their assessments and findings? What will be the assessment and review processes?

Understanding climate change

- What are the magnitudes, rates, and geographic patterns of predicted climate changes?
- What are the causes, mechanisms, and feedbacks involved in climate change (natural and anthropogenic)?
- What is the range of climate's natural variability? How is significant climate change defined?
- How can human activities influence climate and what mechanisms of climate change are beyond man's control?
- Can a reliable early warning system (e.g., indicator species of plants and animals) be fashioned and used to detect the onset of enhanced greenhouse warming?

Impacts of climate change

- Which natural and human systems are the most sensitive to climate? Which are most resilient?
- How will natural and human systems be affected by climate change?
- Which regions are most likely to be negatively affected and which regions could be positively affected?
- Can scientists rule out any possible consequences of future climate changes? What can not be ruled out?
- Can reliable predictions of regional impacts of climate change be made in the next five to ten years? When can further improvements be expected? What resources and tools are needed?

Responding to climate change

- What are the consequences of existing energy and environmental policies for climate change?
- What are the possible response options (mitigation and adaptation) and their associated costs; benefits; level of effectiveness; and the social, economic, and technical barriers to their implementation?
- What are the relative merits of response options and how do they interrelate?
- How does the timing of response options affect the ability to reduce, delay, or avert various effects?
- How do possible initial actions influence future response options?

potential impacts and response options. It is neither necessary nor practical for research to progress sequentially from the climate system, to the impacts, and then to the potential responses in order to provide useful results for decision makers. Much can be done to improve the understanding of impacts without waiting for accurate regional climate predictions. For example, integrated regional and multi-sectoral models—using climate, ecological, demographic, economic, and social data, compiled at the regional level—can provide essential information on potential climate responses, the vulnerability and adaptability of key systems, the extreme ranges of change, and the impacts of climate change on the marketplace.

- Greater Emphasis Needed On Impacts And Responses Research: Information on climate change impacts and response strategies has the greatest potential for assisting decision makers, yet these fields are the least researched. Many of the key questions identified by decision makers involve a significant amount of new socioeconomic, behavioral, and ecological research.
- Funding Alone Is Not Enough: The pace of intellectual and technological advances will affect the progress of research independent of funding levels. Increased communication and coordination among research areas and disciplines—hard sciences, economics, and behavioral sciences—are critical to answering decision makers' questions.
- Global View: An international perspective for research is essential, given the global dimensions of the issue. Even though decision makers may be most concerned with regional and local consequences, developing world issues (such as population and economic development, as well as the pace, quality, and sustainability of development) will be critical.
- Integrated Assessments And Case Studies: Integrated assessments of the causal linkages from emissions through impacts and responses would help structure information for effective use in decision making. Such assessments would incorporate natural and physical sciences, economics, and social factors, including technological change and adaptation. In addition, a coordinated examination of case studies of regional climate variability is needed, based on how societies have responded to climatic variations, using historically documented events. This information would provide valuable insights on how to treat future events.
- Expect the Unexpected: Because of the potential for surprises, decision makers and scientists should frequently re-examine research on possible surprises. Based on this information, contingency plans could be developed to be prepared for dramatic changes in the rate and magnitude of climate change, the seriousness of impacts, and in economic conditions and technological advances.

Tables 15-3, 15-4, and 15-5 provide a brief summary of the potential types of information that research *could* provide to address decision makers' concerns in two, five, and ten years. These time frames were chosen by the decision makers, because of key periods in international negotiations and other activities, as follows:

Table 15-3. Examples of potential research results within one to two years full page.

Understanding climate change
• Comprehensive evaluation of climate change indicators with attributed cause
• Synthesis of existing data to identify significant trends
• Case studies of past climate change

Impacts of climate change
• Preliminary vulnerability assessment of systems and regions most sensitive to climate change
• Simulation models that include ecological, meteorological, demographical, and economic factors
• Preliminary development of environmental change scenarios for regional assessments
• Case studies of impacts of past climate changes

Responding to climate change
• Assessment of land-use patterns and the sources and levels of greenhouse gas emissions
• Inventory of actions that imply relatively long term commitments to emissions at particular levels
• Incorporation and analysis of existing environmental laws and regulations into current economic models, and analysis of their economic and environmental impacts
• Information on potential mitigation and adaptation options
• Catalogue of existing programs addressing climate change
• Methods to evaluate effectiveness/consequences of various policy instruments for implementing mitigation and adaptation strategies
• Economic analyses to evaluate the costs and benefits of response options
• Preliminary assessment of adaptation potential
• Information on interrelationships between U.S. and international activities
• Review of reconstructive ecology
• Case studies of human responses to past climate changes

- 1993 (coinciding with follow-up actions to the U.N. Conference on Environment and Development)
- 1996 (coinciding with the Third World Climate Conference)
- 2000–2005 (coinciding with dates proposed by some in the international community for initial targets and timetables for greenhouse gas emissions stabilization/reduction)
- 2025–2030 (coinciding with dates referred to by some scientists for a potential doubling of greenhouse gas concentrations)

These tables provide an educated estimate by scientists of the potentially available information for time frames of interest to decision makers. They were developed without regard to financial or other resource constraints. Further-

Table 15-4. Examples of potential research results within five years.

Understanding climate change

- Higher spacial resolution GCMs
- Improved understanding of cloud-radiation feedback
- Improved understanding of atmosphere-ocean interaction
- Evaluation of impacts of other climate forcing mechanisms (e.g., ozone and sulfate aerosols)
- General geographic distribution of climate changes

Impacts of climate change

- Information on basic processes influencing changes in systems
- Preliminary baseline monitoring of present state of ecosystems
- Better understanding of how to incorporate technological change and adaptation into impacts analyses
- Regional-scale, future-oriented research on regional vulnerabilities
- Procedures to evaluate validity of existing and future models
- Evaluation of resource conflicts
- Assessment of factors affecting environmental carrying capacity in various situations
- Integrated system studies for various systems and sectors
- Progress towards integrated, interdisciplinary research and assessment
- Continued synthesis of paleoecological information
- Impact assessments for developing countries
- Examination of effects of international trade and aid on food and other resources

Responding to climate change

- Refinement of economic analyses to evaluate costs and benefits of response options
- Initial catalogue of species and sites for use in adaptation analyses
- Preliminary assessment of techniques to aid species migration
- Assessment of technical and institutional responses to major resource shortages

more, the lists suggest what research could do, and *not* what currently planned efforts will do. For more details, refer to the specific research areas discussed in *Phase 2* of the full report.

C. Lessons In Communication

i. Two Communities

Decision makers and researchers inhabit separate domains; to a significant extent, they comprise separate professional cultures. Yet, in the case of the environmental sciences, the two communities are mutually dependent.

Table 15-5. Examples of potential research results within ten years.

Understanding climate change

- Well-defined climate change scenarios to evaluate regional effects
- Improved understanding of magnitudes and rates of change
- More information on atmosphere-ocean-ice interactions and cliamte feedbacks
- Better understanding of climate responses to changing concentrations of greenhouse gases and other mechanisms of climate change
- Ongoing improvements in evaluating predictions of geographic patterns' the role of land-surface processes to predict local and regional climate change near the ground; and the influence of climate change on the biosphere
- Improved weather forecasts

Impacts of climate change

- More defined baseline information on the present state of ecosystems
- Development of a global change social science research program
- Materially closed ecosystem studies to determine effects of climate changes
- Better understanding of species and system response to CO_2 enrichment
- Better understanding of potential climate change impacts
- Further refinement of assessments

Responding to climate change

- Better understanding of social and economic aspects of global change and greenhouse warming
- Information on social behavior and social structures
- Further refinement of assessments

Decision makers rely on the researchers for information, data analyses, and prognoses, while researchers rely on the decision makers for funding, guidance, and ultimately, a *raison d'etre.*

ii. Communications Gap

In light of this interdependence, logic would seem to dictate close, continuing, and frequent communication between the two communities. But this has been far from the case. The normal pattern has been that decision makers and researchers confront each other only twice during the life span of a project or study; at the outset, when the contract is negotiated and months, even years later, when the project is delivered. In the interim, the original decision maker who contracted for the study may have moved on and the ultimate recipient may have had little exposure to, or even knowledge of, the project. This state of affairs is hardly of trivial significance: large sums are expended on research, and the issues being examined are complex and consequential. The policy stakes are considerable.

It would obviously be naive to expect to achieve anything close to an ideal decision maker/researcher relationship. The research community is relatively large in size, broadly scattered geographically, and divided into a myriad of disciplines and subdisciplines. Moreover, many researchers often regard decision makers as uninformed, politically motivated participants in policy-oriented research. The decision makers, for their part, are often preoccupied with day-to-day matters, and are content to leave tedious data gathering and complex analyses to others. As a result too much policy-relevant research is conducted with too little guidance or progress monitoring on the part of the funders. Much can be said for a situation in which the scientists proceed without "interference," but inevitably questions will arise with regard to timeliness, relevance, and usefulness. After all, the issues under investigation are too important for decision makers to default on their duties to provide adequate guidance and oversight, and for researchers to ignore relevance as one criterion for good science and responsible research.

iii. Closing the Gap

Discussions during the *Joint Climate Project* with representatives of both communities (summarized in preceding sections) provided ample evidence that both decision makers and researchers are uncomfortable with the present situation. Both are anxious to develop and sustain a productive dialogue. Both would like to increase the effectiveness of the research community in the decision-making process. Both agree that a two way bridge must be developed to span the communications gap between the two communities.

What emerged from the series of meetings, forums, and workshops was a need and readiness on the part of high-ranking decision makers to meet with principal investigators and their collaborators during the course of a project and to discuss results, methodology, and data with responsible researchers upon the completion of a project. Researchers expressed their desire for more guidance and for frequent contact with the requesters and ultimate consumers of the information their projects produce.

But to truly close this gap, to construct a bridge between the two communities, will take more than wistful expressions and lofty pronouncements. There is no substitute for sustained effort and innovative institutional arrangements. The decision makers and researchers who participated in the project agreed that greater attention must be paid to the development of systematic communications processes. In particular, both sides need to recognize the following points.

- Not An Either/Or Decision: Decision makers' choices are not simply between pursuing research or implementing response strategies. Rather, the challenge is to define the appropriate levels of each over time. Researchers need to provide a broad array of information to address the complex and interacting decisions on global climate change. Decision makers, for their part, need to recognize the long time scales involved in research and, thus, the importance of continuity of funding and program goals.

- Global Climate Change In A Relative Risk Context: Prediction of changes in mean global temperatures does not give an adequate picture of the societal risk that can be related to everyday experiences. Assessing climate change in a relative risk context is difficult, but extremely important. Since the public tends to respond to perceived crises, assigning relative risk would assist decision makers in distinguishing between verifiable, serious threats and possibly misplaced public concern. Given that risk is a function of both the probability and the magnitude of the expected consequences, better data on possible impacts are critical to better estimates of societal risk.
- Urgent Need For Education: A concerted effort is needed to educate decision makers on the facts and uncertainties of global climate change. Considering that public concern is often the impetus for formulating policy, scientists need to communicate technical information to the public more effectively, as well as more frequently. In addition, scientists need to learn more about the decision making process and the types of information most useful for policy. Frequent, two-way communication between decision makers and researchers is fundamental if research is to play an effective role in the decision-making process.
- Managing Uncertainty: There are more ways to manage uncertainties than simply trying to reduce them. For example, building resilient institutions and methodologies would provide a flexible response to any future changes in climate, albeit at potentially significant costs. Contingency plans could allow decision makers to prepare for possible climate outcomes through R&D on response technologies, without needing to deploy them. Furthermore, decision makers and researchers should strive to understand and communicate more effectively the risks of climate change.
- Research Does Not Always Provide The Answer: Decision makers need to realize that additional research actually could increase the amount of uncertainty in some areas. Researchers should inquire about how much certainty decision makers are requiring to take a specific action. To this end, uncertainties that do not matter for decision making should be so identified.
- Develop An Ongoing Assessment Process For Research: To improve communication and better inform decision makers, research efforts should include an iterative assessment process. These assessments would not only help to identify the relevant questions, but also serve to structure the research results and, thus, facilitate clearer communication between the two communities. Furthermore, the assessment process would provide valuable input to the planning of policy-relevant research.

IV. CONTINUING THE PROCESS

The *Joint Climate Project* represents a first step in determining how researchers can assist U.S. decision makers over the coming years and decades, thereby helping to bridge the communication gap between these two communities. A more frequent and systematic two-way dialogue will be needed between decision makers and researchers in order for research to effectively inform the decision-making process.

Discussions with decision makers and researchers during the project revealed that both communities are very interested in developing and sustaining a productive dialogue. Both would like to increase the effectiveness of the research community in assisting the decision-making process. Decision makers suggested meetings with principal investigators and their collaborators during the course of a project to discuss results, methodology, and data, as well as with the researchers upon the completion of a project. Researchers expressed their desire for more guidance and for frequent contact with the requesters and ultimate consumers of the information their projects produce.

Given the debate over how to address potential climate change, representatives from industry, environmental organizations, and other groups also said that they would welcome the chance to participate in a consensus-building effort on areas of agreement for the policy options. Researchers can then focus on areas of disagreement to attempt to enlighten the debate. Similarly, scientists could pursue more frequent communication among and within their various disciplines-especially between those concerned with climate change prediction and those focused on impacts and human responses. In this way, researchers can identify areas of agreement and divergence in order to communicate more effectively with decision makers.

The project's goal was to identify national-level decision maker's broad questions and to determine the general response of the research community. This project is an initial step in the ongoing, iterative exchange needed between science and policy. Further efforts will be required to update these initial results as both science and policy continue to evolve. These national-level findings need to be supplemented with similar efforts on the local, regional, and international levels. Similarly, the *Joint Climate Project* approach could be applied to individual sectors, such as agriculture and energy, to yield significant details on the research needed to address their specific concerns. In any case, the process used by this project serves as a model for improving the process whereby research and policy are integrated and accommodated in research programs addressing societal needs.

NOTES

[1] The *Joint Climate Project* was a federal-private sector effort involving the Electric Power Research Institute (EPRI), Environmental Protection Agency (EPA), Department of the Interior (DOI), Department of Energy (DOE), Department of Agriculture (USDA), Forest Service (USFS), and the National Climate Program Office (NCPO). Full copies of the report (No. TR-100772) on which this chapter is based are available from: EPRI Distribution Center, 207 Coggins Drive, P.O. Box 23205, Pleasant Hill, CA 94523.

[2] Details of these and other questions can be found in the full report of this study (Science and Policy Associates, Inc., 1992).

GLOSSARY

Absorption capability – The ability of a greenhouse gas to change the global solar balance by trapping infrared radiation. Also known as the instantaneous forcing weight.

Adaptation strategies – Strategies implemented to minimize the negative effects and maximize the positive effects of climate change.

Aerobic – Oxygen rich.

Agronomic yield – Crop yield based on total acres of crop planted (in contrast to statistical yield which is based on total acres of crop actually harvested).

Albedo – The surface reflectivity of the globe. Affects the amount of solar radiation being radiated back into space without being absorbed by the earth's climate system.

Anaerobic decay – Decomposition of organic matter by anaerobes-bacteria which thrive in oxygen poor environments, a source of methane.

Anthropogenic emissions – Emissions resulting from human activities.

Biosphere – Refers to the zone of the earth and atmosphere that contains living organisms. The terrestrial biosphere excludes the oceans.

BLS (Basic Linked System) – World food trade model designed at the International Institute for Applied Systems Analysis (IIASA) for food policy studies.

BNF (Biological Nitrogen Fixation) –The process by which some microorganisms assimilate N_2 gas from the atmosphere and convert it to usable soil nitrogen.

C3 Species – Species whose products of photosynthesis are compounds with three carbon atoms. C3 species include wheat, rice, soybean, most horticultural crops, and many weed species.

C4 Species – Species whose products of photosynthesis are compounds of four carbon atoms. C4 species include maize, sorghum, sugarcane, millet, and many pasture, forage, and weed species.

Carbon tax – A tax on fossil fuels based on the individual carbon content of each fuel. Under a carbon tax, coal would be taxed the highest per MBtu, followed by petroleum and then natural gas.

Carbon recycling – Biological sources of carbon engage in carbon recycling. Trees remove CO_2 from the atmosphere during photosynthesis; this CO_2 is returned when the tree dies and decomposes or is burned. Similarly, CO_2 is removed from the atmosphere during the feed growing process; bovines recycle this carbon, returning both CO_2 and CH_4 to the atmosphere.

Cereal production – The production of wheat, rice, maize, millet, sorghum, and other minor grains.

CFCs (Chlorofluorocarbons) – A family of inert gases, including CFC-11, CFC-12, and CFC-113. These gases are of concern for two reasons. First, in the upper stratosphere they result in ozone degradation. Second, they are also potent greenhouse gases. CFCs are currently regulated under the Montreal Protocol on Substances that Deplete the Ozone Layer (1985).

CH_4 (Methane) – A key greenhouse gas with a variety of sources, both natural and human-made.

CO_2 Equivalent Doubling – The radiative effect from all greenhouse gases equivalent to that for a doubling of CO_2.

CO_2 – Carbon dioxide.

CO_2 fertilization – Increased plant productivity due to higher ambient levels of CO_2. Also referred to as CO_2 enrichment or the CO_2 fertilizer effect.

Coefficient of Variation – A measure of variability which is equal to the standard deviation divided by the statistical mean of the variable, expressed in percentage terms.

Comprehensive Approach – An approach that covers all greenhouse gases in aggregate. Each gas would be assigned a weight, based on its effectiveness and atmospheric lifetime, and each country could select its own mixture of greenhouse gas reductions to meet an overall greenhouse gas reduction strategy. It is distinct from an individual gas approach which would require specific reductions for each gas.

CoP – Conference of the Parties to the Framework Convention on Climate Change.

Crop mix – The percentage of land devoted to each crop.

Crop Simulation Model – Mathematical-computer models that simulate the growth process of the plant from the time it is planted to when it matures. These models require climatic inputs (daily minimum and maximum temperatures, precipitation, solar radiation) and agronomic inputs (soil characteristics, variety of plant).

Cropping intensity – The number of crops grown every year on the same land.

Cultivar – Variety of plant species.

DCV (Detrended Coefficient of Variation) – The standard deviation from the linear trend divided by the average of the original series, in percentage terms.

Diurnal cycle – Range of daily maximum minus minimum temperatures.

Doubling of CO_2 – This phrase is used in many ways, but generally means a doubling of atmospheric CO_2 over pre-industrial levels (from 280 to 560 ppmv). Some climate modelers use this same phrase to mean a doubling over their assumed starting levels (from 330 to 660 ppmv).

Downregulation – A downward shift in photosynthesis. The process by which the large increases in photosynthesis observed in response to high CO_2 exposure in the short term are not maintained in the long run.

Drought stress – Stress on a plant due to insufficient root water uptake from the soil. This stress may significantly reduce crop yields. Drought stress is also referred to as water stress.

El Niño – Irregular changes in the ocean currents off the west coast of South America that result in prolonged increases in sea surface temperatures along the coast of Peru and in the equatorial eastern Pacific Ocean. El Niño has been linked to distant atmospheric features having diverse effects, such as the Indian monsoon, shrimp production in Louisiana, and wildland fires in the U.S.

ENSO – El Niño/Southern Oscillations. See El Niño.

Enteric fermentation – The intestinal fermentation which occurs in ruminant animals such as cows; it is a major biological source of methane.

EPIC (Erosion Productivity Impact Calculator) – A crop growth simulation model developed by the U.S. Department of Agriculture.

Equilibrium Response Experiment – A GCM or O/AGCM model run that involves the instantaneous doubling of the atmospheric CO_2 concentration followed by simulation of the climate response over a period of 20–50 years.

Evapotranspiration – Combined water loss from land area due to evaporation of water from soils and transpirational water loss from plants.

FAO – United Nations Food and Agriculture Organization.

FCCC – Framework Convention on Climate Change. Opened for signature at the United Nations Conference on Environment and Development (UNCED), June 1992.

Field time – The amount of time, based on weather conditions, that farmers can perform various operations in the field.

GEF (The Global Environmental Facility) – An organization created in 1990 as a three year pilot program to provide investment and assistance for developing country projects dealing with global warming, biodiversity, international waters, or ozone depletion. It is jointly operated by the World Bank, UNDP, and UNEP. It is named as the "interim" funding agency for the Framework Convention.

General Circulation Models (GCMs) – Large scale computer models used to predict the response of the climate system to a CO_2 equivalent doubling. A/O GCM refers to a coupled ocean-atmosphere GCM.

GFDL – Geophysical Fluid Dynamics Laboratory.

GHG – Greenhouse gas(es), includes carbon dioxide, methane, nitrous oxide, CFCs and other halogenated species, water vapor, ozone, and other trace gases.

Gigatons Carbon (Gt C) – Billion tons of carbon. 1 Gt C = 3.67 Gt CO_2.

GISS – Goddard Institute of Space Studies.

Grain Moisture Content – The moisture level of a crop when it is harvested. Some crops, such as corn, require drying if the moisture level is too high at harvest.

Greenhouse Effect – An atmospheric process by which greenhouse gases (such as CO_2, CH_4, N_2O, and CFCs) affect the global energy balance. Shortwave radiation from the sun that reaches the earth and is re-emitted as long wave infrared radiation is partially absorbed by greenhouse gases (GHGs). In the absence of GHGs the earth's average temperature would be -18°C rather than 15°C.

GWP (Global Warming Potential) – Some greenhouse gases are more effective, on a unit basis, of affecting, or "forcing," the climate system. The GWP combines the capacity of a gas to absorb infrared radiation and its residence time in the atmosphere with a time frame of analysis, then expresses the result relative to CO_2.

Halogenated Species – A key group of greenhouse gases which include chlorofluorocarbons (CFCs), hydrochlorofluorocarbons (HCFCs), halon, methyl chloroform, and carbon tetrachloride.

INC – Intergovernmental Negotiating Committee.

Income elasticity – The expected percentage change in the quantity demand for a good given a one percent change in income. An income elasticity of demand for electricity of 1.0 implies that a one percent increase in income will result in a one percent increase in demand for electricity.

Indeterminate crop species – Crop varieties, like tomatoes, which continue producing new leaves, flowers, and fruits throughout the life cycle.

Industrial-related emissions of fossil fuels – Gases that are produced as a result of a human-induced, non-agricultural process. An industrial-related freeze of emissions would cover CO_2 and CH_4 emissions from burning fossil fuels.

Infrared Radiation – Long wave radiation emitted from the earth. Some of this infrared radiation is absorbed by the greenhouse gases, resulting in warming at the earth's surface and the lower atmosphere.

Instrumental record – Observed weather statistics.

IPCC – Intergovernmental Panel on Climate Change.

Linear programming – Technique that allocates scarce resources in an optimal way given a decision maker's ultimate objective.

Maize – A grain commonly referred to as corn in the U.S.

Market Clearing – The economic condition of supply equalling demand.

MEPs – Marketable emission permits.

Metric Ton – 1000 kilograms (kg). 1 metric ton = 1.1 U.S. (or short) ton.

MINK Study – A study of the likely effects of increasing temperatures on the agricultural economy of the Missouri, Iowa, Nebraska, and Kansas region.

MMT – Million metric tons.

Montreal Protocol on Substances that Deplete the Ozone Layer – An international agreement signed in 1985 which mandates the phaseout of CFCs and halons.

N_2O – Nitrous Oxide

Natural variability of the climate system – That portion of the total variability of climate which is solely due to the internal dynamics of the atmosphere and ocean and has nothing to do with changes in greenhouse gas concentrations or changes in other external factors known to influence climate.

NGO – Non-governmental organization.

NH_3 – Ammonia.

NOAA – National Oceanic and Atmospheric Administration.

NO_x – Nitrogen oxides—NO and NO_2.

O/AGCM – A fully coupled ocean-atmospheric general circulation model. Generally, the atmosphere and oceans are divided into a number of discrete layers, with each layer consisting of a two-dimensional grid of thousands of points. The model then solves equations for the transport of heat, momentum, moisture (in the atmosphere), and salinity (in the ocean) on this three-dimensional grid. The typical resolution is 4° latitude by 5° longitude.

O_3 (Ozone) – In the stratosphere, ozone protects the earth from ultraviolet radiation. In the troposphere, it contributes to smog and is considered a pollutant.

OH (Hydroxyl radical) – Hydroxyl molecules react with CH_4, resulting in the release of CO_2, CO, and H_2O. This is the main sink for CH_4. Uncertainties in the future concentrations of OH lead to uncertainties in predictions of the lifetimes of CH_4. Hydroxyls are also important in the chemistry of tropospheric ozone creation and the dissolution of partially halogenated chemicals in the troposphere.

OMB – Office of Management and Budget.

Output multipliers – Calculated values which indicate how a change in one activity level will affect the economy as a whole.

Paleoclimatic data – Climatic data that can be inferred from such sources as tree rings, ice cores, and layered lake sediments, providing information about climate variability on time scales of centuries to thousands of years.

Persistence – The characteristic inertia of weather and climate which leads to the occurrence of dry or wet spells, and warm or cold periods.

PET (Potential Evapotranspiration) – The maximum amount of soil evaporation and transpiration from a well irrigated crop for a given set of environmental conditions.

Photolysis – Decomposition by solar radiation. N_2O and O_3 are removed from the stratosphere by photolysis.

Photorespiration – A biochemical process occurring in the light in which oxygen is absorbed by plant leaves and CO_2 is released into the atmosphere.

Photosynthesis – The process in which CO_2 enters the plant through small openings in the leaves called stomates, is captured or "fixed" by photosynthetic enzymes, and is then converted to carbohydrates.

ppbv – parts per billion by volume.

ppmv – parts per million by volume.

pptv – parts per trillion by volume.

Preindustrial levels – The atmospheric concentrations of various greenhouse gases present prior to the start of the Industrial Revolution (approximately 1750–1800).

Price elasticities – The expected percentage change in quantity demand for a good given a one percent change in price. A price elasticity of demand for electricity of -0.5 implies that a one percent increase in price will result in a half percent decrease in demand for electricity.

Radiative forcing – Changes in the global balance of incoming solar radiation and outgoing infrared radiation caused by a radiative forcing agent, such as clouds, surface albedo, and greenhouse gases. This results in changes in the global climate.

RI (Rainfall Index) – The national average of total annual precipitation weighted for each station by its long term normal values.

RUE (Radiation use efficiency) – The amount of biomass produced per unit of incoming light energy.

SO_2 – Sulfur Dioxide.

SOC – Soil Organic Carbon.

SST – Sea surface temperature.

Stabilization of Greenhouse Gas Concentrations – The amount by which anthropogenic emissions would have to be reduced in order to stabilize concentrations at current day levels. These are: 60–80% for CO_2, 15–20% for CH_4, 70–80% for N_2O, 70–75% for CFC-11, and 75–85% for CFC-12 (IPCC, 1990).

Statistical yield – The yield from those areas actually harvested. May be significantly higher than the agronomic yield since in poor years, only a fraction of the planted area may actually be harvested. The use of statistical yield will reduce the actual variability of yields.

Stochastic weather generator – A program which generates weather values for daily precipitate, temperatures, and solar radiation based on observed historical patterns.

Stomatal openings – Small openings in the leaves by which CO_2 enters.

Stratosphere – The upper layer of the atmosphere, extending from the troposphere to about 50 km from the earth's surface.

Temperature change commitment – The eventual equilibrium temperature predicted to occur due to a given level of greenhouse gases. The temperature change is not immediate because of the huge thermal mass of the oceans.

Teragram (Tg) – One million metric tones; 10^{12} grams.

Transient Response Experiments – In contrast to the equilibrium response experiments, the atmospheric concentrations of the greenhouses gases are increased gradually over some emissions scenario in a GCM run.

Transpiration – The escape of water vapor from the leaf.

Troposphere – The lower atmosphere, to an altitude of about 8 km at the poles, about 12 km at mid-latitudes, and 16 km in the tropics.

UKMO – United Kingdom Meteorological Office.

UNCED – United Nations Conference on Environment and Development. Held in Rio de Janeiro, June 1992. Also referred to as the Earth Summit.

UV radiation (Ultraviolet radiation) – A type of short wave radiation that is damaging to plants and animals, including humans. The amount of UV radiation that reaches the earth is dependent on the amount of stratospheric ozone. An increase in UV radiation due to a decrease in stratospheric ozone will pose a direct threat to human health (increased cataracts, immune suppression, and skin cancers) and will have a negative impact on plant yields for many species.

Vernalization – The requirement of some temperate cereal crops, such as winter wheat, for a period of low winter temperatures to initiate or accelerate the flowering process.

VS – Volatile Solids.

Wm^{-1} – Watts per meter.

WUE (Water Use Efficiency) – The ratio of crop biomass accumulation or yield to the amount of water used in evapotranspiration.

INDEX